DEVELOPMENTS IN THE THEORY AND PRACTICE OF CYBERCARTOGRAPHY

MODERN CARTOGRAPHY SERIES, VOLUME 5

DEVELOPMENTS IN THE THEORY AND PRACTICE OF CYBERCARTOGRAPHY

APPLICATIONS AND INDIGENOUS MAPPING

SECOND EDITION

Edited By

D. R. FRASER TAYLOR

Geomatics and Cartographic Research Centre (GCRC), Carleton University, Ottawa, Ontario, Canada

Associate Editor

TRACEY P. LAURIAULT

National Institute for Regional and Spatial Analysis (NIRSA), National University of Ireland at Maynooth, Maynooth, Co. Kildare, Republic of Ireland; Member of the Geomatics and Cartographic Research Centre (GCRC), Carleton University, Ottawa, Ontario, Canada

ELSEVIER

AMSTERDAM • BOSTON • HEIDELBERG • LONDON
NEW YORK • OXFORD • PARIS • SAN DIEGO
SAN FRANCISCO • SINGAPORE • SYDNEY • TOKYO

Elsevier
Radarweg 29, PO Box 211, 1000 AE Amsterdam, The Netherlands
The Boulevard, Langford Lane, Kidlington, Oxford, OX5 1GB, UK
225 Wyman Street, Waltham, MA 02451, USA

Second edition 2014

British Library Cataloguing in Publication Data
A catalogue record for this book is available from the British Library

Library of Congress Cataloging-in-Publication Data
A catalog record for this book is available from the Library of Congress

ISBN: 978-0-444-62713-1
ISSN: 1363-0814

Contents

5. Further Developments in the Theory and Practice of Cybercartography: Exploring Web 2.0 and Participatory Software for Building Geolocated Narratives

WILLIAM CARTWRIGHT

6. Tile-Based Mapping with Opacity

ANDREW CLOUSTON AND MICHAEL P. PETERSON

7. Advances in Location-Based Services

GEORG GARTNER

8. Inclusive Cartography: Theoretical and Applied Issues in Brazil

REGINA ARAUJO DE ALMEIDA

9. The Nunaliit Cybercartographic Atlas Framework

AMOS HAYES, PETER L. PULSIFER, J.P. FISET

10. Interactive Audiovisual Design for Cartography: Survey, Prospects, and Example

GLENN BRAUEN

11. A Spatial Typology of Cinematographic Narratives

SÉBASTIEN CAQUARD AND DANIEL NAUD

The Atlases described in this book are available on the web site of the Geomatics and Cartographic Research Centre at Carleton University in Ottawa, Canada http://gcrc.carleton.ca. *The contents are the property of the respective partners, contributors, and sources. Details are available within the atlases themselves.*

The companion website for this book can be found at http://booksite.elsevier.com/9780444627131/

Preface

The book on *Developments in the Theory and Practice of Cybercartography: Applications and Indigenous Mapping* is a substantial revision of the *Cybercartography: Theory and Practice* book published in 2005. This edition is much more than just an update as it contains entirely new material especially, although by no means exclusively, relating to the mapping of indigenous and traditional knowledge. Cybercartography, in both theory and practice, has advanced very substantially since 2005 and this book captures some of these important developments. As in the first edition, there is a major contribution from the research group at Centro de Investigación en Geografía y Geomática 'Ing. Jorge L. Tamayo', A.C (CentroGEO) in Mexico City. There has been a very fruitful cooperation between CentroGEO and the Geomatics and Cartographic Research Centre (GCRC) at Carleton University for over a decade. CentroGEO's focus is on the concept of geocybernetics, a related concept to cybercartography. GCRC is the home of many of the authors of chapters in this volume. The atlases discussed were created by the research team working in the Centre.

Much of the research presented in this book has been carried out in cooperation with a number of Inuit and First Nations communities in Canada. Substantial support for the research reported on here has been provided by the Social Sciences and Humanities Research Council of Canada and the Federal Government of Canada's International Polar Year.

The GCRC and CentroGEO research form the core of the book but is supplemented by complementary work being carried out by research associates in Australia, Austria, Brazil, New Zealand and the United States. Of special interest is the contribution of Dr Teresa Scassa and her colleagues in the Faculty of Law's Centre for Law, Technology and Society as well as the Canadian Internet Public Policy Clinic at the University of Ottawa. Legal and ethical issues in the use and management of geographic information are of growing importance both in Canada and internationally and the partnership established between GCRC and the Faculty of Law led to important new findings in this field. Cybercartography is by its nature multidisciplinary and the addition of legal scholars has been especially beneficial.

Cybercartography, and it main products, cybercartographic atlases including the Nunaliit Cybercartographic Atlas Framework used to create them, has a major role to play in the emerging Web 3.0 era. The holistic nature of cybercartographic theory and the use of location to integrate all kinds of information and present this in interactive, multimedia and multisensory ways give the cybercartographic approach particular appeal. Cybercartography is advancing because of an iterative interaction between theory and practice; people and technology and the chapters in this book present current thinking and point to prospects for the future.

D. R. Fraser Taylor.
July 2013.

Acknowledgements

Many individuals have contributed to the publication of this book and these are too numerous to identify by name. A major contribution to the book has been made by the faculty, postdoctoral fellows, graduate students, and technical and administrative staff of the Geomatics and Cartographic Research Centre at Carleton University, which is the major centre for research on cybercartography. Many of these individuals have made contributions to individual chapters in this book and all are contributing to the ongoing research in the Centre. Special mention should be made for the contribution of Dr Tracey Lauriault, who is Associate Editor of the book. The Centre benefits from having a number of research associates based both in Canada and overseas and again, several of these have contributed chapters to this book including Professors William Cartwright, Georg Gartner, Michael Peterson, Sebastien Caquard and Regina Araujo Almeida, as well as Dr Peter Pulsifer.

The ongoing research cooperation between the Geomatics and Cartographic Research Centre and CentroGeo in Mexico City is once again reflected in the chapters of this edition as it was in the first edition of this book. Special thanks in this respect are due to Dr Carmen Reyes and the Director of CentroGEO and Dr Margarita Paras.

Research collaboration with the Faculty of Law at the University of Ottawa and a talented team led by Dr Teresa Scassa has resulted in increased understanding of the legal and ethical aspects of mapping traditional knowledge and the research reported in several chapters of this book is ongoing.

Our partnerships with indigenous communities and individuals in Canada have been central to the chapters dealing with the mapping of indigenous knowledge. Several of the chapters are co-authored with individuals from these groups and all have benefited from the ongoing collaboration. These include Nunavut Arctic College, the Nunavut Research Institute, the Kitikmeot Heritage Society, the Gwich'in Social and Cultural Institute, the Inuit Heritage Trust, the Inuit Tapiriit Kanatami and a number of Nunavut communities and Elders including Arctic Bay, Cambridge Bay, Pangnirtung, Clyde River, Iqaluit, Cape Dorset and Igloolik. So we would like to acknowledge the importance of Elders and individual community members of the Anishinaabe peoples of the Northern Great Lakes Region of Ontario.

None of our research would have been possible without the generous support of a number of organizations including the International Polar Year Program of the Federal Government of Canada, the Social Science and Humanities Research Council of Canada, the Government of Nunavut, the Department of Fisheries and Oceans Canada, the Kitikmeot Heritage Society, the Inuit Heritage Trust, the Natural Science and Engineering Research Council of Canada, Aboriginal Affairs and Northern Development Canada and the Gwich'in Social and Cultural Institute.

The production of this book has been a team effort and the contributions of all involved are greatly appreciated.

Editor's Biography

Dr D. R. Fraser Taylor is a distinguished research professor and Director of the Geomatics and Cartographic Research Centre at Carleton University, Ottawa, Canada. He has been recognized as one of the world's leading cartographers and a pioneer in the introduction of the use of the computer in cartography. He has served as the president of the International Cartographic Association from 1987 to 1995. Also, in 2008, he was elected a fellow of the Royal Society of Canada in recognition of his achievements. He was awarded the Carl Mannerfelt Gold Medal in August 2013. This highest award of the International Cartographic Association honours cartographers of outstanding merit who have made significant contributions of an original nature to the field of cartography.

He produced two of the world's first computer atlases in 1970. His many publications continue to have a major impact on the field. In 1997, he introduced the innovative new paradigm of cybercartography. He and his team are creating a whole new genre of online multimedia and multisensory atlases including several in cooperation with indigenous communities. He has also published several influential contributions to development studies and many of his publications deal with the relationship between cartography and development in both a national and an international context.

Dr Tracey P. Lauriault is a postdoctoral researcher at the National Institute for Regional and Spatial Analysis (NIRSA), National University of Ireland, Maynooth, Ireland, working on the Programmable City Project. She was a postdoctoral fellow and a graduate student at the Geomatics and Cartographic Research Centre from 2002 to 2013. Her dissertation topic was *Data, Infrastructures and Geographical Imaginations*. She was the fellow assigned to the SSHRC Partnership Development Project entitled *Mapping the Legal and Policy Boundaries of Digital Cartography*, as part of the GCRC's research stream of Law, Society and Cybercartography. She was the research leader for the Pilot Atlas of the Risk of Homelessness funded by HRSDC and was part of the project management team for the Cybercartography and the New Economy Project responsible for collaboration, transdisciplinary research and olfactory cartography. She was the GCRC researcher for the International Research on Permanent Authentic Records in Electronic Systems (InterPARES) 2. She continues to participate in activities and represents the GCRC on topics related to the access to and preservation of data and data policy.

List of Contributors

Alestine Andre Gwich'in Social and Cultural Institute, Yellowknife, Fort McPherson and Tsiigehtchic, Northwest Territories, Canada

Claudio Aporta Marine Affairs Program, Dalhousie University, Halifax, Nova Scotia, Canada

Regina Araujo de Almeida Department of Geography, University of São Paulo, São Paulo, Brazil

Kristi Benson GIS/Heritage Affiliate, Gwich'in Social and Cultural Institute, Santa Clara, Manitoba, Canada; MDT Communications, Ottawa, Ontario, Canada

Glenn Brauen Geomatics and Cartographic Research Centre (GCRC), Carleton University, Ottawa, Canada

María del Carmen Reyes Centro de Investigación en Geografía y Geomática Ing. J. L. Tamayo A.C. (CentroGeo), México

Del Carry MDT Communications, Ottawa, Ontario, Canada

William Cartwright School of Mathematical and Geospatial Sciences, RMIT University, Melbourne, Victoria, Australia

Sébastien Caquard Department of Geography, Planning and Environment, Concordia University, Montreal, Quebec, Canada

Andrew Clouston Geography, Environment and Earth Sciences, University of Victoria, Wellington, New Zealand

Cindy Cowan Nunavut Arctic College, Iqaluit, Nunavut, Canada

Kim Crockatt Kitikmeot Heritage Society

Timothy Di Leo Browne School of Canadian Studies, Carleton University, Ottawa, Ontario, Canada

Nate J. Engler Geomatics and Cartographic Research Centre (GCRC), Carleton University, Ottawa, Ontario, Canada

Sheena Ellison Institute for Comparative Studies in Literature, Art and Culture, Carleton University, Ottawa, Ontario, Canada

William Firth Gwich'in Social and Cultural Institute, Yellowknife, Fort McPherson and Tsiigehtchic, Northwest Territories, Canada

J.P. Fiset Class One Technologies Inc., Ottawa, Ontario, Canada

Georg Gartner Vienna University of Technology, Vienna, Austria

Amos Hayes Geomatics and Cartographic Research Centre (GCRC), Carleton University, Ottawa, Ontario, Canada

Darren Keith Kitikmeot Heritage Society

Ingrid Kritsch Gwich'in Social and Cultural Institute, Yellowknife, Fort McPherson and Tsiigehtchic, Northwest Territories, Canada

Tracey P. Lauriault National Institute for Regional and Spatial Analysis (NIRSA), National University of Ireland at Maynooth, Maynooth, Co. Kildare, Republic of Ireland; Member of the Geomatics and Cartographic Research Centre (GCRC), Carleton University, Ottawa, Ontario, Canada

Gita J. Ljubicic Department of Geography and Environmental Studies, Carleton University, Ottawa, Ontario, Canada

Alejandra A. López-Caloca Centro de Investigación en Geografía y Geomática Ing. J. L. Tamayo A.C. (CentroGeo), México

Fernando López-Caloca Centro de Investigación en Geografía y Geomática Ing. J. L. Tamayo A.C. (CentroGeo), México

Daniel Naud Département de Géographie, Université de Montréal, Montréal, Quebec, Canada

Michael P. Peterson Department of Geography/Geology, University of Nebraska at Omaha, Omaha, USA

Carol Payne Art History, Carleton University, Ottawa, Ontario, Canada

Peter L. Pulsifer National Snow and Ice Data Centre (NSIDC), University of Colorado, Boulder, USA

Stephanie Pyne Geomatics and Cartographic Research Centre (GCRC), Carleton University, Ottawa, Ontario, Canada

Rodolfo Sánchez-Sandoval Centro de Investigación en Geografía y Geomática Ing. J. L. Tamayo A.C. (CentroGeo), México

Teresa Scassa The Centre for Law, Technology and Society, University of Ottawa, Ottawa, Ontario, Canada

Sharon Snowshoe Gwich'in Social and Cultural Institute, Yellowknife, Fort McPherson and Tsiigehtchic, Northwest Territories, Canada

Carmelle Sullivan Department of Geography and Environmental Studies, Carleton University, Ottawa, Ontario, Canada

D. R. Fraser Taylor Geomatics and Cartographic Research Centre (GCRC), Carleton University, Ottawa, Ontario, Canada

CHAPTER

1

Some Recent Developments in the Theory and Practice of Cybercartography: Applications in Indigenous Mapping: An Introduction

D. R. Fraser Taylor

Geomatics and Cartographic Research Centre (GCRC), Carleton University, Ottawa, Ontario, Canada

OUTLINE

Developments in the Theory and Practice of Cybercartography, Second Edition, ISSN 1363-0814
http://dx.doi.org/10.1016/B978-0-444-62713-1.00001-5

1.1 INTRODUCTION

This book is a substantial update of *Cybercartography: Theory and Practice* published in late 2005. In the last paragraph of that book the following statement appeared. '…the chapters in this book suggest that we are at a breakthrough point in the development of cartography and that the paradigm of cybercartography is well worth further exploration. There are, however, many questions still to be answered, and much further research is required if cybercartography is to reach its full potential' (Taylor, 2005:558).

Several challenges and directions were identified for future research:

- The need for more practice;
- and for more rigorous theory;
- design challenges;
- relationships with the arts and humanities;
- the utility of cybercartography;
- the need for multisensory research; and
- the challenges of preservation and archiving.

Since the first edition of this book was published substantial changes have taken place in both the theory and practice of cybercartography and many of these are the results of extensive new practice, especially in cooperation with indigenous communities in Canada. The interaction between theory and practice is a major facet of cybercartography and practice creates a new theory that in turn leads to improved practice. Cybercartography is essentially an iterative process and is holistic in nature. This chapter will outline some of the new theoretical and applied directions of cybercartography since the publication of the first edition.

1.2 THE ELEMENTS OF CYBERCARTOGRAPHY

In the first volume, seven elements of cybercartography were identified as follows:

- Cybercartography is multisensory using vision, hearing, touch, and eventually, smell and taste;
- uses multimedia formats and new telecommunications technologies, such as the World Wide Web;
- is highly interactive and engages the user in new ways;
- is applied to a wide range of topics of interest to the society, not only to location finding and the physical environment;
- is not a stand-alone product like the traditional map, but part of an information/analytical package;
- is compiled by teams of individuals from different disciplines; and involves new research partnerships among academia, government, civil society, and the private sector (Taylor, 2005:3).

In general terms, these seven elements are still in place but have been refined, modified, and expanded around six key ideas. These six key ideas, which are a reflection of the original seven elements, are:

- Individuals use all of their senses while observing what is around them: cybercartography is therefore exploring the possibilities of using all five senses in its representations in order to make cybercartographic atlases as reflective as possible of sensory realities.
- Individuals have different learning preferences and prefer teaching and learning materials in different formats. Cybercartographic atlases have great potential in both formal and informal education and they provide the same information in multiple formats allowing users the freedom to choose which format or combination of formats and modalities they wish to use.
- Educational theory suggests that individuals learn best when they are actively rather than passively involved. This applies both in formal and informal learning situations. Engaging the user requires carefully thought out interactive engagement strategies including the design of effective user interfaces.
- The social media revolution has given people the power to create their own narrative and cartography is challenged to respond to individual and community needs in ways that previously did not exist. The Nunaliit Cybercartographic Atlas Framework is a software platform that provides a mechanism for people to enter their own data into cybercartographic atlases. Cybercartography provides a means for people to tell their own stories as part of a holistic information package. The Framework is open source, provides a metadata structure for the information, and is designed with an interface that does not require special knowledge in order to enter information.
- Many topics of interest to society are complex and the same set of 'facts' on topics such as climate change are open to a variety of representations. Even when there may be an agreement on the facts, there can often a wide variety of interpretations. There are often no simple 'right' or 'wrong' answers to many questions. Cybercartography allows the presentation of different ontologies or narratives on the same topic without privileging one over another. The user can consider the various narratives presented and have a greater understanding of the complexities and uncertainties surrounding many topics. Traditionally, the map was an authoritative source. Cybermaps are much more nuanced.
- Traditional cartography was supply-driven. National mapping agencies supplied the definitive and authoritative maps, which the public used. Technological change allowed for a much more demand-driven approach and cybercartography takes this one step further and empowers individuals and communities to create their own maps including the choice of what to represent and what not to represent. Individuals are new 'prosumers' rather than simply 'cybercartography is consumers' and, as a result, democratizing mapping in new ways. Indigenous people, for example, have often been largely 'invisible' on maps or have been represented by others. Cybercartography gives voice to indigenous people and other community groups both literally and metaphorically.

1.3 DEFINITION OF CYBERCARTOGRAPHY

In 2003 cybercartography was formally defined as '...the organization, presentation, analysis, and communication of spatially referenced information on a wide variety of topics of interest and use to society in an interactive, dynamic, multimedia, multisensory, and multidisciplinary format' (Taylor, 2003:406). This captures some of the elements of cybercartography but has to be read with the major elements and ideas described in Section 1.1 and 1.2 in mind.

It is the combination of these elements and ideas in a holistic manner and the iterative interaction between theory and practice, which defines cybercartography as well as the processes by which cybercartographic atlases are created, which sets cybercartography apart from other approaches. Pyne has recently described cybercartography as a distinctive critical cartographic approach and as 'a set of concepts and tools…that provides an effective atlas building framework for approaching complex social, political and economic phenomena…' (Pyne, in press).

And it has also been more simply described as 'the application of geographic information processing to the analysis of topics of interest to society and the display of the results in ways that people can readily understand' (Taylor, 2013:4). Cybercartography will continue to work in an iterative fashion as a result of the interaction between applications, technology, and theory. Each new practical application brings new theoretical insights as well as new technological developments and the results of each application are used as building blocks for the next.

1.4 NEW PRACTICE

The first edition of the book reported on the application of cybercartography under the Cybercartography and the New Economy Project (Taylor, 2005:7). As part of that project, two cybercartographic atlases – the *Cybercartographic Atlas of Antarctica* and a *Cybercartographic Atlas of Canada's Trade with the World* – were created. These were, however, both primarily supply and technologically-driven atlases, which contain some of the elements of cybercartography described earlier in the chapter but by no means all. The user was involved in their creation but primarily from the human–computer interaction aspects of use and usability and the research reported on in the book included substantial input by human factors and cognitive psychologists. This volume is based largely on an entirely new practice developed since 2007 in cooperation with indigenous groups in Canada hence the sub-title of the book 'Applications in Indigenous Mapping'.

A Social Sciences and Humanities Research Council of Canada (SSHRC) major grant, helped fund the research on cybercartography upon which much of this book was based. SSHRC has also supported the research described in this second edition, especially in relation to the creation of the *Lake Huron Treaty Atlas* (Chapter 17) and the work on Views from the North (Chapter 13). This has been supplemented by substantial support from the Government of Canada. International Polar Year (IPY), which provided major funding of over $1 million for the Inuit Sea Ice Use and Occupancy Project, which created the *Inuit Siku (sea ice) Atlas* (Chapter 14) as well as support for the development of the Nunaliit Cybercartographic Atlas Framework (Chapter 9) and related data-management issues. Both the *Kitikmeot Place Names Atlas* (Chapter 15) and the *Gwich'in Atlas* (Chapter 16) have been developed using funding from the Kitikmeot Heritage Society and the Gwich'in Social and Cultural Institute. The *Arctic Bay Atlas* (Chapter 20) was developed using funding from the Inuit Heritage Trust and Nunavut Arctic College after an initial investment by the Inukshuk Wireless Foundation. Although the amounts involved were not at the same scale as SSHRC and IPY funding, they were especially significant as indigenous communities and institutions were investing their own funds in atlas development.

This new practice has required a consideration of traditional knowledge (TK) and its representation in cybercartographic form. We have worked mainly with Inuit communities and with First Nation groups such as the Anishinaabe and the Gwich'in. The atlases that are created as a result of our partnership with these communities are described in several of the subsequent chapters of this book but the interaction with indigenous people has had a profound effect on the thinking surrounding cybercartography.

1.4.1 The Nature of TK

Oguamaman, in his recent book on traditional knowledge, defines traditional knowledge as '...an aspect of ecological management and environmental stewardship, sustainable development, economic empowerment, self determination, human rights, culture, arts, craft, music, songs, dance and diverse creative repertoire: religion, lifestyles and innumerable aspects of social processes that undergird a people's worldview' (Oguamamam, 2011:46).

The nature of cybercartography lends itself well to the representation of this multifaceted concept but in creating cybercartographic atlases with indigenous groups a major challenge is not only to effectively represent the TK involved but also to consider the processes involved. Kitchen and Dodge have argued that in cartography the processes by which the map is created (ontogenesis) is as important as the ontology, the map itself (Kitchen and Dodge, 2007). Kitchen and Dodge do not explicitly consider legal and ethical aspects of the processes involved but at the Geomatics and Cartographic Research Centre (GCRC) we have found these to be of central importance to our work. The importance of process is central to the creation of products such as the *Inuit Siku (sea ice) Atlas* (Chapter 14), the *Views from the North Atlas* (Chapter 13), the *Kitikmeot Place Names Atlas* (Chapter 15), the *Gwich'in Atlas of Place Names* (Chapter 16), the *Arctic Bay Atlas* (Chapter 20), and the *Lake Huron Treaty Atlas* (Chapter 17). The creations of these are fully described in the chapters listed above. The equally important legal and ethical issues and processes in atlas creation are explicitly considered in the Chapters 18 and 19.

A key element is understanding 'the people's worldview' mentioned in Oguamaman's definition above. This is not easy to articulate as it involves '...understanding of the human place in relation to the universe [and] ...encompasses spiritual relationships, relationships with the natural environment and the use of natural resources, relationships between people; and is reflected in language, social organizations, values, institutions and laws' (Legat, 1991:1). The task is made even more difficult by the fact that descriptions are often written by outsiders rather by the indigenous people themselves. For the Inuit, the term Inuit Qaujimajatuqangit (IQ) or the Inuit 'way of knowing' was developed in 1997 and Shirley Talalik, an Inuit herself, described IQ as follows:

> *Inuit Qaujimajatuqangit* (IQ) is the term used to describe Inuit epistemology or the Indigenous knowledge of the Inuit. The term translates directly "that which Inuit have always known to be true." Like other Indigenous knowledge systems, *Inuit Qaujimajatuqangit* was formally adopted by the Government of Nunavut; however the descriptors used to capture the essence of *Inuit Qaujimajatuqangit* are recognized as being consistent with Inuit worldview as it is described in various Inuit circumpolar jurisdictions. Inuit Elders in Nunavut have identified a framework for *Inuit Qaujimajatuqangit* which is grounded in four big laws or *maligait*. All cultural beliefs and values are associated with the implementation of these *maligait*, ultimately contributing to 'living a good life' which is described as the purpose of being ***(Tagalik, 2001 as quoted in Sullivan, 2013).***

The major principles of IQ are:

- *Inuuqatigiitsiarniq*: the concept of respecting others, building positive relationship, and caring for others;
- *Tunnganarniq*: the concept of fostering good spirit by being open, welcoming, and inclusive;

- *Piliriqatigiingniq*: to develop collaborative relationships and working together for the common good;
- *Avatimik Kamattiarniq*: to show environment stewardship;
- *Pilimmaksarniq*: to be empowered and built capacity through knowledge and skills acquisition;
- *Qanuqtuurungnarniq*: to be resourceful and seek solutions through creativity, adaptability, and flexibility;
- *Aajiiqatigiingniq*: consensus decision-making; and
- *Pijitsirarniq*: to contribute to the common good through serving and leadership (McGregor, 2010:173–174).

The approach to TK of First Nations people is similar to that of the Inuit in terms of the deep relationship between humans and the environment and the TK Working Group of the Northwest Territories described it as follows:

> ...knowledge that derives from, or is rooted in the traditional way of life of Aboriginal people. Traditional knowledge is the accumulated knowledge and understanding of the human place in relation to the universe. This encompasses spiritual relationships with the natural environment and the use of natural resources, relationships between people; and is reflected in language, social organization, values, institutions and laws ***(Legat, 1991:1).***

1.4.2 Cybercartography and TK

Cybercartographic atlases have the technology to represent the multifaceted nature of TK and in the atlases described in subsequent chapters of this book almost all of the aspects of TK appear in a variety of ways and forms. The processes by which the atlases are created are empowering and often involve an explicitly post-colonial, decolonizing approach, for example the *Lake Huron Treaty Atlas* (Chapter 17) and the *Arctic Bay Atlas* (Chapter 20). Indigenous people use the atlases as one of the means of reclaiming their cultural heritage, for example the *Kitikmeot Atlas* (Chapter 15) and *The Gwich'in Atlas* (Chapter 16). The reclamation of cultural heritage through the use of indigenous names of course predated the atlases but the atlases have been adopted by indigenous peoples as a way to further exert their cultural ownership and expression. All of these atlases involve the inclusion of narratives, often by Elders in their own language. These narratives capture the essence of an 'indigenous way of knowing' and both preserve and represent this concern in new ways. To the Inuit, for example, a journey is not simply represented on a map by a line indicating origin and destination. Each journey is a narrative and even where the route taken is the same the narratives are different (Aporta, 2009). The *Inuit Siku (sea ice) Atlas* (Chapter 14), for example, presents the narratives of several Elders and communities and each Elder and community is identified by name. It is possible to look at the collectivity of routes but at the same time to isolate each narrative by individual Elder. This is important because the quality and authenticity of information to Inuit communities is determined not by locational accuracy or the data quality elements usually used in mapping but by who provided the information. We are dealing here with 'living metadata' (Chapter 14), which is quite different from the metadata usually associated with digital maps.

One way of ensuring that a 'people's way of knowing' is adequately reflected is to give the power to the people concerned to choose what they want to represent. In the case of the

Gwich'in Social and Cultural Institute and the Kitikmeot Heritage Society, the content of the atlases is determined by these organizations. In the case of the *Atlas of Arctic Bay*, the same principle applies and often the choice of content encompasses things that would not have been chosen by outsiders. For example, the inclusion by Arctic Bay youth of a rap video entitled 'Don't call me Eskimo'! (http://www.youtube.com/watch?v=tS8RZcKQwBA).

No mapping technology can claim to represent TK in all of its complexity but cybercartography does this better than most, especially when the communities concerned take full ownership of the atlases. Bonny and Berkes (2008) considered this issue some years ago and although technologies have changed, many of their arguments remain valid. They were writing before cybercartography had been used to represent TK. The advantage of the Nunaliit Cybercartographic Atlas Framework is that although it has been developed to enable the online representation of any form of knowledge, its flexible database structure allows it to deliver information in a variety of forms including CD-ROMs and print. This ability allows for broader distribution of the atlases in situations where a user cannot view them online, which is sometimes the case where a bandwidth is limited (see Chapter 9).

1.5 NEW THEORY

In the first edition of this book, the theory of cybercartography was not well developed and certainly was not nearly so well described as the practice. The same applied to the related concept of geocybernetics being developed in parallel by the research group at Centro de Investigación en Geografía y Geomática 'Ing. Jorge L. Tamayo', A.C (CentroGEO) in Mexico City, which was described in the first edition of the book. The Mexican research group has contributed two chapters (Chapters 2 and 3) to this book. Although the CentroGEO research team argues that geocybernetics encompasses cybercartography, there are striking similarities between the two concepts. As with cybercartography, geocybernetics was initially intuitive and empirical rather than theoretical as outlined in Chapters 2 and 3 but CentroGEO researchers have recognized the need to build a transdisciplinary theoretical framework. They are building new linkages between the hard and the social sciences by using holistic and transdisciplinary approaches. Geocybernetics uses Norbert Wiener's concept of social cybernetics in terms of interaction with society and attempts to integrate community knowledge with specialist knowledge to form a 'combined knowledge base'. 'This triggers the creation of a network of stories describing the complex relationships involved' (Chapter 2). They argue that geocybernetics encompasses cybercartography as well as other concepts such as the geomatics prototype, the Reyes method, and emerging knowledge networks with common theoretical building blocks such as chaos theory, complexity theory, and metamathematics. These are important theoretical developments that, like cybercartography, are based on the interaction between theory and practice and recognize the importance of narratives. CentroGEO is building on the experience of over 60 empirical projects, especially the key visual and; verbal expressions of these projects. Whether geocybernetics encompasses cybercartography or the other way around is open to debate but both concepts are based on the interaction between theory and practice and consider environmental and societal issues although in a very different physical and sociocultural context. They also use different language and theoretical constructs to describe this interaction with society.

The theoretical constructs, which have developed, that underpin cybercartography owe much to the interaction with indigenous peoples described earlier. This interaction has helped to clarify thinking about cybercartography in new ways that are quite different from the theoretical approaches described in the first edition of the book.

The macrotheory underlying cybercartography is a reflection of the author's earlier work in Africa, which argued for a 'development-from-within' approach, building on the knowledge and wisdom of local people as a key element in socioeconomic development (Taylor and Mackenzie, 1992). Cybercartography also recognizes the importance of earlier work on traditional mapping described by Woodward and Lewis in their impressive volume on this topic in the *History of Cartography* series (Woodward and Lewis, 1998).

Woodward and Lewis (1998:1) argue that 'maps are more than ...ever-improving representations of the geographical world' and point out that maps have to be thought of as part of a cognitive system as well as a social construct in addition to being part of material culture. In cybercartography all three aspects, the material, the cognitive, and especially the map as a social construct, are included and of course, all three overlap. Cybercartographic atlases are much more than maps. As Brian Harley and David Woodward observed 'there has probably always been a mapping impulse in human consciousness and the mapping experience – including the cognitive mapping of space – indubitably existed long before the physical artifacts we now call maps (Harley and Woodward, 1987:1). Cybercartography encompasses what Woodward and Lewis call 'performance cartography' and 'material cartography' as they include elements such as song, dance, art and speech and as a result, cybermaps are more often 'more interesting than the territory' as Houellebecq (2010) has argued about maps in general.

1.5.1 Cybercartography and Critical Cartography

The interaction with indigenous people has led cybercartography into the theoretical field of critical cartography, especially in relation to cybercartography as a social construct (Taylor and Pyne, 2010). 'Any definition that ignores either the function of maps or their role as social constructs fails to account for the fact that maps are far more than wayfinding devices' (Woodward and Lewis, 1998:6). Cybercartography can, and often does, present a variety of different viewpoints but the atlases dealing with TK and indigenous people described in this volume privilege one voice over all others, that of the indigenous people whose perspectives and cognitive values drive the content of all of the atlases. They give voice, literally and figuratively to both the Inuit and the First Nations people involved.

All of the atlases make extensive use of narrative and, as outlined earlier, explicitly consider the processes by which these atlases are created. Both narratives and ontogenesis are important elements of critical cartography. The *Lake Huron Atlas*, the *Atlas of Arctic Bay*, the *Inuit Siku (sea ice) Atlas* and *the Views from the North Atlas* are all examples of this.

As Caquard has observed, 'overall the critical turn in cartography has drastically modified the relations between maps and narratives in two ways: by deconstructing and exposing the meta-narratives embedded in maps, and by ensuring maps are compelling forms of storytelling' (Caquard, 2013:2). In this volume Caquard provides a spatial typology of cartographic narratives as these relate to cinema.

Cybercartographic atlases are compelling forms of storytelling that are particularly appropriate for Inuit society, which is an oral rather than a literary society. In the cases of atlases involving the Inuit, the telling of the stories by the Elders is also an important means of preserving the valuable knowledge they possess and passing it on to future generations in both formal and informal education processes (Chapters 20 and 21).

1.5.2 Cybercartography and Volunteered Geographic Information

Volunteered Geographic Information (VGI) has been described as 'arguably the most significant change in the whole history of cartography…' (Perkins, 2013). Cybercartographic atlases make extensive use of VGI and in the case of the atlases being created with indigenous people the use of VGI is central. Sui et al. (2012) give an extensive description of VGI and in Chapter 4 Engler, Scassa, and Taylor discuss this in relation to cybercartography. To be useful in the creation of cybercartographic atlases, VGI needs structure and this is provided by the Nunaliit Cybercartographic Atlas Framework, which is described by Hayes and others in Chapter 9. This is a documented oriented framework that replaces earlier relation-based and schematic approach. This framework has been specially designed to allow the easy ingestion of all kinds of VGI. The Nunaliit platform can be used by individuals with little knowledge of geographic information processing and is very easy to use. A prototype Ipad application has been designed to allow for the collection of multiple forms of information including videos, photographs, interviews, etc. to be collected in the field and then uploaded directly into the system (Chapter 9). This tool is ideal for collecting all kinds of VGI and the Atlas Framework also includes metadata giving an indication of the provenance and quality of the VGI entered into the system. In addition, those who enter the data can autonomously decide how they want those data represented and who can use and access them. Intellectual property rights are technologically embedded into the data collection and representation process and aspects of this are discussed in Chapters 18 and 19.

1.5.3 Cybercartography and the Individual

Das Gupta (2013:10) argues that 'the Internet world has put the individual firmly centre stage in terms of interactive information gathering, storage and delivery. The year 2013 should see the emergence of the individual as the focus for geospatial information in a similar fashion.' Mobile devices, such as cell phones, are now ubiquitous and can be used both an input devices for VGI as discussed in Section 1.5.4 and as individually user-centred devices on which cybermaps can be displayed. Gartner, in Chapter 7, discusses advances in location-based services, including the provision of maps although, as Clouston and Peterson observe in Chapter 6 'the slow dissemination of a map to a computer or mobile device is the single major limitation to their use'. Rapid advances are being made and as Hellmis, the Vice-President of the mobile phone maker Nokia observes 'we will redefine mapmaking through computation cartography. This means maps are created on demand to meet a specific purpose, to help govern an individual action, to help answer a personal question' (Hellmis, 2013). For the individual, 'maps have recently become the main interface for accruing data over the internet' (Ron, 2008). In Chapter 5, Cartwright gives an interesting example of how geolocational narratives can be built using social software facilitated through Web 2.0 and use interactive digital media to tell an individual story.

Although many of the cybercartographic atlases described in this volume are community driven, these are collections of individual inputs, often but not always, made by individual Elders who, as in the case of the *Inuit Siku (sea ice) Atlas* (Chapter 14), for example, are identified by name and tell their own individual stories.

1.5.4 The Holistic Nature of Cybercartographic Theory

There are many strands to cybercartographic theory but it is the holistic nature of cybercartography, which is it major characteristic. Cognitive, material, and social constructs are woven together in innovative ways and the results communicated through cybercartographic atlases.

1.6 NEW DESIGN CHALLENGES

Each cybercartographic atlas provides design challenges but all atlases have the same issues as far as use and usability are concerned. In the first edition of this book, the human–computer interaction factors were fully discussed (Lindgaard et al., 2005; Roberts et al., 2005; Tribovitch et al., 2005) and these are not further discussed in this volume, although this continues to inform the practice of creating atlases. There is no doubt that the multisensory and multimedia nature of cybercartographic atlases are engaging but the jury is still out as to whether those using the atlases retain information in long-term memory. The issue here is the possibility of cognitive overload. A balance between simplicity and complexity is not easy to find. User testing is critical but difficult given the nature of online atlases. Determining who the users are is challenging, even when users are also knowledge contributors, as is the case with many of the atlases that represent data contributed by indigenous collaborators.

One area where the GCRC has made considerable progress is the design of user interfaces. These have been much improved and research on this topic continues (Chapter 9). In terms of the design of the atlases dealing with TK, we have to take into account use limitations facing users. To serve communities in the North, for example, the current limitation on band width must always be kept in mind, which meant that the design of the *Inuit Siku (sea ice) Atlas* (Chapter 14) was much more basic than existing technology allows.

One of the better designed atlases is the *Atlas of the Risk of Homelessness* where the limitations outlined above were not present. The 21 variables used to construct the Atlas at a variety of different scales from the city, metropolitan region and the nation posed interesting design problems. One of the more innovative solutions was the development of 'graphomap' (Chapter 12). This is a combination of geographic position presented in schematic form with an interactive means of expressing quantitative values in a comparable form including change over time and space.

Another major design challenge is the integration of the different multisensory and multimedia components of cybercartography. We have made real progress in design terms with sound and audiovisual media as illustrated in Chapter 10 but very limited progress with the others such as olfaction.

In design terms, a major effort has been concentrated on improving the technical design of the Nunaliit Cybercartographic Atlas Framework. This is described in Chapter 9. The technical design has been completely revised to a schema-less, documented platform.

1.7 RELATIONSHIPS WITH ART AND THE HUMANITIES

Given the importance of the cognitive aspects of cybercartography outlined earlier and the inclusion of art, literature, and other aspects, it is perhaps fair to say that mainstream material cartography has failed to effectively include research on cognitive maps, whereas in the arts and the humanities, research has expanded considerably. Caquard and Naud, in their chapter on cinematographic narratives (Chapter 11), touch on this but as Caquard points out (Caquard, 2013) much work has been done in fiction (Roberts, 2012), in cinema (Conley, 2007), art (Pezzuto, 2011), music (Long and Collins, 2012), and poetry (Horowitz, 2001). There are many challenges involved in including these aspects in cybercartography but cybercartography has much to learn from the substantial and long-term research being carried out on cognitive 'mapping' in the arts and humanities.

1.8 MULTISENSORY RESEARCH

We have made considerable progress with new forms of visualization and augmented reality as is illustrated in several chapters in this book. We have also made major strides in the integration of sound and audiovisual media into the design of our atlases as is discussed in Chapter 10. The use of touch and feel has primarily been focussed on use of maps by the blind, which was fully discussed by Araujo de Almeida and Tsuji (2005) in the first edition of the book and some recent developments are discussed by Araujo de Almeida in this volume (Chapter 8). Research on the effective integration of smell into cybercartographic atlases has been discussed by Lauriault and Lindgaard (2006) but as yet has not been integrated into any of our atlases despite the early promise outlined in the first edition of this book. There are two reasons for this. The first is technical. There is as yet no commercially viable olfaction output device that can be used with maps on the market. There are scent diffusers but these are limited to use in specific sectors such as the perfume industry and wine industry or site specific installations and have proven to have limited applicability to cybercartography. Many of the scent diffusers examined also went out of production due to limited marked demand. The potential remains but implementation is a challenge. Adding smell is relatively easy in technical terms but effectively integrating it into design is much more difficult. Brauen (Chapter 10) argues that most uses of sound in mapping, of which there are many, are independent of the map rather than being an integral part of the design.

1.9 PRESERVATION AND ARCHIVING

The need for more effective preservation and archiving maps is the topic of the chapter by Lauriault and Taylor (Chapter 21). The map has been a fundamental facet of the memory of societies from all over the world for millennia. The map has appeared in a variety of forms the main, but by no means the only one, being paper.

The general challenges of archiving and preserving digital maps are substantive as outlined by Lauriault and Taylor (2012) but interactive multimedia cybercartographic atlases pose special problems that as yet have not been fully resolved while the lack of dedicated institutional support towards the creation of trusted digital repositories is another.

In the Internet era, websites are expanding exponentially but many of these are neither effectively maintained nor updated and often disappear without notice. In almost every chapter in the book, reference is made to websites with URLs in order to support an argument or illustrate a point. This is also briefly discussed by Brauen (Chapter 10) as many audiovisual maps examined in the last decade have fallen into disrepair or have disappeared completely. We have no way of knowing if all of the URLs or the new audiovisual maps examined will be accessible by the time this book is published.

The atlases dealing with TK are one means of preserving that knowledge and are especially important because they preserve this in context, not in isolation. In many cases, it is the first time that the aural knowledge shared and represented in the atlases has been documented. TK is not a set of artifacts but is 'a way of knowing" as outlined earlier in this chapter and individual elements lose much of their meaning when removed from their cultural setting.

1.10 LEGAL AND ETHICAL ISSUES

Chapters 4, 18, and 19 deal with important legal and ethical issues, especially as those relate to TK. The notion that researchers working with indigenous people must engage in ethical practices is not new but this takes on new urgency when digital distribution of information is the norm. What does 'informed consent' mean, for example when that information is to be distributed on the Web? When TK is part of a cybercartographic atlas that is accessible by anyone, anywhere, and at any time, what are the most appropriate ethical practices? These ethical questions are considered in Chapters 18 and 19 and suggestions made on best practices.

Cybercartography offers a rich potential for the mapping of TK as outlined earlier in this chapter but that knowledge is much more than a collection of 'artifacts' that can be placed on a map. It forms part of a knowledge system that is often fundamentally different from dominant western systems. Chapter 19 looks at the legal issues involved in protecting community rights to knowledge and fostering a culture of respect, not only for discrete pieces of knowledge but also for the integrity of the knowledge system from which it emanates. Existing law, such as that of copyright, intellectual property, and data ownership are of limited value as these are based on individual rights and not collective or community-centred rights. What is required are entirely new ways of looking at legal and ethical issues including a greater consideration of 'soft law' solutions.

1.11 EDUCATION

Chapter 20 discusses the great potential of cybercartography in both formal and informal education. Many of the characteristics of cybercartography lend themselves to improved teaching and learning in a variety of different settings. In theoretical terms, cybercartography supports the multiple intelligence and interactive learning theories of Howard Gartner (2011). In an applied sense, the Cybercartographic Atlas of Arctic Bay (Chapter 20) and the *Inuit Siku (sea ice) Atlas* (Chapter 14) are both being used to improve education in the North. The *Atlas of Arctic Bay* is being used as a central part of a new course for community educators at Nunavut Arctic College and components of the *Sea Ice Atlas* are being introduced into the high school curriculum in Nunavut (Chapter 20). In educational terms the atlases provide a new way of introducing TK into the teaching and learning process, which is both interactive and engaging and, as is argued in Chapter 20, the processes by which these atlases are created with the involvement with Inuit communities is in itself a very important educational experience, which is of equal importance to the artifacts themselves. The atlases also provide an important resource for all those, outside of the communities themselves, who are interested in learning more about TK in Canada's North. All cybercartographic atlases present knowledge in a form that people can readily understand making them powerful educational tools.

1.12 CONCLUSION

Cybercartography has come a long way since the first edition of this book was published. Indeed, this volume contains over 90% new content and is, in essence, a completely new book. Many of the same authors from the first edition have contributed entirely new chapters. These include William Cartwright, Georg Gartner, and Michael Peterson (with Andrew Clouston), three of the world's leading cartographers in the field of mapping on the Internet. The research group at CentroGEO in Mexico City, which made an important contribution to the first edition, have substantially updated their work on the concept of geocybernetics and presented a new theory based on extensive practice. Regina Araujo has updated her work on mapping for the blind and added a new description of participatory indigenous mapping in the Brazilian Amazon, while Sebastien Caquard has described his most recent work on cinematographic cartography. The extensive work on mapping with the Inuit and First Nations is entirely new as is the important topic of the legal and ethical issues involved in mapping TK. Several of the chapters on mapping the TK have been co-authored with writers from the Kitikmeot Heritage Society, the Gwich'in Social and Cultural Institute, and Nunavut Arctic College. This input has been of great value to the content of the chapters concerned and is an indication of the valuable partnerships on which the creation of many our cybercartographic atlases depend as well as the importance of the processes by which these atlases are built. Inuit and First Nations' input to the atlases through the Elders, youth, and other community members, is critical and the quality, and acceptability and authenticity of the information, is dependent on this input. The narratives they present are the core of many of our atlases.

Cybercartography and the Nunaliit Cybercartographic Atlas Framework can be applied to a wide variety of topic in addition to the mapping of TK. In this volume, for example, there is a description of the *Pilot Atlas of the Risk of Homelessness*. There are also common issues regardless of what application area is involved such as archiving and preservation. The legal and ethical issues, for example, apply to a much wider field than traditional the knowledge context in which we discuss them.

Sir Tim Berners-Lee (2009), widely recognized as the originator of the World Wide Web, identified two major challenges for the future of the Web. The first is linking datasets on different topics, both quantitative and qualitative and generating new and more useful combined information. The second is finding new innovative ways of displaying that information in ways that people can be more readily understood. Cybercartography, using location as a linking mechanism, does both and is very much a technology for the Web 3.0 era.

References

Aporta, C., 2009. The trail is here: inuit and their Pan-Arctic network of routes. Home Ecology 57, 131–146.
Araujo de Almeida, R., Tsuji, B., 2005. Interactive mapping for people who are blind or visually impaired. In: Taylor, D.R.F. (Ed.), Cybercartography: Theory and Practice, Elsevier, Amsterdam, pp. 411–432.
Berners-Lee, T., 2009. Notes on Design and Linked Data. Available at: http://www.w3.org/design issues/linked data.html.
Bonny, E., Berkes, F., 2008. Community traditional environmental knowledge: addressing the diversity of knowledge, audiences and media types. Polar Record 49/230, 243–253.
Caquard, S., 2013. Cartography 1: mapping narrative cartography. Progress in Human Cartography 37/1, 135–144.
Conley, T., 2007. Cartographic Cinema. University of Michigan, Ann Arbor.
Gartner, H., 2011. Frames of Mind: The Theory of Multiple Intelligences, sixth ed. Basic Books, New York.
Das Gupta, A., 2013. The world we want in 2013. Geospatial World 3/6, 8–10.
Harley, B., Woodward, D., 1987. The map and the development of the history of cartography. The History of Cartography 1, 1–42.
Hellmis, C., 2013. Location brings a new Dawn in map making. Geospatial World 3/6, 92–93.
Houellebecq, M., 2010. La carte et le territoire. Flamarrion, Paris.
Horowitz, H., 2001. Wordmaps. In: Dear, M., Ketchum, S.L., Richardson, D. (Eds.), Geohumanities, Routledge, London/New York, pp. 144–159.
Kitchen, R., Dodge, M., 2007. Rethinking maps. Progress in Human Geography 31/3, 331–344.
Lauriault, T.P., Lindgaard, G., 2006. Scented cartography: exploring possibilities. Cartographica 41/1, 72–92.
Lindgaard, G., Brown, A., Bronsther, A., 2005. Interface design challenges in virtual space. In: Fraser Taylor, D.R. (Ed.), Cybercartography: Theory and Practice, Elsevier, Amsterdam, pp. 211–230.
Lauriault, T.P., Taylor, D.R.F., 2012. The map as a fundamental source in the mapping of the world. In: Duranti, L., Shaffer, E. (Eds.), Conference Proceedings, UNESCO Memory of the World in the Digital Age: Digitalization and Preservation, Vancouver, Available from: http://www.ciscra.org/docs/Lauriault_Taylor_Map.pdf.
Legat, A., 1991. Report of the Traditional Knowledge Working Group. Government of the Northwest Territories, Yellowknife 23.
Long, P., Collins, J., 2012. Mapping the soundscapes of popular music heritage. In: Roberts, L. (Ed.), Mapping Cultures: Place, Practice, Performance, Palgrave/McMillan, Basingstoke, pp. 144–159.
McGregor, H., 2010. Inuit Education and Schools in the Eastern Arctic. UBC Press, Vancouver.
Oguamamam, C., 2011. Intellectual Property in Global Governance. Routledge, London/New York.
Perkins, C., 2013. "Review of crowdsourcing geographic Knowledge: Volunteered geographic information (VGI) in theory and practice" by Sui, D., Elwood, S. and M. Goodchild. Cartographica 48/1, 76–78.
Pezzuto, D., 2011. Leonardo's Val di Chiana Map in the Mona Lisa. Cartographica 46/3, 149–159.
Pyne, S., in press. Spatializing history: the cybercartographic atlas of the Lake Huron Treaty process. In: This is Indian Land: The Robinson Huron Treaties of 1850. UBC Press, Vancouver.
Roberts, L. (Ed.), 2012. Mapping Cultures: Place, Practice, Performance, Palgrave/McMillan, Basingstoke.
Roberts, S., Purush, A., Lindgaard, G., 2005. Cognitive theories and aids to support navigation of multimedia information space. In: Fraser Taylor, D.R. (Ed.), Cybercartography: Theory and Practice, Elsevier, Amsterdam, pp. 231–256.
Ron, L., 2008. Google Maps = Google on Maps, Presentation to the Where 2.0 Conference 12–14 May, Burlingame, California.
Sui, D., Elwood, S., Goodchild, M., 2012. Crowdsourcing Geographic Knowledge: Volunteered Geographic Knowledge (VGI) in Theory and Practice. Springer, Dordrecht.
Sullivan, C., 2013. Integrating Culturally Relevant Learning in Nunavut High Schools: Student and Educator Perspectives from Pangnirtung, Nunavut and Ottawa. Unpublished Master's Thesis, Carleton University, Ottawa.
Tagalik, S., 2001. Inuit Qaujimajatuqangit: The Role of Indigenous Knowledge in Supporting Wellness in Inuit Communities in Nunavut. National Collaboratory Centre for Aboriginal Health Fact Sheet University of Northern British Columbia, Prince George.
Taylor, D.R.F., Mackenzie, F., 1992. Development from within: Survival in Rural Africa. Routledge, London.
Taylor, D.R.F., Pyne, S., 2010. The History and Development of the Theory and Practice of Cybercartography. International Journal of Digital Earth 3 (1), 1–14.
Taylor, D.R.F., 2003. The concept of cybercartography. In: Peterson, M.P. (Ed.), Maps and the Internet, Elsevier, Amsterdam, pp. 405–420.

Taylor, D.R.F., 2005. Cybercartography: Theory and Practice. Elsevier, Amsterdam.
Taylor, D.R.F., 2013. Cybercartography is charting a new route. Geospatial World 3/6, 50–53.
Tribovitch, S., Lindgaard, G., Dillon, R.F., 2005. Cybercartography: a multimedia approach. In: Taylor, D.R.F. (Ed.), Cybercartography: Theory and Practice, Elsevier, Amsterdam, pp. 254–284.
Woodward, D., Lewis, M.L., 1998. Cartography in the traditional African. American, Arctic and Pacific societies. The History of Cartography, vol. 2. Book 3. University of Chicago Press, Chicago.

CHAPTER

2

From Cybercartography to the Paradigm of Geocybernetics: A Formal Perspective

Fernando López-Caloca, Rodolfo Sánchez-Sandoval, María del Carmen Reyes, Alejandra A. López-Caloca

Centro de Investigación en Geografía y Geomática Ing. J. L. Tamayo A. C. (CentroGeo), México

OUTLINE

2.1 INTRODUCTION

Nearly two decades of research has involved empirical, theoretical, and methodological work in cybercartography. Initially, Taylor proposed the concept of cybercartography as a line of research, which included modeling, design, technological development, and the

Developments in the Theory and Practice of Cybercartography, Second Edition, ISSN 1363-0814
http://dx.doi.org/10.1016/B978-0-444-62713-1.00002-7

intertwining of these with society (Taylor, 2003) and further developed in chapter 1 of this volume. Later, Reyes established the methodological and conceptual bases for the construction of a theoretical framework (Reyes, 2005, 67) and left the door open for the formalization of the semantics and syntax of its metalanguage. Reyes also proposed the concept of geocybernetics as a more comprehensive avenue of research (Reyes et al., 2006).

To understand and advance this new geocybernetics paradigm, we need to clarify certain aspects that distinguish cybercartography. According to Reyes, topology is to geometry what algebra is to arithmetic. Likewise, we can analogously say what geocybernetics is to cybercartography. Although this chapter does not focus on algebra and topology, we can adopt a similar formal approach in terms of the use of a symbolic language to help to generalize and facilitate the construction of abstract structures that enable organization of human thought. For example, in our case, we can begin to explore a conceptual language and to study rules to manipulate geocybernetic expressions that involve stories such as those described in geocybernetics and enable the construction of formal structures that help to establish and advance this new paradigm.

Similar to cybercartography, geocybernetics also has its own body of knowledge; that is, it has its own theoretical framework and building blocks (Reyes and Paras, 2012). This new conceptual framework – called geocybernetics – leads to the inclusion of preexisting paradigms, combining quantitative and qualitative methods under the cybernetic, complex, and chaotic vision stemming from the structure, functioning and behaviour of living and social systems interacting in space-time.

The path to undertake in the formalization of the cognitive process implicit in the body of knowledge of geocybernetics became evident after several decades of empirical work (Reyes and Paras, 2012). This chapter is organized according to metatheory, informal theory, and, finally, formal systems, which are schemas that have been followed by scientists for many centuries. Informal theory is presented using the Reyes method, geocybernetics, and the emerging knowledge network as background. The metatheory adopted for our purpose is the current scientific building blocks that scientists have developed over at least the last 20 centuries. Included as references and presented in the introduction to the third section of the chapter are elements of logic proposed by Whitehead and Russell (Nagel and Newman, 2001, xii) and Gödel (Nagel and Newman, 2001, xiii) and Kleene (2009) the contributions of metamathematics and the main conceptual notions of high-profile physicists such as Heisenberg (1990) and Einstein (1982).

The formalization of emerging knowledge can require the collaboration of many scientists over a considerable amount of time. In Section 2.4, the authors share their vision of four ideas to support the attempt to establish a formal system for geocybernetics. Proposed as driving forces in the effort to formalize geocybernetics are: conceptual languages, an analogy with the accepted metamathematics framework, scientific qualitative prose, and the exploration of visual language concepts. Finally, the authors comment on research topics that would facilitate the advancement of this avenue of investigation.

2.1.1 Informal Theory

Experience gained from many applications with different focuses (social, political, economic, and ecological, among others) shows us that a new spatial-temporal framework is needed to establish organizational forms that guide behaviour and permit the structural coupling of teams of transdisciplinary groups during different stages of design, modeling, and process development. Based on this new framework, the societal actors and teams of specialists make their knowledge about a territory explicit and a nontechnical, conceptual level of communications is established among those involved.

The spatial-conceptual framework (networks, spatial dissemination or regionalization, among others) has been adopted based on both qualitative and quantitative perspectives. Using the spatial dynamic of social and natural processes in the territory as a reference, the networks can be seen as a constantly changing structures in which local and global properties are observed. Thus, the network as a whole is a global property, whereas the average number of lines that cross at a vertex constitute a local property (Hofstadter, 1999, 371–372). In addition, spatial dissemination should be understood as the propagation of something (reproduction, pandemics, knowledge, waves or particles, among others) in space-time. In the case of regionalization, this should lead to a new approach that goes beyond the appropriation of space. This should involve the evolutionary regionalization of the territory, which takes into account the spatial dynamics resulting from social construction and the whims of nature, as well as other unforeseeable factors. It is important to consider the territory from a spatial-temporal, scientific perspective that leads us to an evolutionary regionalization that explains the story of the natural manner in which living beings describe how they maintain their livelihood in a dynamic environment by performing their activities in space-time (Kauffman, 2000, 104) (Hilhorst, 1976, 51). In summary, a regionalization is needed that tells us how living beings evolve in space-time to pursue their livelihood in a changing environment that is continually being constructed. Undoubtedly, this symbiosis between biology and geography enables building conceptual bridges, points of contact for conversations, and the intertwining of biological concepts such as adaptation and learning with geographic concepts about networks, spatial dissemination, and evolutionary regionalization, in which cumulative and propagative work and organization occurs in space-time (Kauffman, 2000, 104). Supporting these transdisciplinary concepts are those from physics, such as the case of intertwining in Maxwell's Demon, the second law of thermodynamics and Shannon's quantification of information (Mitchell, 2009, 54).

Concepts of cumulative and propagative work and organization proposed by Kauffman open new avenues to study the enormously complicated interaction among interwoven, recursive social, and natural phenomena. This tells us that space-time not only provides context but also plays a relevant or determinant role in the way in which social and natural phenomena are organized. Space-time falls within a propagative organization; it is part of the organization and cannot be separated from it. This leads us to the idea that this spatiotemporal analysis perspective provides us with concepts and terminology to build conceptual bridges or points of contact for conversation among the different fields of knowledge.

Nevertheless, as part of all these biological–geographical processes, we also find concepts from Wiener's science of cybernetics regarding control (positive and negative feedback) and communications (Wiener, 1954), in addition to complexity and chaos theory, such as dissipative structure (Prigogine, 1997, 66), autopoiesis (Margulis, and Sagan, 1995, 17), complex adaptive systems (Holland, 1995, 4), self-organization (Mitchell, 2009, 40), nonlinearity and attractors (Mainzer, 2007, 1–4), among others.

2.1.2 The Role of Knowledge Management

After more than two decades of empirical research, it can be seen that when seeking a solution based on societal demand, a battery of quantitative models is not necessarily sufficient, even though they provide a good way to describe objects and actions and are a powerful analytical tool. Experience tells us that in order to address societal problems, an entire working methodology is needed, or a strategy that takes into account the different aspects (knowledge, methodologies and procedures) that lead to a geocybernetic solution. Due to the complexity of their organization, we need new methods to describe the processes that interact in a territory (Minsky, 1988, 105). The Strabo method (Luscombe, 1986; Luscombe and Reyes, 2004) and the Reyes method (Lopez, 2011, 117–129) are among those that incorporate knowledge about the territory in the search for solutions to societal demands.

In the particular case of the Reyes method (which we will discuss below), in which geocybernetic solutions are designed through the modeling process, not only is explicit and formal scientific knowledge valuable, but the creativity and tacit knowledge of the societal claimant is also recognized. Figure 2.1 shows a model summarizing the criteria proposed by Reyes. On one side, we can see the scientific knowledge of the transdisciplinary team that is going to interact in the search for a solution, whose knowledge is essentially formal and explicit. On the other side is the societal actor or claimant, whose knowledge can be tacit and empirical (knowledge based on experience, practice, analogy, etc.). The knowledge of the societal claimant should be made explicit with the support of geocybernetic specialists, taking into account how the societal actor sees the territory, as well as the role of political, administrative, and operational circumstances. All the knowledge should be intertwined to create a common *knowledge base* that, in the form of concepts, synthesizes the knowledge of the societal

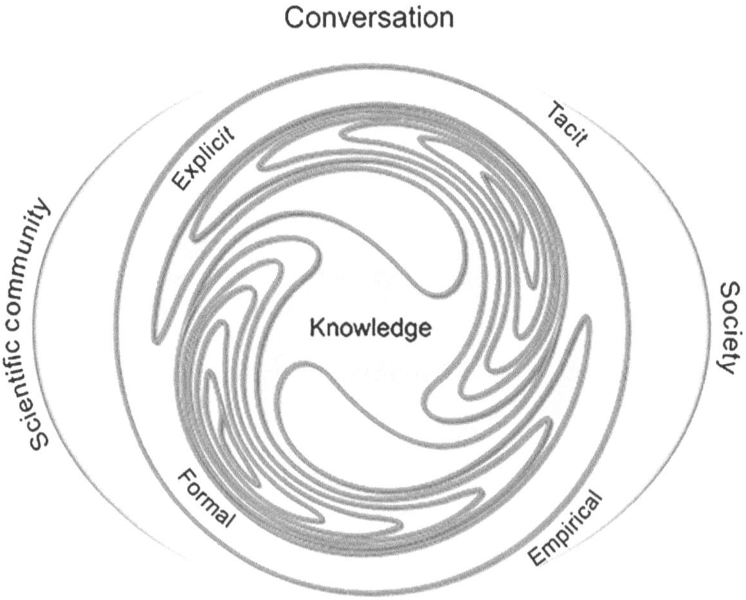

FIGURE 2.1 Knowledge based approach.

claimant and the specialists. These fundamental and concise concepts should serve as the knowledge base to trigger the construction of a network of stories describing the complex relationships involved in the social and natural processes that interact in territories with new models and new emergent concepts. This continues successively until attaining a level at which the concepts derived from the stories of the network become mathematical, physical, and statistical quantitative models or metaphors in the absence of deduction (Kauffman, 2000, 135). This semantic network, as a whole, is called an emergent knowledge network and should reflect the holistic view of the territory previously described. Nevertheless, this does not mean it is the result of deduction, but rather, as Kleene explains, the semantic network results from intuitive reasoning and its justification must be based on the conceptual meaning of preceding models and territorial evidence.

Societal claimants seek comprehensive solutions based on the guiding principles they establish, rather than a mere collection of partial solutions that do not lead to a coordinated, holistic, and complete solution. Although disciplinarity involves a diversity of concepts from different approaches or realms of knowledge, and even though these concepts have a scientific basis, it is not possible to expect to resolve a problem in a comprehensive manner with isolated concepts lacking logical unity. Rather, they need to be combined and adapted to the proposed problem (Polya, 1973, 157). To compensate for this deficiency, a methodology is needed that leads to establishing *a system of the greatest conceivable unity, and of the greatest poverty of concepts of the logical foundations, which are still compatible with the observations made by our senses* (Einstein, 1982, 294). What is critical is the objective of representing the multitude of concepts pertaining to the knowledge base, the most relevantly possible, fundamental concepts and relationships, which in turn can be freely chosen and can provide a transdisciplinary perspective. Einstein explains that *The liberty of choice, however, is of a special kind; it is not in any way similar to the liberty of a writer of fiction. Rather, it is similar to that of a man engaged in solving a well designed word puzzle. He may, it is true, propose any word as the solution; but, there is only one word which really solves the puzzle in all its parts* (Einstein, 1982, 294–295).

This knowledge base constitutes a metasynthesis of concepts and is different than ordinary concepts in that it contains the elements needed to resolve the societal claimant's problem. Therefore, it is possible to obtain a metasynthesis of concepts that enables performing deductive, heuristic, inductive, or analogical reasoning (Polya, 1973, 113), and which, at the same time, when intertwined with the accumulated experience of the territory, enables selecting the next concepts to be applied to obtain a geocybernetic solution of the problem. Thus, the resolution of conflicts or improvement of a strategy to resolve existing conflicts (among different disciplines) is proposed (Jackson, 1990, 147). While the above is a simple sentence, it has many implications, since it requires experience, creativity, intuition, imagination, and sensitivity in order to organize, manage, and apply what is learned and known (Minsky, 1988, 80). Not only do the experts in geocybernetics – based on the metasynthesis of concepts and knowing when and how to use their concepts, with the territory as their reference – broaden and enrich the knowledge base by constructing the stories of the networks and conglomerates of existing and emergent concepts, but this broadened base itself leads to a geocybernetic solution. The above forms are part of the essence of modeling based on knowledge (Reyes and Paras, 2012).

Let us briefly summarize. The metasynthesis of concepts, in which their potential to trigger each other is found, constitutes the model for the activity of the territory, albeit in the real universe or in alternative universes (Hofstadter, 1999, 360). This flexibility in modeling the activity of the territory based on the metasynthesis of concepts affects the resolution of conflicts and ambiguities by representing various models and their relationships with both the observers (from different disciplines) and the phenomena they represent (Heylighen and Joslyn, 2001). This helps us to get closer to understanding their structure, functioning, and behaviour. An analogy that illustrates the construction of an emergent knowledge network based on the metasynthesis of concepts is the generation of a well-designed crossword puzzle. Any story can be proposed as a solution, but there is only one family of stories that can actually solve the crossword puzzle in every way.

2.1.3 The Reyes Method and Knowledge Management

In essence, the Reyes method allows, through a collaborative knowledge-building process, the design of geocybernetic solutions. This method was developed by Carmen Reyes and is the result of over three decades as a researcher and of work as an advisor to different social organizations, companies, and governmental institutions in Mexico and Latin America. One can identify three main components of the method: methodological aspects, design processes, and the emergence of knowledge networks as shown in Figure 2.2.

From a methodological perspective, the Reyes method is a conceptual guide. And though it does not exactly say what has to be done, as in a recipe, it indicates what has to be considered, what has to be questioned first, what has to be reflected upon, and that is what helps the process to resolve the problem. That is, it focuses and directs the reasoning needed to solve the problem, to know *how* to do what is mentioned in books that discuss *what* but do not say *how*. Colloquially, one would wonder: How should the problem be dealt with? The Reyes method involves in the methodological component three conceptual approaches: qualitative research, requirement analysis, and a geographic model (Reyes, 2005, 88). While this distinction is made for academic understanding, in practice these are interwoven and it is sometimes difficult to know in which approach one is engaged, since the boundaries are blurred. Qualitative research focuses on processes to insert geocybernetic solutions into the environment and the requirement analysis component on the systemic functionality of the solution. Qualitative research considers the political, organizational, cultural, and social context in which a geocybernetic solution is to be inserted. This process must be conducted based on coupling the proposed solution with individual, community, organizational, and institutional processes. Qualitative research is employed in order to recreate, through conversations, how the societal claimant performs the institutional and organizational processes in his or her social and political environment, thereby formalizing his or her knowledge and, when

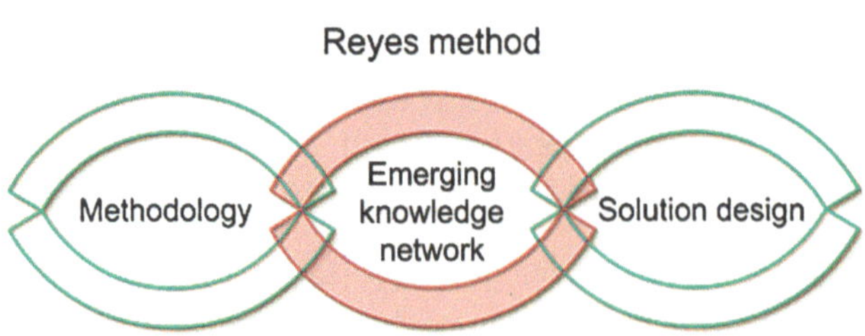

FIGURE 2.2 Components of the Reyes method.

applicable, making it explicit. The requirement analysis takes into account the strategies and functional models of the organization. Similar to qualitative research, the geomatics specialist must have conversations with the claimants and the users of the solution, and dialogue with them, resolving questions such as: what is it that you would like, exactly?, what satisfactory solutions are required?, among others.

Both analyses are usually complex and difficult and are not achieved in a single session but rather through a series of sessions. A key element of the process depends on the willingness of the interviewees (societal actors and claimants) to cooperate and on their detailed explanations of their models, in order to identify the ideas or concepts behind the monitoring (internal and external) of their vision of the territory or how they think about it, the politics of its management, their effectiveness or intentionality in decision-making about the territory, among others. When needed, geocybernetic specialists responsible for engaging in the conversations will have to guide the stakeholders in the search for their models and identify the key concepts underlying their different management and functional models.

Later, these institutional and organizational management models are presented to the other geocybernetic specialists so they can be structurally coupled with the societal actors.

When all the knowledge models – those of the societal actor as well as the specialist – have been made explicit, then the search for the metasynthesis of the concepts begins, initially named the geographic model.

While constructing the geographic model, conversation is crucial and includes cybernetics and its positive and negative feedback processes (Wiener, 1954, 33). In addition, the agents (societal actors, claimants and specialists) learn and adapt based on their experiences (Wiener, 1954, 48) and establish a propagative and cumulative organization of knowledge (Kauffman, 2000, 105). The participants interact through the recursive dynamics of conversation, by expressing and sharing their preferences. This is where arguments are made, such as why they think a particular option is better than another one. These arguments can convince others that the option presented is better, or bring out counter-arguments. In the best of cases, the arguments and counter-arguments presented will highlight the most important implications of the different options, combining the best of those previously considered. Through this process of recombining, a new option or brilliant idea may also emerge that represents an abrupt and significant transformation in the perspective of the agents and a sudden reorganization in the way in which the problem is conceived (Polya, 1973, 159). This argument is very similar to the metaphor of genetics which, as Holland (1995, 65) says, in a system based on rules, 'Strong rules represent knowledge won. Under competition, strong rules usually determine the agent's actions, so they are the core of the agent's internal model.'

Reyes considers 'conversation' the mechanism for resolving problems. Her method proposes various conditions and heuristics that can lead to the social structural coupling that produces a geographic model, resulting from the act of seeing with a more comprehensive perspective. This process can be seen as a game formulated as follows:

- Setting of the game:
  - Each of the specialists and societal actors and claimants has a knowledge model.
  - The dialogue or conversation among them should be focused on the territory.
  - The specialists and the actors must cooperate and have empathy.

- Heuristics of the Reyes method:
  - The tacit knowledge models of the actors and the specialists should be made explicit based on the territory.
  - Communication among everyone (specialists and actors) should be established based on a conceptual level as opposed to a technical one.

Normally, during these processes of social coupling we forget the role played by logic. Nevertheless, this is key to *formulate and organize the ideas* (Byers, 2007, 194) or concepts. *If we want to be able to communicate at all, we have to adopt some common base, and it pretty well has to include logic* (Hofstadter, 1999, 99). Nonetheless, the conditions and heuristics in the Reyes model are essential to coordinating the conversation process which, through interaction and negotiation, seeks to reach agreements that will enable designing a geographical model. The problem involved in the joint work of specialists and actors involves managing the interdependencies among the activities of the agents. Therefore, a mechanism is needed to coordinate the activities and interactions involved in the different participations of the agents during the negotiation process (Wooldridge, 2009, 162).

Wooldridge (2009, 315) explains the negotiations techniques traditionally used when seeking consensus, where any negotiations scenarios would, in general, include four different components: (1) a set of negotiations, (2) a protocol, (3) a set of strategies, and (4) a rule to determine when an agreement has been reached and what it establishes. In terms of the Reyes method, these four parameters lead to a very rich and complex environment for analysis. Negotiation is generally the product of a series of rounds of conversations, with a feedback process involved in each round. The proposals presented by the agents should be extracted from the set of negotiations and must be legal, according to that defined by the protocol. If an agreement is reached, according to that which is defined by the rules for agreement, the negotiations are concluded with the negotiated agreement.

Because the negotiations process is self-organizing, there is no agreement mechanism in the Reyes method to facilitate the rounds, during which each agent presents a proposal. In addition, there is no set of negotiations since the societal actor, based on his or her focus, defines how the territory should be viewed in the geocybernetic solution. In the protocol, the legal concepts are those that are made explicit by the societal actor through qualitative research and requirement analysis. In terms of the specialists, the concepts that enter into play are those that take into account the focus of the societal actor. The strategies and negotiations are part of the conversation mechanism through which, as mentioned previously, the agents adapt and learn throughout the entire process. This can therefore be modeled based on the genetics metaphor contained in Holland's (1995, 70) algorithm:

- Reproduction: Select the best established concepts (pertaining to the territory) from among those proposed by the specialists and societal actors. The more useful the concept (when the arguments are stronger due to the obtained knowledge) the more likely it will be chosen and even used several times.
- Recombination: The arguments or statements by the specialists or actors can convince others that their concept is the best or provoke counter-arguments or the development of new statements. In the best of cases, the most outstanding implications of the different concepts will be highlighted or new concepts, innovations, or brilliant ideas will emerge.

- Replacement: The specialists and actors analyse the results, reevaluating their individual concepts and maintaining them if their implications are satisfactory or adopting concepts of other specialists or new ones that emerge from this process. This recursive process produces a succession of geographic models and concludes when the specialists and the actors decide that the model contains the metasynthesis of the concepts that can provide an answer to the social demand.

From a theoretical perspective, the feedback process presented in each round can generate what is called a propagative and cumulative organization (Kauffman, 2000, 81) since there is a recombination of the information received with previously accumulated and stored information. This argument enables us to establish a relationship in which social and natural processes are intertwined, and cybernetics and space-time are an indissoluble part of its own organization. This further supports our fundamental concept that geocybernetics is also a window onto the world, allowing for the observation of the phenomenon of propagative work as well as propagative and cumulative organization, control and communication, learning and adaptation, self-organization, and evolution. The reflection of all this can be observed when applying the Reyes method.

2.1.4 Comments

The concepts, stories, metasynthesis of concepts, geocybernetics, conceptual frameworks, and semantic networks tell us that a theory is needed to organize and integrate all the knowledge constructed over a decade of investigation and to facilitate the emergence of new avenues of research. In this respect, mathematics has always served to support the formalization of scientific knowledge and, thus, by looking within its branches it was found that an appropriate theory could be established using formal systems as an initial approach. Byers (2007, 257) explains, *In mathematics and in science in general, logic is the organizational principle par excellence.* Later, Byers (p. 259) argues, *It is connected to the notions of rationality and reason. In fact, logic is the form in which reason appears in our scientific civilization.* We obviously need to be careful with these formal systems since they only involve deductive reasoning. Nevertheless, their abstract conceptual structure can be utilized, which enables organizing some of the complexity of human reasoning. But we can also broaden the vision by using other reasoning such as creativity, intuition, and analogy, in addition to deductive reasoning. Enderton (2001, xii) explains that deductive reasoning includes not only classical logic but also other models, such as many-valued logic, modal logic, and intuitionistic logic, that represent different selections of the properties of real-life deductions. Also included in mathematics is heuristic (inductive or analogical) reasoning (Polya, 1973, 113) as well as practical, and means and ends reasoning, where desires, beliefs, intentions, empathies, plans, etc. are found (Wooldridge, 2009, 69–75).

2.2 A METATHEORY FOR GEOCYBERNETICS

While each science develops and works with its own concepts, it is not so easy to define what a concept is. Heisenberg (1990, 71) explains that *the essence of abstraction consists in singling out one feature, which, in contrast to all other properties, is consider to be particularly important in this connection*. Therefore, all formation of concepts depends on this abstraction process and presupposes the abilities to recognize similarities. In practice, total similarity among phenomena is never produced. Similarity is only suggested through the process of abstraction, through the singularization of one characteristic while setting aside all others. In addition, Einstein (1982, 292) says, *The connection of the elementary concepts of everyday thinking with complexes of sense experiences can only be comprehended intuitively and it is unadaptable to scientifically logical fixation. The totality of these connections – none of which is expressible in notional terms – is the only thing which differentiates the great building which is science from a logical but empty scheme of concepts.*

Kleene (2009, 51) writes: *Intuition is nothing other than the faculty of considering separately particular concepts and inferences which occur regularly in ordinary thinking*. Thus, the essence of abstraction consists of finding the conceptual units that are behind those ordinary thoughts. According to Kleene (2009, 62), *Metamathematics includes the description or definition of formal systems as well as the investigation of properties of formal systems. In dealing with a particular formal system, we may call the system the object theory, and the metamathematics relating to it its metatheory*. From this perspective, geocybernetics is the object theory, and the metatheory certainly includes metamathematics. Whether there are other elements of the formal systems missing in the metatheory is still an open question. Furthermore, Kleene (2009, 60–61) tells us that we use Euclid's axiomatic-deductive method and Aristotle's formal treatment of logic when formalizing a mathematical theory. Instead of using a natural language, a symbolic language is constructed in order to express the theory, since the words of the natural language are too cumbersome and ambiguous to be adequate. As Kleene explains: *The symbols in a symbolic language will usually correspond to whole words instead of to letters; and sequences of symbols which correspond to sentences will be called 'formulas'*.

2.3 FORMALIZING GEOCYBERNETICS: AN ATTEMPT

Logic as a metatheory and formal system is the backbone of scientific thought, even though the paradigm, which reached its pinnacle with the work of Bertrand Russell and Alfred North Whitehead in *Principia Mathematica* – a belief that all mathematics could be grounded in pure logic (Hofstadter, viii) – was changed later by the work of Gödel. In this section, the authors propose several avenues to explore in an attempt to advance the development of a formal system for geocybernetics. First, the notion of geocybernetics as a conceptual language is explored. Then, a metamathematical way of thinking is considered and, using analogical reasoning, a parallelism is established with geocybernetics. Third, a broader approach to scientific language is adopted, identifying what is named 'qualitative scientific prose' and, finally, some ideas regarding the role of visual language in geocybernetics are discussed.

2.3.1 Geocybernetics as a Conceptual Language

Since we are dealing essentially with knowledge theory and knowledge management in the discourse about geocybernetics, ideas involving 'chunk' concepts (as explained in

Section 2.4), the metasynthesis of concepts, cognitive frameworks, and models are elements of the geocybernetics language. Mendelson (1979, 29) defines a formal theory that can be used to advance the identification of the key elements of the geocybernetic conceptual language. For example, we can (1) adopt the notion of 'chunk' concepts, (2) recognize networks and clusters of 'chunk' concepts as the backbone of the stories, (3) consider the metasynthesis of concepts to be the driving force to establish emergent knowledge networks, and (4) consider the emergent knowledge network to be a contextual restriction for the developing stories.

Regarding the rules of inference, it was mentioned that the construction of stories should take into account the agreements and heuristics proposed by Reyes, the rules of inference used by experts in their different reasoning disciplines and the establishment of consensus or agreement, when required. Thus, a story should be a consequence of the meaning of the metasynthesis of concepts and preceding stories, by virtue of the rules of inference. In addition, each inference should be proven in accordance with the contextual evidence in the territory (Kleene, 2009, 62).

2.3.2 An Analogy with the Language of Metamathematics

In our case, besides using mathematics as a symbolic language, a conceptual language is adopted in order to advance the formalization of geocybernetics, as mentioned in Section 2.4.1. Therefore, analogous to the formal system in metamathematics, formal symbols are identified as the 'chunk' concepts, the formulas are the stories related to societal processes, the axioms are the assumed (adopted) emergent knowledge network, and the formation rules of formulas are based on heuristics, social agreements, and empathies, among others.

Geocybernetic reasoning encompasses deductive reasoning as well as others that have been present in artificial intelligence and in science in general, such as heuristical, intuitive, analogical, empirical, and intentional reasoning. The network and clusters of 'chunk' concepts allow for the construction of stories (i.e. formulas) which are constrained by the semantic network (axiomatic assumptions).

An explicit example, based on a very fruitful empirical background (Reyes and Paras, 2012; Lopez, 2011, 16), can facilitate understanding the use of an analogy that points to a mathematical way of thinking in the development of geocybernetics. In the Cybercartographic Atlas of the Lake of Chapala (Figure 2.3), the 'chunk' concepts are found in the menus of the computer artifact: history, spatial dynamics of the basin, the lake and its surroundings, environmental issues and management aspects, among others. The menus themselves are clusters of 'chunk' concepts and the Atlas is conceived as a network of 'chunk' concepts. In this case, one can identify three main knowledge frameworks in the semantic network: general systems theory, territorial planning theory, and the landscape ecology model.

As mentioned by Lopez (2011) and Trujillo (2009), the stories (formulas) told using the Atlas can be identified by the cognitive environment of each storyteller and the societal context in which it is told.

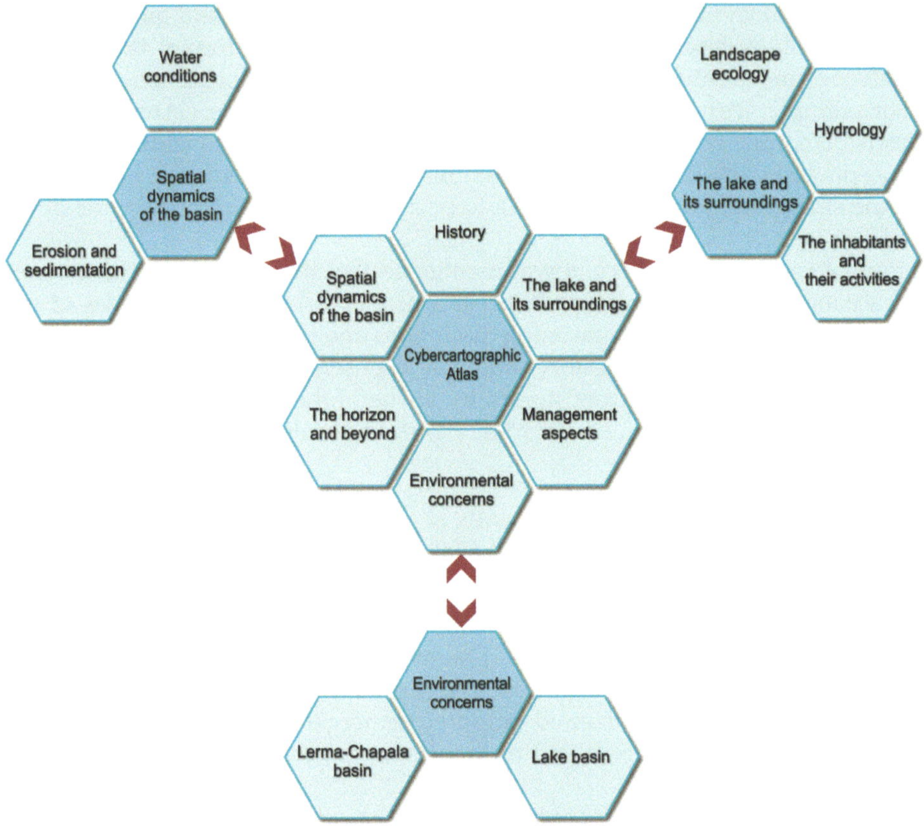

FIGURE 2.3 'Chunk' concepts of the Lake of Chapala artifact.

2.4 QUALITATIVE SCIENTIFIC PROSE AS A FORMAL ELEMENT

Natural languages have always been used in the advancement of science, and in most knowledge domains, qualitative prose has been found to have made substantial and relevant contributions. As mentioned in Reyes (2005, 74), geographers have used natural language for centuries as a means to communicate and represent spatial knowledge. As in cybercartography, in geocybernetics the use of natural language includes both written and verbal expressions. What role does qualitative scientific prose play in geocybernetics?

In the face-to-face transmission of mathematical knowledge, teachers usually introduce topics verbally and using symbolic mathematical language. In a similar manner, in mathematical books, written language is intertwined with the symbolic one. As mentioned before, in geocybernetics, all these good scientific practices are embraced; additionally, certain elements that are the key to its processes are emphasized. For example, the 'conversation' that occurs using the Reyes method to build a conceptual metasynthesis mainly involves verbal and visual languages. Both, the content of the prose and its expression become essential. While over the years we have identified the importance of the 'storyteller' (Trujillo, 2009) in cybercartography, when other components are incorporated in geocybernetics – such as explicit quantitative and qualitative modeling (i.e. simulation, decision trees, prospective, expert systems, etc.) – a 'tailor-made' qualitative prose has to be introduced. Therefore, scientific qualitative prose becomes a key formal component, since the paths of the reasoning used in geocybernetics contain transdisciplinary bridges (implicit and explicit) that connect different knowledge domains (Lopez and Muñoz, 2012).

A brief example could aid us in introducing the need to further explore and formalize qualitative prose in geocybernetics. One of the main issues in a state-level project in Mexico is the design and implementation of a cadastral system. While geomatics has a long tradition addressing these types of tasks, a land inventory alone is seldom of interest to public policy teams unless the territory is an explicit component of their political strategies. As explained in Lopez and Muñoz (2012), it becomes necessary to identify the different knowledge domains and transdisciplinary bridges involved. From an empirical perspective, these challenges have been overcome by the research team on dozens of occasions. The conversation with societal actors advanced properly with the use of verbal prose that incorporate 'chunk' concepts such as territorial public policy, territorial information management, cadastral information systems, and geospatial data infrastructures. In order for the collaborative knowledge network to emerge, the process included conversations using the right language and qualitative prose. Certainly at some point, the actors can interact using more conventional language, for example, tools such as geospatial data structures for the design of geocomputing solutions.

In this regard, several questions remain: Can we differentiate geocybernetic prose from that used in other disciplines? How can someone learn to use qualitative prose as part of the Reyes method? How is this qualitative prose related to the cognitive domains of the disciplines involved? Are there elements of language formalization that are worth studying?

2.5 THE ROLE OF VISUAL LANGUAGE IN GEOCYBERNETICS

In the well-known words of Horace:

> A feebler impress through the ear is made, Than what is by the faithful eye conveyed ***(Byrne, viii)***

Surely we all remember our lessons in mathematics where the teacher uses prose, drawings and formal symbology to explain not only geometry but mathematical analysis, topology, differential equations, arithmetic and so on. In the tradition of science, visual language plays a key role. In this regard, for our purpose, we mention the work of Oliver Byrne (1847), who created a visual language with the intent of facilitating the teachings of Euclidian Geometry. In our case, visual languages are essential to both empirical work in cybercartography and the Reyes method.

Visual languages have been developed for centuries in the academic context. In cybercartography, Reyes et al. (2006) argue: *Cybercartography uses a metalanguage: a holistic language formed by a variety of other languages. In the design of a cybercartographic atlas, the language that most effectively expresses a process, a concept or a problem is selected and used to generate messages to be transmitted through different media. However, the main elements of this metalanguage of cybercartography including formal codes, semantics, and syntax are yet to be developed.*

Cybercartography is a metalanguage that brings us closer to society. It is what we need to connect science and society. More specifically, through this metalanguage, messages are transmitted and feedback processes are achieved, enabling societal actors to take advantage of what cybercartography offers in terms of their own needs.

Cybercartography can be viewed as both a conceptual and visual language. As Reyes (2005, 74–75) mentions: *There is an explicit bridge between spatial representation and mathematics. The realms of geometry in its different expressions (Euclidean, projective, hyperbolic, etc.) as well as topology (algebraic and surface) are examples of ways of encoding and communicating spatial information and knowledge through a symbolic language. In cybercartography, communication takes place using diverse resources including cartographic, linguistic, mathematical, statistical, musical and visual languages.*

One of the main characteristics of the Reyes method is the strong emphasis on the conceptual design of geocybernetic solutions to respond to societal actors. Looking at the empirical work of the last two decades, one can clearly identify the design process that has been supported and expressed through visual language. Visual language usually plays two roles: first, it enables the expert designer to establish a systemic order to the ideas derived from the conversations; and second, it is used as a communication and conversation tool to convince both scientific and societal stakeholders to invest in the proposed solutions. As an exercise to initiate a systematic analysis of the empirical geocybernetic visual language, the authors are exploring, in a backward, step-by-step fashion, the key visual elements used by past projects.

In cybercartography, our point of departure is probably the most robust visual language – cartography itself. Although as a metalanguage there are still open questions to be explored, empirically, cybercartography has been the cornerstone in geocybernetics. Nevertheless, in the process of applying the Reyes method, the team has had to create its own visual language, beyond cartography, in order to create theoretical transdisciplinary bridges as well as to have effective conversations with societal actors.

In terms of design, the extensive work of George Stiny (2006) came to our attention. Several of the main ideas he poses have also been present in the course of our research. These include the need to use a scheme that *includes both visual and verbal expression with shapes and symbols, respectively*, his acknowledgement of a broad understanding of design (Stiny, 2006, 7–8) and his emphasis on designing as a practical way of reasoning, in which seeing is key.

Last but not least, design is in the essence of geocybernetics. It can be explicitly found in the Reyes method, for example, where the team discovered that visual language was a key to the conversation process with societal actors. Just as Stiny states that *metaphors are good heuristics* (Stiny, 2006, 311) we state that design is an essential element in geocybernetics. Stiny compares design to calculating, and in our case, the question of whether geocybernetic design is a particular case of Stiny's shape grammars still has to be explored.

2.6 FINAL COMMENTS

As mentioned in the introduction of this chapter, our main aim is to attempt to formalize the empirical and informal knowledge that has emerged from research in geocybernetics over at least the past three decades. Why is it important to take this step? The authors are aware that in scientific work, formalization processes facilitate the communication of findings and allow for showing the consistency of theoretical frameworks.

Since the scientific management model explicitly and fully adopted by the research team over the last 13 years emphasizes interaction with societal actors as a key element in the process, many of the results obtained are immersed in contextual environments with their own cultural and ideological characteristics. It has become apparent that communicating results with societal actors who have similar contexts is much easier than with those whose contexts more greatly differ. As an example, we could mention workshops held during 2012 in which the Reyes method was recently applied in the State of Puebla and with several government offices in the Philippines. The messages were very well understood by both audiences, collaborative knowledge processes were triggered, and the results turned out to be an excellent platform for further collaboration.

A more formal framework that refers to well-established scientific ideas and uses analogical reasoning has allowed the team to identify at least four lines of research that could aid in this attempt to formalize geocybernetics.

Even though analogical reasoning was not explicitly and fully acknowledged in the twentieth century – when the formalization of mathematics advanced quite significantly – and rational, deductive reasoning was taken as the point of departure for initial developments in metamathematics, analogical reasoning was present and certainly has had a significant impact in the development of science. In fact, several of us clearly identify having solved mathematical problems by analogy when we were very young. In this respect, we were happy to find the following thoughts of Minsky (2006) regarding analogical reasoning.

> This kind of thinking is important because it helps us to deal with new situations – and in fact, that is almost always the case, because no two situations are ever the same, and this means that we're always making analogies. For example, if the problem that you are facing now reminds you of one that you solved in the past, then you may be able to use that knowledge to solve your problem by using a procedure like this: The problem that I am working on reminds me of a similar one that I solved in the past, but the method that was successful then does not quite work on the problem that I am facing now. However, if I can describe the differences between that old problem and this new one, those differences might help me to change that old method so that it will work for me now.

The team is very enthusiastic about this research project and we can identify several benefits provided by its outcome: the opportunity to pose relevant questions from the perspective of a formal system that is better-recognized in the current scientific mainstream; the possibility of communicating the findings and results of the research team to the scientific community around the world; sharing findings and results with societal actors interested in solving problems similar to those with which we have had success; and finally, building new paths among the so-called hard sciences and the social sciences by using the holistic, collaborative and transdisciplinary approach proposed by geocybernetics.

Acknowledgements

The original vision of D. R. Fraser Taylor opened the avenue of research of cybercartography. The team of authors has explored and promoted a new paradigm, geocybernetics. Each one has contributed substantially from its own scientific background in a transdisciplinary and collaborative manner. Special thanks to Rafael García for his creative contribution in the design of the figures, to Jorge Dávila for his comments on the text and to Tracey P. Lauriault for aiding in the revision of the chapter.

References

Byers, W., 2007. How Mathematicians Think. Princeton University Press, Princeton, New Jersey.

Byrne, O., 1847. The First Six Books of the Elements of Euclid in Which Coloured Diagrams and Symbols Are Used Instead of Letters for the Greater Ease of Learners. W. Pickering, London Facsimile of the first edition, Taschen: Los Angeles.

Enderton, H., 2001. A Mathematical Introduction to Logic. Harcourt/Academic Press, Los Angeles, CA.

Einstein, A., 1982. Ideas and Opinions. Three Rivers Press, New York, NY.

Heisenberg, W., 1990. Across the Frontiers. Ox Bow Press, Woodbridge, CT.

Heylighen, F., Joslyn, C., 2001. Cybernetics and second-order cybernetics. In: Meyers, R.A. (Ed.), Encyclopedia of Physical Science & Technology, third ed. Academic Press, New York.

Hilhorst, J., 1976. La Teoría del Desarrollo Regional: Un Intento de Síntesis. Ensayos sobre planificación regional del desarrollo, ILPES Instituto Latinoamericano de Planificación Regional Económica y Social, Mexico, pp. 51–69 Siglo XXI Eds.

Hofstadter, D., 1999. Gödel, Escher, Bach: an Eternal Golden Braid. Basic Books. Inc, New York.

Holland, J., 1995. Hidden Order: How Adaptation Builds Complexity. Basic Books. Inc, New York.

Jackson, P., 1990. Introduction to Expert Systems, second ed. Addison Wesley, Reading, Mass.

Kauffman, S., 2000. Investigations. Oxford University Press, New York.

Kleene, S., 2009. Introduction to Metamathematics. Ishi Press International, New York.

López, F., 2011. Un Aporte Teórico: El Prototipo Geomático. Tesis de Doctorado en Geomática, Centro de Investigación en Geografía y Geomática Ing. J.L. Tamayo, México City.

López, A., Muñoz, E., 2012. Aportaciones desde la geocibernética y la percepción remota a la política pública de áreas verdes urbanas. GEOcibernética: I+G+S (Año 1, Número 1). Available at: www.geocybernetics.org.

Luscombe, B.W., Reyes, C., 2004. Building consensus in environmental decision making: a methodology integrating GIS tools and structured communication. Paper Read at the 29th Annual Conference of the National Association of Environmental Professionals (NAEP), 25–28 April Portland, OR.

Luscombe, B.W., 1986. Spatial Data Handling in Data-Poor Environments. PhD dissertation. Simon Fraser University, Burnaby, BC, Canada.

Margulis, L., Sagan, D., 1995. What Is Life? University of California Press, Berkeley & Los Angeles, CA.

Mainzer, K., 2007. Thinking in Complexity: The Computational Dynamics of Matter, Mind, and Mankind. Springer, New York.

Mendelson, E., 1979. Introduction to Mathematical Logic. D. Van Nostrand Co, New York, NY.

Minsky, M., 2006. The Emotion Machine: Commonsense Thinking, Artificial Intelligence, and the Future of the Human Mind. Simon & Schuster, New York.

Minsky, M., 1988. The Society of Mind. Simon & Schuster, New York.

Mitchell, M., 2009. Complexity: A Guided Tour. Oxford University Press, New York.

Nagel, E., Newman, J.R., 2001. Gödel's Proof. New York University Press, New York and London.

Polya, G., 1973. How to Solve It: A New Aspect of Mathematical Method. Princeton University Press, Princeton, NJ.

Prigogine, I., 1997. The End of Certainty: Time, Chaos, and the New Laws of Nature. Free Press, New York.

Reyes, M.C., 2005. Cybercartography from a modeling perspective. In: Taylor, D.R.F. (Ed.), Cybercartography: Theory and Practice, Elsevier, Amsterdam.

Reyes, M.C., Taylor, D.R.F., Martínez, E., López Caloca, F., 2006. Geo-cybernetics: a new avenue of research in geomatics? Cartographica Special Issue on Cybercartography 41/1, 7–20.

Reyes, M.C., Parás, M., 2012. Geocybernetics: a pathway from empiricism to cognitive frameworks. GEOcibernética: I+G+S (Año 1, Número 1). Available at: www.geocybernetics.org.

Stiny, G., 2006. Shape: Talking About Seeing and Doing. MIT Press, Cambridge, MA.

Taylor, D.R.F., 2003. The concept of cybercartography. In: Peterson, M.J. (Ed.), Maps and the Internet, Elsevier, Amsterdam.

Trujillo, A.J., 2009. Cibercartografia Web: Hacia la construccion de un modelo colaborativo de conocimiento cibercartografico. Tesis de Maestría en Geomática, Centro de Investigación en Geografía y Geomática Ing. J. L. Tamayo, México City.

Wiener, N., 1954. The Human Use of Human Beings: Cybernetics and Society. Da Capo Press, Lexington, KY.

Wooldridge, M., 2009. An Introduction to Multiagent Systems. John Wiley & Sons, Ltd, Chichester, West Sussex, UK.

CHAPTER

3

Geocybernetics as a Tool for the Development of Transdisciplinary Frameworks

María del Carmen Reyes, Alejandra A. López-Caloca, Fernando López-Caloca, Rodolfo Sánchez-Sandoval

Centro de Investigación en Geografía y Geomática Ing. J. L. Tamayo A. C. (CentroGeo), México

OUTLINE

3.1 INTRODUCTION

A large part of the initiatives involved in geocybernetic investigations have focused on studying the building of transdisciplinary bridges between the empirical and knowledge domains. This also applies to the creation of several cybercartographic atlases described in later chapters in this volume. The creation and meaning of these bridges emerged from the demands of society and cutting-edge science. This chapter contributes to explore the process

Developments in the Theory and Practice of Cybercartography, Second Edition, ISSN 1363-0814
http://dx.doi.org/10.1016/B978-0-444-62713-1.00003-9

of using analogical reasoning to build transdisciplinary bridges between the boundaries of different knowledge domains. The geocybernetic approach is transdisciplinary and creates conceptual metasynthesis (see Chapter 2) in the process to construct geoknowledge and communication in relationship and interaction with society.

A transdisciplinary approach closes various discipline gaps in the construction of knowledge through communications processes. Communication creates a transverse axis on which actors travel as they provide their explicit knowledge frameworks for the evolution of ideas and actions that respond to societal problems.

Reasoning is a key to resolving conflicts at the level of 'chunk' concepts (see Chapter 2) in the modeling of a geocybernetic solution. For this, we use different ways of thinking, such as logic, which plays a role, but is not the only element. As (Minsky 2006, 33) explains, *The concept of rational thinking is incomplete – because logic can only help us to draw conclusions from the assumptions that we happen to make – but logic, alone, says nothing about which assumptions we ought to make*. Therefore, not all themes are equally capable of being formalized – such as medicine, economy, politics, management, among others – even though since the last century they have used brilliant mathematics and technology. This is because some themes are more dependent on a knowledge base acquired through in-depth experience (expertise: to know what one knows and to know how and when to use it (Jackson, 1990, 10)) and progressive adaptation to the environment (see Reyes method, Chapter 2). Thus, other reasoning methods are necessary to address these sciences and disciplines, one of which can come from finding useful analogies (Minsky, 2006, 33), that is, analogical reasoning.

This chapter explores the role of transdisciplinary and analogical reasoning in geocybernetics. It presents a transdisciplinary empirical and theoretical exercise in geocybernetics, points out some methodological aspects, and provides final comments from the research team.

3.2 THE ROLE OF TRANSDISCIPLINARY AND ANALOGICAL REASONING IN GEOCYBERNETICS

After over two decades of developing diverse applications in geocybernetics, it can be stated that societal actors and claimants as well as specialists from different disciplines who participate in a transdisciplinary manner in the process of modeling the territory at a conceptual level do not use simple concepts such as those commonly used to describe objects or landmarks. Rather, they use 'chunk' concepts such as education, employment, social security, health, pollution, transportation, management, among others, to attempt to describe, through semantic networks (see Chapter 2), the evolutionary dynamic of the territory (Minsky, 2006, 2).

The use of these 'chunk' concepts by specialists and societal actors enables them, in each of their disciplines, to single out the concepts and inferences that occur in ordinary thinking (Kleene, 2009, 51). These simplifications reduce situations to their conceptual structure and assume their abstract essence by summarizing, in the form of 'chunk' concepts a number of processes which, at the lowest level, are reduced to conceptual units. This allows a focus on what is essential and to understand and model social and natural processes that interact in the territory.

A characteristic of a transdisciplinary approach is that each discipline as well as the societal actor can have different points of view and perceptions about the territory. Thus, all the 'chunk' concepts can assume different meanings and appear to be unarticulated with no points of contact, even though they sometimes use the same names. Some of the challenges of transdisciplinarity were described by Lauriault and Taylor in the first edition of this book (Lauriault and Taylor, 2005). In addition, transdisciplinarity needs to be treated as a social process requiring conversation or dialogue, in which cooperation, coordination, and negotiation can be established, as well as consensus, agreements, and commitments in decision making and the resolution of conflicts. In geocybernetics, it has been possible to overcome this difficulty by using the Reyes method, as explained in Chapter 2.

In a transdisciplinary approach, analogies can help to establish points of contact, or conceptual bridges, in the gaps among specialists from different disciplines and the societal actors. In a territorial context in which the complexity of relationships and functionalities appear, analogies could trigger the key ideas for the design of geocybernetic solutions through a natural mapping as Hofstadter (2008, 150) explains:

> ... a remark made with the aim of talking about situation A can also implicitly apply to situation B, even if there was no intention of talking about B and B was never mentioned at all. All it takes is that there be an easy analogy – an unforced mapping that reveals both situations to have essentially the same central structure or conceptual core – and then the extra meaning is there to be read, whether one chooses to read it or not.

The experience of over two decades of applying the Reyes method can be supported by Hofstadter's argument (2008, 153) ... *analogy has force in proportion to its precision and its visibility*. The territory comes into play regarding precision and evidence by restricting, coupling, and organizing the gaps presented among disciplines and societal actors (López, 2011, 147). For example, a public policy maker can identify by analogy between his/her 'chunk' concepts and those of the specialists key ideas to design the geocybernetic solution. This is possible because it is within the territory that we find the space-time in which we perform our daily activities, and we pursue our livelihood economically, socially, politically, and culturally. Thus, space-time models provide a good perspective and evidence by indicating what, where, and when, we can perform these activities. On the other hand, the georeferenced data in the computational models that store and organize geospatial information (cartography, spatial maps, satellite images, etc.) provide precision. Therefore, these spatial-temporal models and georeferenced data can enable reasoning through analogy to reach, in a certain way, a good level of 'mathematical precision' (Polya, 1973, 37–38).

3.2.1 The Role of Analogical Reasoning in Geocybernetics

After more than a decade of presenting geocybernetic solutions to societal actors, it became evident that they are able to discern how the Reyes method offers successful societal actions such as territorial public policies, political interactions, or management models. Even though the solutions presented do not always respond to the specific problems that the actors had in mind in the initial conversations. As a result, for years, the research team have been asking themselves: why have these empirical exercises been so successful? why the participation of the societal actors allowed the emergence of the conceptual metasynthesis? It is clear that a natural map or transition phase, as described by Hofstadter is in the essence of the Reyes method. Nonetheless, the underlying question, what is the structure or conceptual nucleus that establishes the natural map or transition phase, so that the analogy can be established? The answer to this question is explored in the following sections.

3.3 A TRANSDISCIPLINARY EXERCISE WITHIN GEOCYBERNETICS

In this section, the role of transdisciplinary reasoning in geocybernetics is further explored by pointing out explicit bridges among disciplines and outlining the paths that have allowed for the design and implementation of geocybernetic solutions.

To understand the transdisciplinary processes that have been conducted in geocybernetic exercises involving interaction with society, a metacognitive ('thinking about thinking') strategy is shown in Figure 3.1, as an example. This is a conceptual guide and considers different levels, as described below.

It is important to mention that the point of departure is an initial effort to make explicit the transdisciplinary processes in geocybernetics (López and Muñoz, 2012), which was based on an empirical exercise undertaken by the research team.

3.3.1 Society

We acknowledge that society as a whole needs knowledge and geospatial information to support a wide variety of activities that include, among others, better articulation of public policies, improvement of ecosystemic or environmental services that provide support, regulation, cultural provision, and understanding and natural conditions and processes, as well

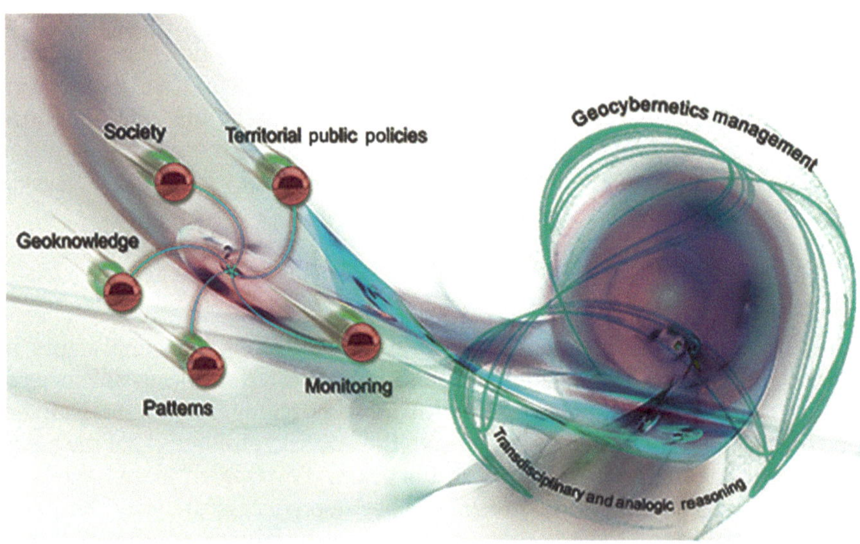

FIGURE 3.1 The role of geocybernetic solutions immersed in a transdisciplinary model.

as activities that address problems such as poverty, land use and ownership, sustainability, deforestation, food, health, sustainable food, public safety, risks and vulnerability, business investments, etc. Societal demands can be viewed as a transversal axis in the process.

3.3.2 Territorial Public Policies

Public policy expresses the goals, decisions, and actions adopted by a government for political, social, and economic management. There is an increasing need for policies to be formulated by establishing close connections with academics, the responsible politicians, and society. In the process to develop and implement public policies with a scientific approach, Sabatier, 1988 has focused his studies on the Advocacy Coalition Framework and Policy-Oriented Learning. In addition, approaching the territorial dimension from a theoretical framework has been a key element in providing geocybernetic solutions for public policies in over 60 projects developed based on this schema.

3.3.3 Geocybernetic Management

Geocybernetics management establishes strategies developed to resolve problems in the territory, in which the geospatial contexts of societal processes are identified. The development and the proposed solutions are based on the consideration of conceptual elements through communication, conversation, spatial language, and the construction of knowledge models, as well as transdisciplinarity. This process of development using the Reyes method, in which a semantic network (Chapter 2) emerges, shows the conceptual metasynthesis underlying the processes derived from the society and, in this case, territorial public policies.

'Chunk' concepts emerged during the conversation process among the specialists and societal actors, such as territorial, public policies, and space-based planning (inventories, diagnostics, prospective and management). The significance of each of these 'chunk' concepts is not necessarily the same for all participants. They could take on different meanings not only because of the different academic backgrounds and experiences but also because of the current intentions behind the project they have undertaken.

At one point, a conceptual metasynthesis emerges. One or more public policy makers expressed interest in the dynamics of the territory in order to comply with their constituents in the management of natural resources. One of the specialists showed examples of the possibility of using visual language to follow temporal-spatial changes in the study region. For the specialist in satellite image processing, it became clear that proposing a monitoring plan relevant to the phenomena of interest to the societal actor was useful.

The transdisciplinary bridge in the form of a conceptual metasynthesis also emerged. From then on, the conversations became much clearer and the scope, objectives, and goals of the collaborative project were much easier to establish and agree upon. 'Monitoring' as a conceptual metasynthesis became a key word in the prose of the conversations among the specialists and stakeholders, as well as in the contracts. This is a very significant step since one always wants to make sure that the results satisfy the claimant, albeit for financial reasons, prestige, or academic interests.

As mentioned in Chapter 2, three of the most relevant languages used by geocybernetics played a role in the above-mentioned process: conceptual, visual, and qualitative prose.

As shown in Figure 3.1, the geocybernetic management behaves as an attractor, and the transdisciplinary and analogic reasoning of the actors and specialists occurs in this process. The societal actor had to make explicit the similarity between his/her purposes and the 'chunk' concepts presented by the specialists through the conversations, and the specialist had to deal with at least two different semantics due to the different type of languages adopted.

3.3.4 Monitoring

An investigative strategy is proposed that incorporates the key concepts in geocybernetics, such as control (regulation), monitoring (supervision), spatial-temporal variability, feedback and learning, all of which are concentrated and converge in geoknowledge.

Control and monitoring are reciprocally connected in geocybernetics. They facilitate self-regulation during processes and events as well as establish strengths and opportunities, determine weaknesses in the territory and project scenarios. Monitoring or follow-up processes are projected in time and used to evaluate the current success of goals and the degree to which they have been met. They also define trends and guide the ongoing activities. Monitoring integrates concepts of space and time, spatial dynamics, and change.

To illustrate the concept of change, Galton (2000, 207) mentions that:

> The things that go on in time can be broadly classified as states, process, and events, and approximate definitions of these are as follows: a state is characterized by absence of change, a process is characterized by change as it is actually occurring; an event is a chunk of change picked out as an individual from the ongoing flux (is an episode of change with a beginning and end, considered as a completed whole).

Monitoring makes it possible to experience feedback during the event or process, generate future expectations about solutions, analyse the degree of understanding of the problem, analyse the connectivity between both new and previous informations as well as with the other events, understand the observed resulting patterns to establish standards with which to compare future patterns, and outline tasks, goals, and new strategies or actions.

Monitoring involves two visions of the territory. One is the management (global or macro) vision, based on which the individual contributions of human agents can be observed, with their activities at the macro-state level of society, representing the emergence of the cultural, political, economic, and ecological order, among others. This order strongly influences the individuals in a society, guiding their activities and activating and deactivating their attitudes and abilities. This type of feedback is typical of dynamic complex systems (Mainzer, 2007, 424). The other vision is operational, which involves specific and timely problems in the territory as a response to the management process.

Spatial dynamics is a concept that is increasingly emerging in solutions to problems presented to researchers by societal actors. This involves the spatial-temporal vision, which is related to how certain institutions, companies and organizations, among others, observe, model, and organize by not only monitoring the external world but also by monitoring their own needs and internal processes which, in turn, reflect the external situation around them. As a result, their policies and operational and management models should be kept up-to-date through feedback cycles, depending on the spatial dynamics of their social processes as well as internal and external natural processes that interact in space-time in such a way that, in the end, they reflect the external world (Mainzer, 2007, 310; Hofstadter, 1999, 324).

In this case, that which is external not only involves contiguity but also functional connectivity. Understanding small fluctuations of change is valuable to the design of solutions to problems posed by society. For example, knowledge of the danger of occurrence and recurrence for exposed systems and social groups enables identifying phenomena involving risk and vulnerability and utilizing the capacities to engage in planning and decision making. For example, in the case of land use and ground-cover changes, tasks related to follow-up, surveillance, and modeling help to identify the driving forces of these processes, in order to address the impacts of changes in forest cover, land degradation, intensification of agriculture, etc. The projection of future changes requires understanding and integrating trends occurring in these processes through time-series analyses of land change processes and feedback, as well as incorporating assumed scenarios.

3.3.5 Patterns

There is a need for geospatial information in the development of projects for society. This includes tasks that range from obtaining field data to the generation of products, services, and information systems. The observation of the state of the phenomena being posed serves as an initial diagnostic and, therefore, the identification of geospatial patterns in the place being observed provides an understanding of that place in order to better model its functional and structural behaviour. Patterns are a valuable tool to describe and reuse empirical knowledge. Byers (2007, 200) refers to the notion of patterns or regularities, as Keith Devlin explains:

> What the mathematician does is examine abstract "patterns"—numerical patterns, patterns of shape, patterns of motion, patterns of behavior, voting patterns in a population, patterns of repeating chance events, and so on. Those patterns can be either real or imagined, visual or mental, static or dynamic, qualitative or quantitative, purely utilitarian or of little more than recreational interest. They can arise from the world around us, from the depths of space and time or from the workings of the human mind.

Technological innovation has facilitated our understanding of the planet. Observation of the Earth continues to generate data about the territory and has emphasized the identification of linear and nonlinear patterns, some of which take into account time-series, with their nearly real-time acquisition or access. To a certain degree, using models to extract and analyse patterns can facilitate the various tasks involved in understanding processes between the society and those occurring in the territory. For example, both the spatial and time distribution of land patterns have variability. They are not passive and static observations of events. In fact, the identification and definition of this space-time variability will determine the way in which the pattern can be found.

As an example of the importance of the above, consider the power of identifying vulnerability patterns in a likely dangerous space with social-spatial patterns in order to obtain risk scenarios that are useful to the planning of scenarios to reduce the impact of floods, earthquakes, or landslides in an urban or rural context. Another example closely related to space-time is a crop land. Changes in observations of crop lands vary according to the changes in the spatial (type of crop, daily observation) and temporal (planting rotation patterns) extensions of physiological changes in the crops, which are complex and dynamic systems.

Satellite images are one technological instrument used to identify patterns, whose strengths make it possible to explore different applications that respond to the needs of different realms of society. The data provided by satellite images are simple observations of the Earth and are the input for an interpretation process. The basic principles for extracting information from the image and translating it into an understandable representation are put into practice through semantic and cognitive understanding (Datcu and Klaus, 2005). Semantics refers primarily to the type of algorithm used to extract the information, and the extracted information is added (merged) to define components of objects and contextual structures. In terms of the cognitive part, interpretation, the scene is understood to be a set of objects in a certain context and its meaning.

3.3.6 Geoknowledge

Transdisciplinary reasoning has increasingly contributed to enable researchers from diverse scientific fields to meet the challenge of advancing the scientific understanding of geospatial information through conceptual metasynthesis. Geospatial information can also be obtained from experience based knowledge as well as from theoretical models.

Regarding the information–geoknowledge interface one can observe relevant processes such as:

- Express geospatial and spatial-temporal information. Express spatial-temporal concepts related to a phenomenon and the social or cultural processes related with it.
- Communicate geospatial information and tools to represent it. People code knowledge when generates forms of representation, for example, through speaking, writing, diagrams, and mathematical notations. The representation of geoknowledge generally looks for adequate and explicit symbolic models.
- Arguments related to spatial-temporal information. During the reasoning process, it is necessary to make inferences based on spatial-temporal information. Integration tasks are performed when multiple pieces may or may not be related to each other. They are therefore aligned or organized, and lacking information may even be completed.

In general terms, the objective of the contributions of geoknowledge is to generate research results that are accessible in a community context and focused on the society, and are reflected in easy-to-access educational subjects, complex geomatic solutions, cybercartography (atlas), geospatial expressions, geoinformatic artifacts i.e. geocybernetic solutions. All of these applications are designed to support learning about and understanding the problem posed by the society, in the communication between persons and machines, to facilitate the interpretation, analysis, evaluation, and inference of the problem. They are also designed to develop skills for high-level thinking and critical thinking, skills to participate in investigations and collaborative investigations and to help and facilitate performing transdisciplinary cognitive tasks in order to understand and learn about complex subjects.

Geoknowledge makes it possible to confront situations or problems in society. It is a complex process that intertwines the activities of society, institutions, and academia. The economic, political, and social actors are participants and builders in the territory. The building of bridges throughout the process makes it possible to create important networks in order to provide direction so as to guide action in the territory. In other words, geoknowledge is part of the semantic or emergent knowledge networks (Chapter 2).

3.4 METHODOLOGICAL CONSIDERATIONS

Once and again in the dozens of projects where a geocybernetic approach has been adopted, similar processes as the ones described in the previous section have been identified by the team of researchers. Can we identify certain patterns that would allow to draft a method to build transdisciplinary frameworks as the one described? Should we look for replicability or each case is unique?

Adopting the notion that a methodology is essentially a conceptual guide to pursue a purpose and following a heuristical reasoning we could say that the transdisciplinary bridges are not necessarily unique and that there is a high dependence on the cognitive frameworks of all the participants. It is well known in science that researchers in similar contextual cognitive environments often obtain similar results and/or propose the same solutions.

The essence of geocybernetics points more to complex processes, where emergence and autorganization are the driving forces in the transdisciplinary and analogic reasoning involved. A more detailed study of the empirical results regarding the transdisciplinary bridges that have emerged throughout the years will allow the identification of key conceptual elements and patterns.

3.5 FINAL COMMENTS

As the reader can perceive, Chapters 2 and 3 of this book are part of the same avenue of research. In Chapter 2 an attempt on the formalization of geocybernetics as a scientific realm is presented and, we hope, this will facilitate the communication of results to other scientists and to potential societal actors interested in specific topics. On the other hand, this Chapter focuses on a theoretical exploration of a method to build transdisciplinary frameworks in geocybernetics using analogical reasoning.

Reviewing the avenues of research proposed in Chapter 4 of *Cybercartography: Theory and Practice* published in 2005, one can see that overall the investigation in the topic has advanced considerably. For example, exploration towards artificial intelligence was pointed out, geocybernetics emerged encompassing cybercartography and complex solutions in geomatics, the Reyes method is a refinement of the methodological issues previously proposed and a new examination of the role of mathematics allowed initial steps towards formalization.

Acknowledgements

Each one of the authors has contributed substantially from its own scientific background in a transdisciplinary and collaborative manner. Special thanks to Rafael García for his creative contribution in the design of the figures and to Tracy Lauriault for aiding in the revision of the chapter.

References

Byers, W., 2007. How Mathematicians Think: Using Ambiguity, Contradiction, and Paradox to Create Mathematics. Princeton University Press.

Datcu, M., Klaus, S., 2005. Human centered concepts for exploration and understanding of images. IEEE Transactions on Geoscience and Remote Sensing 43/3, 601–609.

Galton, A., 2000. Qualitative Spatial Change. Oxford University Press.
Hofstadter, D., 1999. Gödel, Escher, Bach: An Eternal Golden Braid. Basic Books. Inc, New York, NY.
Hofstadter, D., 2008. I am a Strange Loop. Basic Books. Inc, New York, NY.
Jackson, P., 1990. Introduction to Expert Systems, second ed. Addison Wesley, Reading, Mass.
Kleene, S., 2009. Introduction to Metamathematics. Ishi Press International, New York, NY.
Lauriault, T.P., Taylor, D.R.F., 2005. Cybercartography and the new economy: collaborative research in action. In: Taylor, D.R.F. (Ed.), Cybercartography: Theory and Practice, Elsevier, Amsterdam, pp. 181–210.
López-Caloca, A.A., Muñoz-Goncen, E., 2012. Aportaciones desde la geocibernética y la percepción remota a la política pública de áreas verdes urbanas. GeoCibernética innovando en geomática para la sociedad 1 (1). Available at: http://www.geocibernetica.org/journal/index.php/diciembre-2012.
López, F., 2011. Un Aporte Teórico: El Prototipo Geomático. Tesis de Doctorado en Geomática, Centro de Investigación en Geografía y Geomática Ing. J.L. Tamayo, México. Biblioteca del CentroGeo.
Mainzer, K., 2007. Thinking in Complexity. The Computational Dynamics of Matter, Mind, and Mankind. Springer-Verlag, Berlin Heidelberg.
Minsky, M., 2006. The emotion machine: commonsense thinking, artificial intelligence, and the future of the human mind. Simon & Schuster Paperbacks, New York.
Polya, G., 1973. How To Solve It: A New Aspect of Mathematical Method. Princeton University Press, Princeton, NJ.
Sabatier, P., 1988. An advocacy coalition framework of policy change and the role of policy-oriented learning therein. Policy Sciences 21, 129–168.

CHAPTER

4

Cybercartography and Volunteered Geographic Information

Nate J. Engler[1], Teresa Scassa[2], D. R. Fraser Taylor[1]

[1]Geomatics and Cartographic Research Centre (GCRC), Carleton University, Ottawa, Ontario, Canada, [2]The Centre for Law, Technology and Society, University of Ottawa, Ottawa, Ontario, Canada

OUTLINE

4.1 INTRODUCTION

The increasingly place-based nature of discourse and information sharing over the Internet by both experts (cartographers and geographic specialists) and nonexperts ('neogeographers') is well into its second decade (Kuhn, 2007; Turner, 2006). As a result, the map rapidly is becoming a unifying framework for online sharing of data and information (Tsou, 2011). This turn towards a 'cartographic Web' represents a general broadening of the nature of geographic information collection, distribution, and use, from an activity traditionally dominated by a relatively small community of expert users of scientifically-oriented geographic information systems (GIS) software, to one that embraces diverse communities of geographic information produsers (a portmanteau of *producer* and *user*) (Budhathoki et al., 2008) and prosumers (a portmanteau of *producer* and *consumer*) (Budhathoki et al., 2008; Hardy, 2008; Toffler, 1980).

Developments in the Theory and Practice of Cybercartography, Second Edition, ISSN 1363-0814
http://dx.doi.org/10.1016/B978-0-444-62713-1.00004-0

The collaborative and assertive approach of these communities towards geographic information has prompted numerous and fundamental changes to the ways in which maps are produced and consumed, with *cartographic interactivity* (e.g. the customization, population, and dissemination of digital maps online through the actions of multiple participants) being among the most significant of these new paradigm shifts (Roth, 2013). In this paradigm, the 'cartographic Web' represents a natural spatial extension of the prevailing archetype of social media and the Web 2.0 model (Sui and Goodchild, 2011).

This interactive approach to cartography also is generating new forms of digital spatial information. Prior to the advent of platforms such as OpenStreetMap (OSM), access to usable and current digital geographic information was locked behind expensive paywalls and encumbered by strict licencing policies. This substantially constrained the availability and use of much of the spatial data resources that had been collected and formatted by public sector governmental agencies as well as by private sector corporations (Johnson, 1995).

In the new era of cartographic interactivity however, new forms of geographic information are generated directly and indirectly as manifestations of individuals' daily routines, such as commuting in a global positioning systems (GPSs)-equipped automobile, or capturing multimedia content of a local event with locationally-aware smart phones or other hand-held devices (Goodchild and Glennon, 2010). This information is used to populate collaborative road atlases (e.g. OSM) to aid in the reporting, monitoring, and relief coordination of natural disasters (e.g. Twitter) (Goodchild and Glennon, 2010), to coordinate and respond to riots, protests, and other political gatherings (e.g. Ushahidi) (Liu and Palen, 2010), to augment existing repositories of location-based multimedia (e.g. Flickr) (Köbben et al., 2012), and to share information among hobbyist communities (e.g. the Annual Christmas Bird Count of the Audubon Society) (Elwood, 2008).

Indeed, a growing percentage of individuals are now equipped with locationally-aware devices such as digital cameras, smart phones, tablets, notebooks, and laptops equipped with digital compasses and accelerometers (Yin and Carswell, 2011). In effect, these human 'agents' are able to capture passively or actively geographic information about their immediate surroundings, along with the means to measure environmental variables or augment the information they are collecting with supplemental sociocultural information in the form of embedded tags or comments (Castelein et al., 2010). The products of these activities are collectively labelled volunteered geographic information (VGI); defined as 'geographic information acquired and made available to others through the voluntary activity of individuals or groups, with the intent of providing information about the geographic world' (Elwood et al., 2011, p. 5; Goodchild, 2007a).

On a global scale, VGI is now contributed regularly to location-based services that augment, and may rival, or eventually replace, extant public and private sector Web-based products (Haklay, 2010; Zielstra and Zipf, 2010). Although still in its infancy, the theory and practice of VGI is in many ways parallel to that of another conceptual advance in the geographic sciences, namely cybercartography. Cybercartography is a theoretical construct defined as 'the organization, presentation, analysis, and communication of spatially referenced information on a wide variety of topics of interest and use to society in an interactive, dynamic, multimedia, multisensory, and multidisciplinary format' (Taylor, 2005; Taylor and Pyne, 2010). The latest developments in cybercartography are outlined in Chapter 1.

This chapter will examine both VGI and cybercartography and discuss how cybercartography as a technology, a mode of production, and a unifying framework is ideal for the management, dissemination, and visualization of VGI. The chapter begins with a brief description of

VGI and a review of its nature, content, and modes of production. This is followed by a critical analysis of cybercartography and its potential as a unifying framework for VGI, including opportunities and challenges in both theory and practice. The final section presents a critical look into the current legal aspects of VGI and its relationship to what has been called 'authoritative' information. This examines some of the pertinent legal issues that present particular challenges to the use of VGI in cybercartography.

4.2 VOLUNTEERED GEOGRAPHIC INFORMATION

The prospect of freely available and ubiquitous VGI has been facilitated by technological advances, in both software and hardware, as well as by the prevailing paradigm of the social media movement (Elwood, 2010). Software-level innovations such as virtual globes (e.g. Google Earth and NASA World Wind) reverse the traditional Web navigational path. Instead of seeking and accessing information on the Internet by navigating through a series of hyperlinks, users can browse content by first selecting a particular geographic location and then navigate to the related contextual information. This process has formalized location as a unifying framework for the presentation and visualization of location-based information (McDougall, 2009b).

These innovations also have made possible the liberation of spatial data through processes of bottom-up user creation and dissemination. Online platforms such as Google Maps have also made available, free of charge, public domain earth-observing satellite and aerial imagery, which has engaged citizens in virtual exploration (Schachtman, 2013), enabled the identification and contextualization of real world features and locations, and prompted public exchanges about appropriate ways to label and demarcate places and spaces (BBC News, 2011).

Hardware advances such as the inclusion of GPS receivers and accelerometers in most mobile devices (e.g. cellular phones, smart phones, tablets, touchpads, netbooks, etc.) has made possible both active and passive real-time collection of location-based information (Yin and Carswell, 2011).

Hence, a new landscape (i.e. a Web cartography) is being produced over the Internet. New technologies, including Twitter's GeoAPI (enabling geotagging of microblog posts), Google Earth and other virtual globes, location-based social media applications such as Foursquare and Gowalla, map mash-ups, application programming interfaces that allow users to incorporate and integrate content from multiple web services, and the devices used to connect to these applications, 'entail a deployment of geovisual artefacts to structure a visual experience; a prioritisation of individualised interactive/exploratory ways of knowing; hyper-granular, highly immediate, experiential cartographic representations de-coupled from conventional practices of cartographic abstraction; and approaches to asserting credibility through witnessing, peer verification and transparency' (Elwood and Leszczynski, 2012, p. 12). In effect, ground truthing, initially a quantitative and precise exercise of framing the natural world within the constraints of the point, polyline, and polygon taxonomy of early GIS has become a discursive process of collaboration and assertion by online communities.

Typologies of VGI are still in their infancy; however, several notable contributions have been made. Elwood et al. (2011) base their typology on the *purpose* underlying the creation and sharing of the VGI. They distinguish between allocentric and egocentric VGI where allocentric information is inherently community-oriented and is contributed for the benefit of others, rather than for specific, self-interested purposes. Examples include contributions to Web-based map services such as Wikimapia or OSM, or participation in disaster relief efforts through folksonomic geotagging on Twitter. Egocentric information, on the other hand, refers specifically to the produser's real-time location and/or movements and subsequently provides a means for connecting to others in time and space. Examples include geotagged digital photographs uploaded to Flickr, or the use of Facebook's 'Places' tool, which allows individuals to broadcast their presence at real-world venues and to simultaneously realize the presence of others from their social network (Elwood et al., 2011).

Others have proposed a taxonomy based on the *nature* of the VGI; a distinction between explicitly and implicitly volunteered information (Craglia et al., 2012). The difference can variously be attributed to (1) the format of the content, (2) the mode of production of the content, or (3) the intent of the volunteer in creating the content. For example, the content of explicit VGI contains a specific coordinate pair, an address, or the name of a feature, building, or public space embedded in or attached to the contribution (Hardy, 2008). Similarly the mode of production for explicit VGI suggests that the content is the product of active and engaged sharing by either a volunteer using some subset of their five senses and/or through the deliberate use of their location-aware devices (Craglia et al., 2012). The term participatory sensing has also been proposed for this mode of production (Jiang and McGill, 2010). Finally, with explicit VGI, the content is contributed with a specific purpose in mind – i.e. such as to aid in disaster relief, to display and share a user's location with their social network, or to contribute to, edit, or correct a feature in a community repository.

Implicit VGI on the other hand, refers to content that may include, for example, a natural language tag such as 'near Ottawa' without identifying a specific geographical location in real world coordinates. Another example would be metainformation about the origin or location of production of the information attached to the post, tag, or content. While the content and/or the metacontent may still be parsed to return a coordinate pair that approximates a real world location, this level of detail likely is beyond what the contributor intended to volunteer (Coleman et al., 2009). This represents a deviation from the ideals that underlie VGI as both a product and a process in the sense that while the information is voluntarily provided, it is not geographic information per se, but rather an artefact that contains geographic footprints (Stefanidis et al., 2013).

Stefanidis et al. (2013) have phrased this as 'ambient geospatial information' and other terms such as 'opportunistic' sensing (Jiang and McGill, 2010) have also been employed. At the extreme, harvesting of implicit VGI can represent potential harm to the contributor, such as through identity theft, and occurs through processes such as unauthorised capture of tracking data from personal mobile devices (Emspak, 2011; Scassa and Sattler, 2011; Thurm and Kane, 2010). Table 4.1 outlines a specific typology based on explicit and implicit VGI within the framework of Twitter contributions (i.e. Tweets).

Thus, the emerging reality is that neogeographers, lacking formal training in geographic or cartographic theory or practice, are creating their own digital maps from their own geospatial information and that of others. This has provoked both optimism and apprehension (Turner, 2006). The most frequently cited benefits associated with VGI centre predominately on the democratisation of spatial data production and consumption (Elwood, 2008). For example, the collaborative and assertive nature of social media enables the task of data production to

TABLE 4.1 Typology of VGI

Contributor's Intent	Geographic Reference	
	Explicit	**Implicit**
Explicit	Volunteered geographic information • GeoTwitter tweets with explicit latitude/longitude information tagged to the tweet ('Raining hard here') with embedded geoJSON: "coordinates":{"coordinates" [−75.14310264, 40.05701649], "type": "Point"}	Volunteered geospatial information • Tweet referencing properties of an identifiable place ('Traffic accident 401 eastbound at Exit 116 near London').
Implicit	Citizen-generated geographic content • Tweets referencing properties of an identifiable place ('Meeting friends at the Royal Ontario Museum')	Citizen-generated geospatial content • Tweets containing a geographic reference in the context of a nongeographic topic ('Meeting friends downtown tonight!')

Source: Adapted from Craglia et al. (2012).

be divided among a community of users (Sunstein, 2009). This realigns the process of geospatial data creation based on personal or community needs rather than corporate or government mandates, priorities, or capabilities (Feick and Deparday, 2010; Goodchild et al., 2007b). Integral to this democratisation are new opportunities for individuals to enhance, validate and in some cases dispute authoritative geographic information with contextualized representations of their own. This may include experiential and local knowledge that otherwise may not be accessible or documented in public databases (e.g. the locations of rare plants, unmarked sites of local historic significance, or places of spiritual significance). The plurality of data and the interactivity inherent in the editing and assertion processes provides new prospects for mapping different perspectives and realities compared to the single perspective that has dominated GIS for decades (Feick and Deparday, 2010). Table 4.2 lists some common benefits and disadvantages associated with VGI.

Research into volunteer motivations has ranged far and wide since the foundation of Wikipedia and similar projects. In the VGI context, produsers participate voluntarily with no expectation or assumption of financial reward. In fact, there is evidence to support the view that financial incentives are anathema to VGI motivations and ultimately erode not only trust in the user but also undermine the sense of collaboration and trust in VGI communities (Cook, 2008). However, participation does require the produser to make some investment of time and/or expertise – either to produce and contribute data, or to edit and verify the validity of a contribution.

The produser may have altruistic intentions, or a desire to participate in the creation of a digital commons to further the common good by providing the underlying data and information from which new understandings of a particular region or neighbourhood can be developed (Coleman et al., 2009). However, no matter the motivation, the extent to which the results of a volunteer's work influence decisions or are in fact used by the community or by those in a position of authority has an overall effect on the volunteer's satisfaction and by extension their ongoing contribution (Gouveia and Fonseca, 2008). Table 4.3 describes and classifies the various motivations for volunteerism from several research studies.

TABLE 4.2 Benefits and Disadvantages of VGI

Type	Benefit	Disadvantage
Data	• Support the development of early warning systems • Data collection for areas where no commercial or governmental interest exists • Big data – heterogeneous and multiple data sources for features • Rapid update of contextual information in times of change – i.e. construction, infrastructure, political upheaval	• Data credibility may be low • Metadata are scarce • Due to the volunteers' lack of specific knowledge and training VGI is not always comparable to data collected by others • Unpredictable level of volunteer commitment
Logistics	• The participation of volunteers can be a cost-effective method to maintain data collection activities • Augment the geographic area and the time period being monitored	• Public participation may require training or close supervision of volunteer work • Requires the organization of training materials
Social Impacts	• More informed and educated public concerning environmental problems and scientific methods • Volunteer monitoring promotes cooperation instead of the traditional 'us versus them' approach	• The Not in my Backyard (NIMBY) principle, may affect citizen participation, distorting in some cases the goals of the volunteer initiative • Vandalism

Source: Adapted from Gouveia and Fonseca (2008).

TABLE 4.3 Suggested Motivations for Volunteerism

Category	Reason
General Interest	1. Technology and its novelty (McDougall, 2009a) 2. Personal learning, enjoyment (Gouveia and Fonseca, 2008) 3. Convenience – making information available to family and friends (Goodchild, 2007a) 4. Personal satisfaction (Goodchild, 2007a)
Altruism	1. Contributing to the improvement of a fundamental data set that is used widely for the benefit of the public and assisting the deployment of community resources such as emergency services (McDougall, 2009b) 2. Supporting others within a community of users (Flanagin and Metzger, 2008) 3. Because doing so is consistent with an altruistic world view of generalized reciprocity (Flanagin and Metzger, 2008; Gouveia and Fonseca, 2008)
Egotism	1. Studies from social network analysis indicate the ability to build social capital (Gouveia and Fonseca, 2008; McDougall, 2009b; Tulloch, 2008) 2. Self-promotion (Goodchild, 2007a) 3. Reputation (Gouveia and Fonseca, 2008)
Power	1. To achieve desired political outcomes (Flanagin and Metzger, 2008) 2. Sway public processes or political agendas (Gouveia and Fonseca, 2008)

Source: Adapted and expanded from Coleman, Sabone and Nkhwanana (2010).

4.3 CYBERCARTOGRAPHY

Cybercartography was introduced in 1997 (Taylor, 1997), the same year as the launch of two innovative new platforms: Classmates.com (which still records over 800,000 visitors per month[1]) and the now defunct SixDegrees.com. At Classmates.com early users could follow colleagues and classmates, view old yearbooks, and participate in virtual reunions. SixDegrees.com, based on the theory that everyone on earth can be connected to everyone else through a chain of no more than six links, was the first site to allow its users to create profiles, invite friends, organize groups, and survey other user profiles. From these fledgling social networking sites was spawned the mass movement that exploded with the arrival of Facebook.com and that generated the new paradigm of social media: 'a group of Internet-based applications that build on the ideological and technological foundations of Web 2.0, and that allow the creation and exchange of user-generated content' (Kaplan and Haenlein, 2010, p. 61).

Within this new paradigm, cartography has advanced beyond its traditional authoritative roles of place-finding and formalized representations of the physical environment into that of providing a powerful unifying framework for a wide range of cultural topics, socioeconomic concerns, as well as subjects of a hobbyist nature. Indeed, online digital maps are increasingly using multimedia formats that go beyond text and graphics, and integrate video, sound, and touch (Taylor and Pyne, 2010). Hence, both mapping as a process and the map as a product has taken on new meanings in the information era. This is reflected by the shift from map users to map creators, which establishes new forms of democratised teaching and learning.

The underlying theory of cybercartography is built on the precepts of an interactive and facilitated approach to the development and population of maps and atlases. Throughout history, the map has been important as an artefact, however, maps potentially are much more than simple products that indicate location or place. They have been defined as 'graphic representations that facilitate an understanding of things, concepts, processes, or events in the human world' (Harley and Woodward, 1987). In this context the theory of cybercartography requires an approach to content creation that emphasises interactive involvement with the community of practice for whom the atlas is designed as well as the researchers, and technical staff that facilitate the development of the schema, representations, and visualizations based on the needs of the community. Hence, cybercartography is designed to facilitate a bottom-up development.

The technology that enables the theory, namely the *Nunaliit Cybercartographic Framework*, an open source framework developed at the Geomatics and Cartographic Research Centre (GCRC), is a means to an end, not an end in itself (Taylor, 2013). The development of this framework is explored in more detail by Hayes in Chapter 10. In practice, cybercartography is rapidly evolving to approach and consider new social challenges and new topics of interest to society. For example, cybercartography privileges an asserted approach to addressing questions such as those around anthropogenic climate change, as well as the social determinants of health where social, economic, and cultural complexity suggests that there is no simple right or wrong answer.

To appreciate the complexities of these questions, the framework is designed for the inclusion, without prejudice, of multiple perspectives. Each of these perspectives has its own ontological underpinnings that result in different narratives. Often these narratives have particular

[1] http://social-networking.findthebest.com/l/195/Classmates-com.

and specific locational references. For example, the description of a trail to an Inuit elder is not simply the location of that trail but the stories that go with each journey. Each journey along a trail, even if it follows exactly the same route, is a new story (Aporta, 2009).

Of particular importance in this respect is the ability to give a voice to indigenous peoples and to those in demographic groups not privileged by the dominant power structure of their national culture. By facilitating the ability of these marginalized groups to create their own narratives, cybercartography offers a platform where they can speak for themselves both literally and figuratively (Taylor and Pyne, 2010). These themes are more fully developed in subsequent chapters of this volume where a number of cybercartographic atlases involving Inuit and First Nation's peoples are discussed.

4.4 SYNTHESIS OF VGI AND CYBERCARTOGRAPHY

Cybercartography represents a powerful tool for the (1) management, (2) dissemination, and (3) visualization of geographic narratives, or geonarratives, derived from heterogeneously sourced VGI. Thus, cybercartography offers an unprecedented opportunity for rethinking the way location-based information is collected, produced, engaged with, disseminated, and represented over the Internet. Going beyond the current approach of popular VGI-populated platforms such as OSM, Google MapMaker, and Foursquare, cybercartography reinforces the interplay between theory and practice. It also emphasizes the relationships and capabilities engendered between developers, facilitators, and community members in the production of an atlas and describes entirely new processes and results in new products of engaged cartographic interactivity.

The cybercartographic atlas represents a metaphor for both qualitative and quantitative information. While location is a central organizing principle these atlases also are capable of ingesting, cataloguing, and representing nonlocation-based information. Each atlas is based on an unstructured schema-less document database system, namely CouchDB, that greatly improves information management, scalability, and performance. The technical aspects of the software are more fully discussed by Hayes in Chapter 9. The approach represents a new way of thinking about VGI management that is scalable, flexible, and responsive to unforeseen needs and uses that may arise as the atlas matures. In this way, each atlas can respond to the changing and expanding needs of the community developing it, such as: new and unanticipated forms of quantitative or qualitative data; new unexpected relationships between types or kinds of information; and new categories of users, contributors, and consumers of the atlas who for security or other reasons may be limited to some subset of the entire collection contained within the atlas.

Dissemination of the atlas is managed at the level of the community, based on their needs for information sharing and/or protection. In some cases communities and aboriginal groups wish to publicize broadly their information, such as with the Kitikmeot Place Names Atlas, described in Chapter 15. In these instances, public sharing of traditional place names and points of interest to the community translates to provenance – that is ownership and sovereignty over lands that they have occupied for generations. In other cases, communities wish to share their information, such as hunting routes and/or trapping sites, only within their own community or with neighbouring communities. In these cases protection of these resources is critical to ensure the communities continued sustenance needs while simultaneously preventing external depletion or exploitation of these resources by the so-called 'eco-tourism' corporations. The Nunaliit platform is designed specifically to handle multiple layers of information that can be coded to user types such that information presented on the atlas is determined by the status of the user. In many cases this means a simple dichotomy between a public and a private layer; however, more complex systems may be developed as needed by the community.

Geonarratives – oral histories, life histories, and biographies told not only in time, but also in space – are at the heart of the cybercartographic approach. The Nunaliit platform, although rooted in the ontological paradigm of point, (poly)line, and polygon common to GIS software, does not prescribe in advance the visual representation of the VGI contributed by the community members. Hence, symbology, palette, and other aesthetic choices are left to the community, and then incorporated by the researchers and developers. Table 4.4 gives a point-by-point analysis of the synthesis between VGI and cybercartography, based on the seven defined elements of cybercartography (Taylor 1997; Taylor, 2005).

TABLE 4.4 Synthesis between VGI and Cybercartography

Element	VGI Approach	Cybercartographic Atlas
Multisensory	Generates digital picture, audio, and video files, as well as text in the form of comments and tags.	Designed to format and present information for all five senses, including vision, hearing, touch, and eventually smell and taste.
Multimedia	Generates multimedia files such as .jpg, .mp3, and .mpeg, and also GPS data such as waypoints, routes, and tracks in .gpx format.	Designed to ingest all popular file formats via an automated java library that compresses, formats, and makes consistent all input for a common look and feel unique to each atlas.
Interactivity	The use, editing, and sharing of VGI is inherently interactive, and is improved as more contributors come together to contribute more content.	The atlas framework uses a nonschema document database that is scalable, flexible, and adaptable to new and unanticipated data formats and content types.
Integration	The contributions of multiple users combine to form a consensus, asserted truth, about some feature or phenomena of the world.	The nonschema approach allows for easy linking and association of contributions, regardless of content type or format.
User-group	Contributors include anyone with an interest or a desire to participate or make use of the product of others contributions.	Designed for ease of use of the mapping and data contribution interfaces, appealing to the broadest range of users and contributors.
Interdisciplinarity	Contributors are from multiple backgrounds and limited only by desire or interest to participate.	Each atlas is developed by a multidisciplinary team that may include geographers, sociologists, and anthropologists, as well as the community members.
Partnerships	Partnerships develop as users work together to achieve consensus from the volunteered content.	Strong partnerships at the initial stages of atlas development allow for rapid design, development, and use of the atlas by communities, research groups, etc.

4.5 LEGAL ISSUES WITH VGI

The incorporation of VGI into cybercartography projects offers great potential for expanding not only the nature of the content that may be incorporated into an atlas, it also redefines the nature of individual and community participation in any project. At the same time, however, it can raise a number of legal issues that are either less likely to be present in a more top-down project, or that are manifested or managed very differently in more conventional mapping contexts. The traditional, top-down, highly controlled process of the development of 'official' cartographic works has rendered the legal considerations largely invisible. In other words, the conventions of the genre have evolved to ensure that the process for creation and dissemination of cartographic works fits within the parameters of the governing laws. This is much less likely to be the case in the context with both cybercartography and VGI. In each of these contexts (and in their combination) the traditional practices around creation and dissemination may be significantly disrupted. It is thus necessary to give some thought to the impact of legal rules on the design and execution of such projects and the ethical issues involved. These issues such as intellectual property rights, privacy, ethics, and civil liability are more fully explored by Scassa and others in Chapter 18 and by Deleo Brown in Chapter 12. Some observations from the specific perspective of VGI will be given here.

4.6 INTELLECTUAL PROPERTY

In copyright terms, a cybercartographic atlas is a multimedia work; in other words, it is a compilation consisting of multiple different works. While the overall author or authors of the compilation will have a copyright in the atlas itself, where VGI is relied upon, it is likely that multiple individual contributors will have uploaded their own photos, video, or written comments to the compilation. This means that there are likely to be a great number of works embedded in the compilation, each of which may have its own copyright, resulting in a large number of different rights holders (Scassa, 2012). As a result, the creators of a cybercartographic atlas that relies upon VGI must give some attention to how rights in the different contributions will be managed. One strategy is to require contributors to assign copyright in their works to the creators of the atlas, but this may not only be a deterrent to contributions, it may also be counter to the open and collaborative goals of the project. A less invasive strategy may be to require contributors to give the project a nonexclusive licence to reproduce and disseminate the contributed works. The licence should specify if it is to be perpetual or only for a limited span of time; for a project with no foreseeable termination point, a perpetual licence is preferable.

It may be the case that some of the contributed content will be works to which the contributor does not hold the rights, and is thus not in a position to licence them to the creators of the atlas. For example, an individual may upload a photograph taken by a friend (or found elsewhere online), or they may contribute text, audio or video content for which they are not the author. This may leave the creators of the atlas vulnerable to accusations of copyright infringement. The creators of a cybercartographic atlas may, through notices, urge contributors to contribute only content for which they hold rights. The licence agreement discussed above may also stipulate that the contributor warrants that they have rights in any contributed content, and that they will reimburse (hold harmless) the creators of the atlas for any liability or legal costs incurred if infringing content is uploaded. This will only be a realistic option if contributors are required to identify themselves. In many cases, even if complaints are received about specific contributions to the atlas, it may be enough simply to remove the offending content promptly. This may or may not be an issue for the creators of the atlas, depending upon how unique the contribution is, how important it is to the goals of the atlas, or whether there are concerns that the complaints are motivated by a desire to suppress the telling of particular stories via the atlas.

In general terms, copyright law does not protect facts or information – only the original expression of that information (Judge and Scassa, 2010; Vaver, 2011). A map is considered an artistic work that is protected by copyright – it is a visual expression of facts or information. Another mapmaker can create his/her own expression of the same data. The fact/expression dichotomy applies as well to cybercartographic atlases and the individual works (photos, videos, text) of which they are composed. Thus although the specific works are protected by copyright law (for example a photograph), the facts or information that are a part of the work will be considered to be in the public domain. This distinction may prove problematic in contexts where the cybercartographic project incorporates indigenous or traditional knowledge. The lack of IP protection for the knowledge embedded within a recorded interview with an elder, for example, may run counter to expectations. Other normative obligations may apply to the traditional knowledge in such cases. These issues are addressed more specifically in Chapters 12 and 18.

4.7 PRIVACY

There exists a wide range of data protection and privacy laws – even within a single jurisdiction – that provide a level of privacy protection for individuals. A cybercartographic project, which by definition is not confined to national boundaries, is potentially open to privacy complaints from multiple jurisdictions.

Personal information is generally considered to be any information about an identifiable individual (Scassa and Deturbide, 2012). The individual need not actually be identified – it is enough if the information provided can be matched with other available information so as to identify a particular individual. In our information society there are so many available data sources and technological tools for searching and matching data that much information that is not obviously about an identifiable individual may well become so (Scassa, 2010a).

Data-protection laws generally prohibit the collection, use, or disclosure of personal information without the individual's consent. Where an individual is contributing his/her own personal information to a cybercartographic project (for example, information about their own location or activities) consent may be implied – at least for the explicit purposes of the cybercartographic project. Express consent can also form part of a licence agreement where appropriate. However, where an individual is contributing information about other people – for example, by uploading geo-referenced photographs that feature other individuals – the consent of those individuals to the sharing of that information is not apparent.

In many cases, the privacy issues may not be significant. For example, data-protection laws may not apply to noncommercial private – sector activity (Scassa and Deturbide, 2012). There are also exceptions for information collected, used or disclosed for artistic, literary or journalistic purposes, although the full scope of these exceptions in the digital era is not established (Scassa, 2010b). However, if the cybercartographic project is hosted by a public institution such as a government agency or a university, public-sector data-protection laws may apply, and may place constraints upon how personal information is collected, used, or disclosed.

Of course, some projects may inherently raise more privacy issues than others. Where, for example, the goal of a project is to invite contributors to report on illegal, antisocial, or simply controversial activities, there may be more specific privacy risks and potential liability – not just for invasion of privacy, but possibly also for other forms of civil liability discussed below. In such cases, the cybercartographers will need specifically to direct their attention to managing the legal risks associated with the project.

The fact that privacy issues may be inevitable means that cybercartographers will need to turn their attention to what, if any, regime directly applies to them and to how they can best manage the unpredictability of VGI within the context of that regime.

4.8 CIVIL LIABILITY

In addition to intellectual property and privacy issues, the incorporation of VGI into cybercartography projects may also raise other liability issues. For example, volunteers may contribute content about other individuals or their activities that is defamatory. More significantly, perhaps, questions have been raised about the extent to which VGI may create risks of civil liability for faulty map information (Rak et al., 2012; Robson, 2012). Because of the multifaceted nature of VGI, the liability may be not simply for faulty geographic information, but for other flawed information regarding, for example, environmental or disaster-related conditions. While liability may arise with any type of map or information-based product (Chandler and Levitt, 2011), VGI has prompted particular concerns in large part because the reliance on the contributions of volunteers has raised issues about the quality and credibility of the resultant maps. Much research is being done on assessing the quality and credibility of VGI, and on finding ways to improve it (Flanagin and Metzger, 2008; Goodchild, 2008; Haklay et al. 2010). Nevertheless, concerns remain that the de-professionalization of GIS brought about by the neogeography movement will lead to added risk.

The civil liability issues are difficult to assess in this context. In the first place, liability for the harms caused by reliance on faulty information may only arise if it was reasonable for individuals to rely upon the information. Clear disclosure about the limits of the quality of the mapped information, or clear directions as to its fitness (or lack of fitness) for particular uses may substantially reduce risks of liability, and most licences for information-based products also include clauses limiting or excluding liability (Chandler and Levitt, 2011). A cybercartography project may also have entirely different goals from providing information to be relied upon in navigation or other risky activities.

4.9 CONCLUSION

While VGI as a process and cybercartography, in practice, share a suite of commonalities, it seems more fruitful to regard cybercartography as a unifying framework that can maximize the potential of the management, dissemination, and visualization of VGI. Within this model, the interplay between theory and practice, in an open and bottom-up process of cocreation can lead to significant and interesting new ways of looking at the world from the perspectives of multiple communities and peoples.

VGI has already thrown a central pillar of the geographic sciences into question, namely, the notion of accuracy. Through the social media movement the rigid and authoritative accuracy so critical to data formatted for GIS has been replaced with an asserted and collective approach to determining what constitutes an accurate representation of any real world event, feature, location, or sociocultural phenomena. While there is certainly only one type of dimensional accuracy – i.e. placing the label, feature, or boundary at the correct location on the map – this is at the minimum one level of interest and sophistication less than a more holistic approach to 'accuracy' that captures the story, the reality, and the experience of a world view in all its complexity.

For example, in publicly sponsored atlases, such as those often produced by countries such as Canada, toponyms are rigorously vetted and analysed prior to inclusion (see for example the Canadian Geographical Names Data Base, which store names records for populated places and physical features across Canada). In the VGI paradigm however, maps such as those of OSM are often littered with a plethora of colloquial, locally-known, and cherished place names that help to contextualize neighbourhoods and communities in whole new ways.

Of more importance is the role of toponyms in claims upon sovereignty and rights to land. It is the continued habitation of a people and their reciprocal relationship with the land, sea, and ice that gives them their claim to sovereignty over a particular region. In Canada, over a thousand years of occupation by the Inuit has been recorded in toponyms for inhabited places as well as land, sea, and ice features. This history marks the region as a homeland of a particular people, a culture, and country (Christie, 2011). The preservation of the local and Traditional Knowledge is of critical importance to indigenous communities who seek to preserve their cultures and identities while at the same time asserting their own claims to self-determination and self-governance (Cutler, 2011).

Finally, the synthesis of these two conceptual advances, namely VGI and cybercartography, despite the legal concerns described in this chapter, may well form the basis for a cross-sector solution to the two major challenges of the future Web as identified by Tim Berners-Lee as outlined in Chapter 1; namely, linking large data sets on disparate topics (e.g. VGI) and displaying the new information created in innovative ways (i.e. cybercartography).

References

Aporta, C., 2009. The trail as home: inuit and their pan-arctic network of routes. Human Ecology 37 (2), 131–146.

BBC News, 2011. Google to Amend Rio Maps over Brazil Favela Complaints. Available at: http://www.bbc.co.uk/news/world-latin-america-13193503.

Budhathoki, N., Bruce, B., Nedovic-Budic, Z., 2008. Reconceptualizing the role of the user of spatial data infrastructure. GeoJournal 72 (3–4), 149–160.

Castelein, W., Grus, Ł., Crompvoets, J., Bregt, A., 2010. A characterization of volunteered geographic information. Proceedings of the AGILE International Conference on Geographic Information Science, Guimarães, Portugal, pp. 32–39.

Chandler, J., Levitt, K., 2011. Spatial data quality: the duty to warn users of risks associated with using spatial data. Alberta Law Review 9 (1), 79–106.

Christie, G., 2011. Indigeneity and sovereignty in Canada's far north: the arctic and inuit sovereignty. South Atlantic Quarterly 110 (2), 329–346.

Coleman, D., Georgiadou, Y., Labonte, J., 2009. Volunteered geographic information: the nature and motivation of produsers. International Journal of Spatial Data Infrastructures Research 4, 332–358.

Coleman, D., Sabone, B., Nkhwanana, N.J., 2010. Volunteering geographic information to authoritative databases: Linking contributor motivations to program characteristics. Geomatica 64 (1), 27–39.

Cook, S., 2008. The contribution revolution: letting volunteers build your business. Harvard Business Review 86 (10), 60–69.

Craglia, M., Ostermann, F., Spinsanti, L., 2012. Digital earth from vision to practice: making sense of citizen-generated content. International Journal of Digital Earth 5 (5), 398–416.

Cutler, C., 2011. The globalization of international law, indigenous identity, and the new constitutionalism. In: Coleman, W. (Ed.), Property, Territory, Globalization: Struggles over Autonomy, UBC Press, Vancouver, BC, pp. 29–57.

Elwood, S., 2010. Geographic information science: emerging research on the societal implications of the geospatial web. Progress in Human Geography 34 (3), 349–357.

Elwood, S., 2008. Volunteered geographic information: key questions, concepts and methods to guide emerging research and practice. GeoJournal 72 (3–4), 133–135.

Elwood, S., Goodchild, M.F., Sui, D., 2011. Researching volunteered geographic information: spatial data, geographic research, and new social practice. Annals of the Association of American Geographers 102 (3), 571–590.

Elwood, S., Leszczynski, A., 2012. New spatial media, new knowledge politics. Transactions of the Institute of British Geographers10.1111/j.1475-5661.2012.00543.x.

Emspak, J., 2 May 2011. Google sued for tracking users' locations. International Business Times.

Feick, R., Deparday, V., 2010. Evaluating selected visualization methods for exploring VGI. Geomatica 64 (4), 427–437.

Flanagin, A., Metzger, M., 2008. The credibility of volunteered geographic information. GeoJournal 72, 137–148.

Goodchild, M.F., 2008. Spatial accuracy 2.0. Proceedings of Proceedings of the Eighth International Symposium on Spatial Accuracy Assessment in Natural Resources and Environmental Sciences, Shanghai, China, pp. 1–7.

Goodchild, M.F., 2007a. Citizens as voluntary sensors: spatial data infrastructure in the world of web 2.0. International Journal of Spatial Data Infrastructures Research 2, 24–32.

Goodchild, M.F., Glennon, J.A., 2010. Crowdsourcing geographic information for disaster response: a research frontier. International Journal of Digital Earth 3 (3), 231–241.

Goodchild, M.F., Kitchin, R., Perkins, C., 2007b. Citizens as sensors: the world of volunteered geography. GeoJournal 69 (4), 211–221.

Gouveia, C., Fonseca, A., 2008. New approaches to environmental monitoring: the use of ICT to explore volunteered geographic information. GeoJournal 72 (3–4), 185–197.

Haklay, M., 2010. How good is volunteered geographical information? A comparative study of OpenStreetMap and ordnance survey datasets. Environment and Planning B: Planning and Design 37 (4), 682–703.

Haklay, M., Basiouka, S., Antoniou, V., Ather, A., 2010. How many volunteers does it take to map an area well? The validity of Linus' law to volunteered geographic information. The Cartographic Journal 47 (4), 315–322.

Hardy, D., 2008. Discovering behavioral patterns in collective authorship of place-based information. Proceedings of 9th International Conference of the Association of Internet Researchers, Copenhagen, Denmark.

Harley, B., Woodward, D. (Eds.), 1987. The History of Cartography, University of Chicago Press, Chicago, IL.

Jiang, M., McGill, W., 2010. Human-centered sensing for crisis response and management analysis campaigns. Proceedings of Proceedings of the 7th International ISCRAM Conference, Seattle, WA, pp. 1–11.

Johnson, J., 1995. Case studies of dissemination policy in local government GIS agencies. Computers, Environment and Urban Systems 19 (5–6), 373–389.

Judge, E., Scassa, T., 2010. Intellectual property and the licensing of Canadian government geospatial data: an examination of geoconnections' recommendations for best practices and template licences. The Canadian Geographer 54 (3), 366–374.

Kaplan, A., Haenlein, M., 2010. Users of the world, unite! The challenges and opportunities of social media. Business Horizons 53 (1), 59–68.

Köbben, B., Huisman, O., Lin, H., 2012. Combining VGI with viewsheds for photo tag suggestion. In: Gartner, G., Ortag, F., Berlin, D.E. (Eds.), Advances in Location-Based Services, Springer, pp. 181–190.

Kuhn, W., 2007. Volunteered Geographic Information and GIScience. Available at: http://www.ncgia.ucsb.edu/projects/vgi/docs/position/Kuhn_paper.pdf.

Liu, S., Palen, L., 2010. The new cartographers: crisis map mashups and the emergence of nongeographic practice. Cartography and Geographic Information Science 37 (1), 69–90.

McDougall, K., 2009a. The potential of citizen volunteered spatial information for building SDI. Proceedings of the GSDI 11 World Conference on Spatial Data Infrastructure Convergence, Rotterdam, The Netherlands.

McDougall, K., 2009b. Volunteered geographic information for building SDI. Proceedings of Proceedings of the Surveying & Spatial Sciences Institute Biennial Conference, Adelaide, Australia, pp. 645–653.

Rak, A., Coleman, D., Nichols, S., 2012. Legal liability concerns surrounding volunteered geographic information applicable to Canada. In: Rajabifard, A., Coleman, D. (Eds.), Spatially Enabling Government, Industry and Citizens: Research and Development Perspectives, GSDI Association Press, Needham, USA, pp. 125–141.

Robson, E., 2012. Responding to Liability: Evaluating and Reducing Tort Liability for Digital Volunteers. Woodrow Wilson International Center for Scholars, Washington, DC.

Roth, E., 2013. Interactive mapping: what we know and what we need to know. Journal of Spatial Information Science 6, 59–115.

Scassa, T., 2012. Legal issues with volunteered geographic information. The Canadian Geographer 57 (1), 1–10.

Scassa, T., 2010a. Geographic information as personal information. Oxford University Commonwealth Law Journal 10 (2), 185–214.

Scassa, T., 2010b. Journalistic purposes and private sector data protection legislation: blogs, tweets, and information maps. Queen's Law Journal 35, 733–781.

Scassa, T., Deturbide, M., 2012. Electronic Commerce and Internet Law in Canada, second ed. CCH Canadian, Toronto, ON.

Scassa, T., Sattler, A., 2011. Location-based services and privacy. Canadian Journal of Law and Technology 9 (2), 99–134.

Schachtman, N., 2013. What did Google Earth spot in the Chinese desert? Even an Ex-CIA analyst isn't sure. Wired. Available at: http://www.wired.com/dangerroom/2013/01/google-earth-china-hunh/.

Stefanidis, A., Crooks, A., Radzikowski, J., 2013. Harvesting ambient geospatial information from social media feeds. GeoJournal 78 (2), 319–338.

Sui, D., Goodchild, M.F., 2011. The convergence of GIS and social media: challenges for GIScience. International Journal of Geographical Information Science 25 (11), 1737–1748.

Sunstein, C., 2009. Going to Extremes: How like Minds Unite and Divide. Oxford University Press, New York, NY.

Taylor, D.R.F., January 2013. Cybercartography is charting a new route. Geospatial World.

Taylor, D.R.F., 1997. Maps and mapping in the information era. Proceedings of the 18th ICA Conference, Stockholm, Sweden. 1–10.

Taylor, D.R.F., 2005. Cybercartography: Theory and Practice, first ed. Elsevier, Amsterdam.

Taylor, D.R.F., Pyne, S., 2010. The history and development of the theory and practice of cybercartography. International Journal of Digital Earth 3 (1), 2–15.

Thurm, S., Kane, Y.I., 17 December 2010. Your apps are watching you. The Wall Street Journal.

Toffler, A., 1980. The Third Wave. Bantam Books, USA.

Tsou, M., 2011. Revisiting web cartography in the United States: the rise of user-centered design. Cartography and Geographic Information Science 38 (3), 258–259.

Tulloch, D., 2008. Is VGI participation? from vernal pools to video games. GeoJournal 72, 161–171.

Turner, A., 2006. Introduction to Neogeography. O'Reilly, Sebastopol, CA.

Vaver, D., 2011. Intellectual Property Law, second ed. Irwin Law, Toronto, ON.

Yin, J., Carswell, J.D., 2011. Touch2Query Enabled Mobile Devices: A Case Study Using OpenStreetMap and iPhone. Dublin Institute of Technology, Dublin.

Zielstra, D., Zipf, A., 2010. A comparative study of proprietary geodata and volunteered geographic information for Germany. Proceedings of Proceedings of the AGILE International Conference on Geographic Information Science, Guimarães, Portugal.

CHAPTER

5

Further Developments in the Theory and Practice of Cybercartography: Exploring Web 2.0 and Participatory Software for Building Geolocated Narratives

William Cartwright

School of Mathematical and Geospatial Sciences, RMIT University, Melbourne, Victoria, Australia

OUTLINE

Developments in the Theory and Practice of Cybercartography, Second Edition, ISSN 1363-0814
http://dx.doi.org/10.1016/B978-0-444-62713-1.00005-2

5.1 INTRODUCTION

In June 2009, I was undertaking research to ascertain the availability of maps and geographically related information that was produced by France during their involvement in the Gallipoli landings and subsequent battles. This involved visiting the Service Historique de la Defense, Département de l'armée de division des terres, Château de Vincennes in Paris. Archivists Madame Marie-Anne de Villèle and Mademoiselle Ponnou had prepared some documentation for me prior to my arrival and they noted that one particular archive publication, *Inventaire sommaire des archives de la guerre 1914-1918*, Ministère d'état Charge de la défense nationale (Nicot et al., 1972), a general reference to the Service historique de la défense, collection might provide the key to accessing appropriate documentation stored in the archive. The publication provided a comprehensive catalogue to artefacts in the archive up to 1972. This included maps and other geographically related documents. The section of interest to research into the French involvement in the Gallipoli campaign was 'Grandes unités françaises d'orient and commandement des armées alliés en orient – Corps expéditionnaire d'orient (C.E.O.) (22 February to 4 October 1915) and Corps expéditionnaire des Dardanelles (C.E.D) (4 October 1915 to 6 January 1916).'

Recorded in the *Inventaire sommaire des archives de la guerre 1914-1918* was the contents of all of the archive boxes stored at the Service historique de la défense at Château de Vincennes. I was provided with five dust-covered boxes from the archive that held maps and other documents.

One particular box – Box 20N33 – contained official military maps, sketches and reports. But this box also contained one additional document – a commercially produced *Colour map of Europe and Turkey*, folded and reinforced with linen at the folds (Figure 5.1). It did not fit with the other documents.

FIGURE 5.1 Annotated map of Europe and Asia Minor.1. *Source: Service historique de la défense, Département de l'armée de la division des Terres, Château de Vincennes, Paris. Image: William Cartwright. 29 June 2009.*

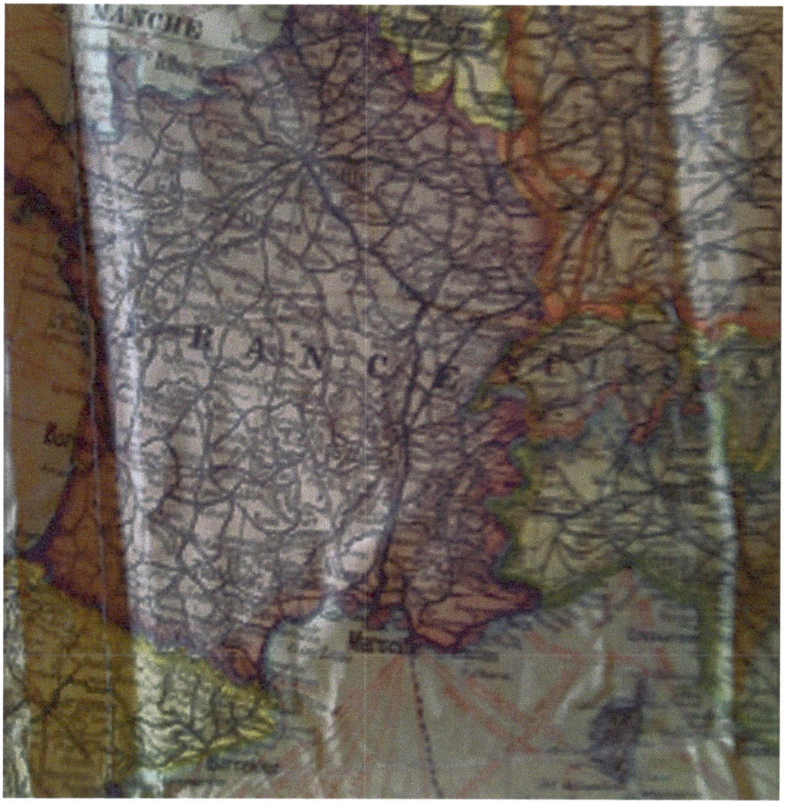

FIGURE 5.2 Start of the journey overland trip to Marseilles [by train?]. *Source: Service historique de la défense, Département de l'armée de la division des Terres, Château de Vincennes, Paris. Image: William Cartwright. 29 June 2009.*

Its possible owner, H. Barrot, had annotated this map with a 'travel line' from Paris to Marseilles to the Dardanelles. (This name was written on the verso.) The map had a pencil line drawn over the shipping lines that were included in the map and it traced his journey from Paris to the Dardanelles. It was a record of Barrot's involvement in the preliminary movement of troops before the landings at Gallipoli. This representation shows part of Barrot's personal geography of the campaign.

The annotations showed the overland trip to Marseilles [by train?] (Figure 5.2), then the sea voyage from Marseilles to Bizerte, Tunisia, and then between Sicily and Malta following regular shipping lanes and then south of Crete, and finally turning to sail north to a point (marked 'E', by hand) to a point just to the east of Lemnos (Figure 5.3).

What appears to be a sketch map (Figure 5.4) is on the verso. This map is yet to be deciphered.

This find at the Service historique de la défense, Château de Vincennes, Paris was the catalyst that led me to become interested in how personal geolocated geographies, here a simple annotated paper map, might better tell human stories behind major events. In this example, the story of the journey of just one person – Barrot – was just one element of the entire World

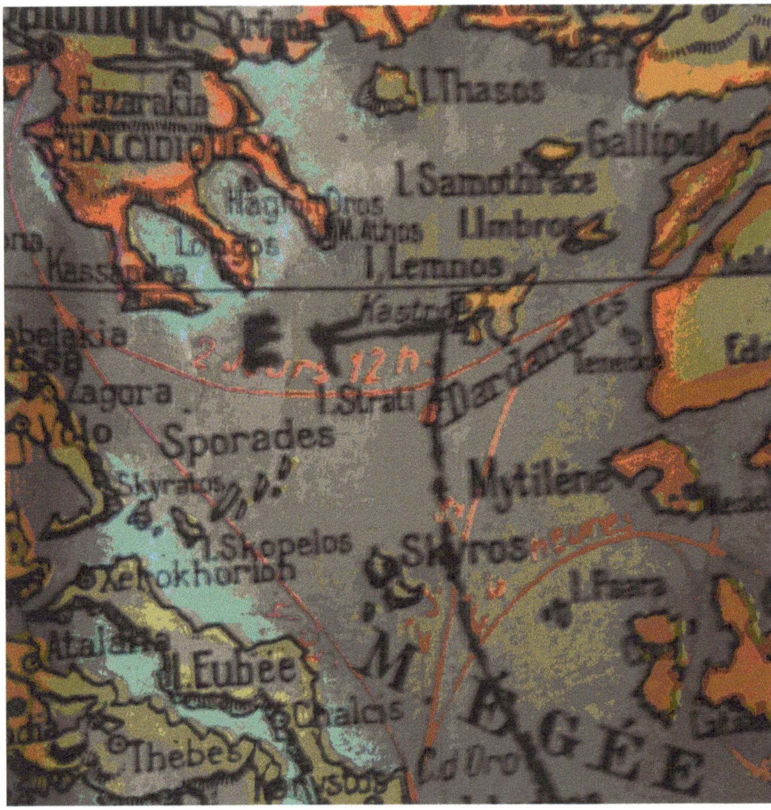

FIGURE 5.3 Turning to sail north to a point (marked 'E', by hand), just to the east of Lemnos. *Source: Service historique de la défense, Département de l'armée de la division des Terres, Château de Vincennes, Paris. Image: William Cartwright. 29 June 2009.*

War I story. Barrot's simple annotation, with just a pencil on a map, provided an insight into the human movement and displacement that wars bring into play.

5.2 NARRATIVES OF 'PERSONAL GEOGRAPHIES'

This narrative of Barrot's personal geography led me to develop a concept about how interactive integrated media might be employed to better communicate an individual's story. Here, in this example, the theatre of war provided the stage upon which Barrot's story is told via a geolocated narrative: the annotated map. This concept provided the vehicle to develop a resource that would enable me to construct a narrative that would realize ideas related to providing personal stories – linked to place – about everyday Australians who volunteered to fight 'for King and Country' far from home in World War I. More particularly, it would look at Australians who fought in the Gallipoli Campaign of 1915 in the Dardanelles.

FIGURE 5.4 Sketch map on verso. Contents to be deciphered. *Source: Service historique de la défense, Département de l'armée de la division des Terres, Château de Vincennes, Paris. Image: William Cartwright. 29 June 2009.*

This project entailed seeking geographic locations from archival resources – letters, photographs, military records, Red Cross reports, diaries, annotated publications and maps. Once a location was determined, this would be the basis of (geo)tagging that resource and using this tag to link it to similar resources and resources that covered a similar geographical area. This process has been called 'geotranscription' by Pyne (2009).

The underpinnings of the methodologies that were used in this project fall under the 'umbrella' of the theories and practices that have been developed by the contributing members of the Cybercartography (Taylor, 1997, 2003; Taylor and Pyne, 2010) research community. The Cybercartography research group at the Geomatics and Cartographic Research Centre at Carleton University developed a number of geonarrative products, based around the idea of 'collective memories', which used (geo)location as the 'glue' that held the various components of the narrative together, as well as facilitating links to other like media and resources. These products included *The Cybercartographic Atlas of Indigenous Perspectives and Knowledge of the Great Lakes Region* (Caquard et al., 2009), *Inuit siku Atlas* (Chapter 14 in this volume) and *The Cybercartographic Atlas of the Lake Huron Treaty Relationship Process* (Pyne, 2009 and Chapter 17 in this volume).

In this project, the data are organized according to (geo)location and contemporary interactive software is used to develop a collaborative resource that can be accessed and contributions made via the Web, more specifically Web 2.0. It presents multiple media resources to construct a Web-delivered geonarrative.

I have a personal interest in developing a methodology for presenting a personal geography of warfare. My great uncle, the brother of my paternal grandfather was killed in the 'Gallipoli' campaign of World War I. His name is John Henry Cartwright. This campaign, which took the lives of many soldiers of the new nation of Australia, is remembered in Australia as when Australia became a nation in its own right – independent of 'mother England'.

Therefore, a pilot project was developed sourcing Web-delivered data and artefacts, from Australian and international on-line archives to explore how a geolocated narrative might be built. It focused initially on one soldier in the Gallipoli campaign – John Henry Cartwright – who enlisted in the 14th Battalion, Australian Imperial Forces (AIF). His Battalion left Australia in December 1914 and he died in Gallipoli in May 1915. This pilot project is outlined later in this chapter.

5.3 GEOLOCATED NARRATIVE

A personal geography can be 'mapped' in different ways using interactive integrated media. It can 'map' not just the physical geography of a terrain or landscape, but also more personal, human geographies as well. Constructing personal narratives of geography using interactive integrated media can enhance and personalise geographic information. Some users may wish to be 'told' a story while they view map or graphical depictions on a screen. The narrative can be presented using multiscreens to deliver digital facsimiles of artefacts that exist at key nodes of a story, or build digital timelines akin to Barrot's marked-up map of his journey from Paris to the Dardanelles, outlined earlier in this paper.

The geolocated narrative therefore falls into the category of oral storytelling, whereby using personal documentation (each having its determinable 'place' – a location, a point in time, a 'located' event or experience) gives authority from lived experience. Authority is given to the information telling due to the fact that, even though the author is absent, documentation and artefacts recording a key event that provide information where a place can be determined can be used to provide the elements of a narrative that, when composed into a comprehensive whole, can provide a rich resource for understanding events and experiences. As in writing, where communicating with words tells the story, delivering a geolocated narrative also needs the input of artistry to build a credible and informative information resource. This has mostly been done using the medium of print. However, digital delivery gains authority and credibility by the ability and skill used to write and author packages and generally mastering the medium. Content is important, but the context is coloured by the authority of the 'teller'.

For geography, narratives can just provide statements of facts, where no embellishment is required and the user only wants to know 'the facts'. These facts can be stand-alone, or supported by expert resources, that provision the reader with rich information relating to the geographical space being explored. It may be a documentary-type video, supported by comprehensive and interactive artefacts. The narrative can 'walk' a user through 'unknown territory', but one that is richly illustrated using artefacts that are key tools for better understanding the story through the provision of background information. Users may construct their own story, or be 'talked' through an area, where they construct their own story, selecting and using support materials and navigation aids that are part of an interactive suite of information. Finally, they may decide that they wish to experience what has happened in someone's life by investigating a 'personal landscape' through an interactive geolocated narrative.

5.4 BUILDING A GEOLOCATED NARRATIVE

An effective design strategy for building a geolocated narrative would be to allow users easy and structured access to comprehensive map and graphics sets and complementary textual, audio-, and video-support materials. Using interactive integrated media, users can aggregate or disaggregate the screen image and select displays, allowing for a display to be 'assembled' by users that best suits their own particular needs. Users can see exactly what they want to see; the system becomes transparent and can be used without the need to consider usage restrictions.

When producing conventional map or atlas products, the terms of scale, symbolisation, classification and generalisation are used to define the amount of information provided and the detail illustrated. It is argued that the same terms can be applied to the geolocated narrative, whereby the stories are geographically referenced and 'scaled' and their contents classified and generalised (i.e. more generalised stories developed). For example, when using scale, the scale chosen determines the amount of detail that a story provides. A very small scale will cover a large area, but only provide general details. At a large scale a much smaller geographical 'footprint' is made, and the story provided will contain much more detailed text. Accordingly, scale determines how much a reader can 'zoom' into a story. Similarly, how information is symbolised dictates the genre of the story. Classification allows for similar stories to be clustered and assembled in a hierarchy. Generalisation is directly related to scale – the smaller the scale the greater the generalisation of the storytelling.

5.5 IDENTIFYING PLACE

For the prototype 'map' to be produced, it was necessary to investigate individual records to identify the actual 'place' where each event in John Henry Cartwright's 'personal geography of warfare' occurred. As noted earlier, the records accessed were mainly from online repositories. The idea was to determine the 'place' where each event occurred – a personal node in an overall 'space'.

French Sociologist Michel de Certeau proposed the spatial practices of everyday life in a chapter of his book, *The Practice of Everyday Life* (1984). In the chapter, entitled 'Spatial Stories', de Certeau distinguished between space and place (1984:117) saw PLACE as 'the order (of whatever kind) in accord with which the elements are distributed in relationships of coexistence'. Here elements are defined by their relative positions to each other, making a place 'an instantaneous configuration of positions' (Pyne, 2009), implying 'an indication of stability' (Pyne, 2009). Therefore, every PLACE so defined, has a unique geographical characteristic.

Therefore, when inspecting the resources identified, the actual 'place' where an event took place was important to ascertain. This could be the name of a town, a street address, the name of an Army Battalion, or a hill somewhere on the Gallipoli Peninsula. Once these places were determined a timeline could be assembled, and the places subsequently mapped – as a timeline. The prototype cartographic interactive artefact would represent place + time, and in so doing represent the personal geography of warfare of John Henry Cartwright.

Like many Australians of his generation, John Henry Cartwright enlisted in the AIF to serve 'King and Country'. He joined the AIF at 31 years of age and he traded his job as a farm hand in Korumburra, rural Victoria, for one where tools of warfare were exchanged for tools of agriculture.

His route to the Gallipoli campaign took him from Korumburra, to the Broadmeadows training camp, to Egypt, Greece and then to the Dardanelles. He arrived in Turkey, on the Gallipoli Peninsula with the 14th Battalion of the AIF and it was here that he met his fate – missing in action 18 August 1915, later reported killed in action.

The Australian War Memorial, in Canberra, has digitized the records of all Australians who served in World War I. It was in these on-line archives that I was able to find the story of John Henry's 'personal geography of warfare' – one that stretched from a small Australian farming and timber town to his final resting place in Turkey. An investigation of these resources found original AIF records, reports from the Red Cross that formally established his demise, to the records of the posthumous awarding of military medals to his family. Some of these records are shown in figures later in this chapter (Figure 5.5).

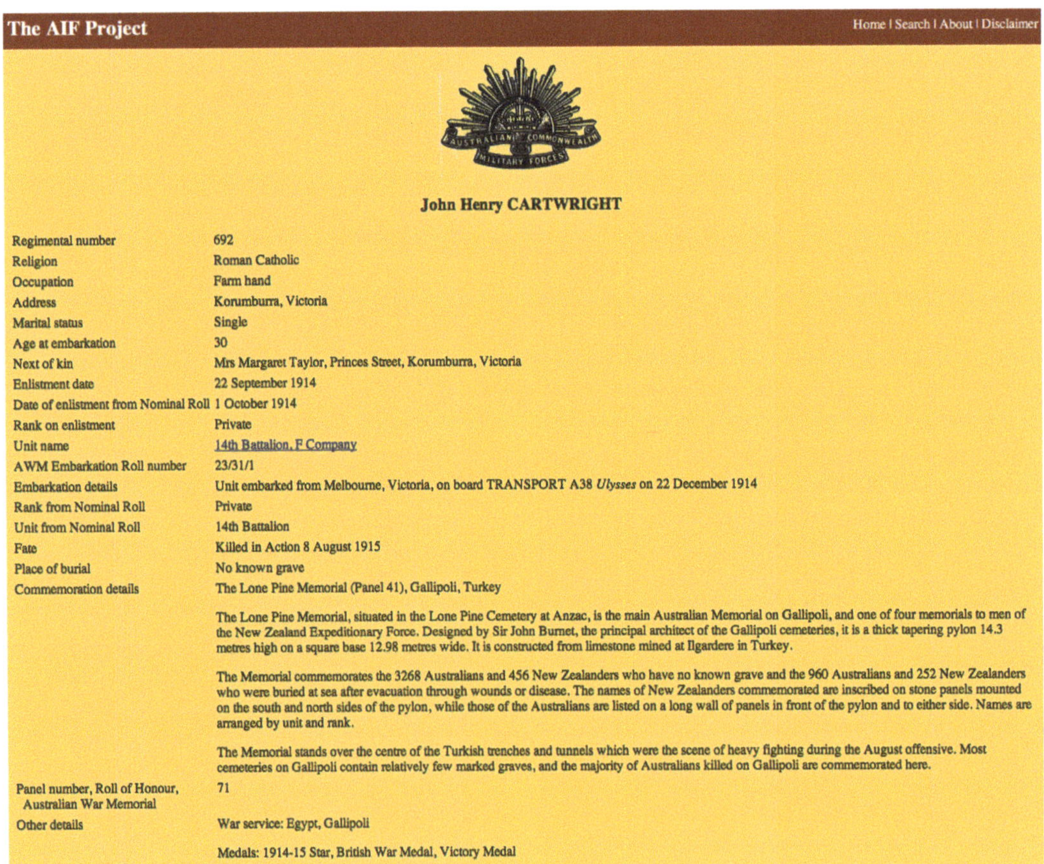
The AIF Project Home | Search | About | Disclaimer

John Henry CARTWRIGHT

Regimental number	692
Religion	Roman Catholic
Occupation	Farm hand
Address	Korumburra, Victoria
Marital status	Single
Age at embarkation	30
Next of kin	Mrs Margaret Taylor, Princes Street, Korumburra, Victoria
Enlistment date	22 September 1914
Date of enlistment from Nominal Roll	1 October 1914
Rank on enlistment	Private
Unit name	14th Battalion, F Company
AWM Embarkation Roll number	23/31/1
Embarkation details	Unit embarked from Melbourne, Victoria, on board TRANSPORT A38 *Ulysses* on 22 December 1914
Rank from Nominal Roll	Private
Unit from Nominal Roll	14th Battalion
Fate	Killed in Action 8 August 1915
Place of burial	No known grave
Commemoration details	The Lone Pine Memorial (Panel 41), Gallipoli, Turkey The Lone Pine Memorial, situated in the Lone Pine Cemetery at Anzac, is the main Australian Memorial on Gallipoli, and one of four memorials to men of the New Zealand Expeditionary Force. Designed by Sir John Burnet, the principal architect of the Gallipoli cemeteries, it is a thick tapering pylon 14.3 metres high on a square base 12.98 metres wide. It is constructed from limestone mined at Ilgardere in Turkey. The Memorial commemorates the 3268 Australians and 456 New Zealanders who have no known grave and the 960 Australians and 252 New Zealanders who were buried at sea after evacuation through wounds or disease. The names of New Zealanders commemorated are inscribed on stone panels mounted on the south and north sides of the pylon, while those of the Australians are listed on a long wall of panels in front of the pylon and to either side. Names are arranged by unit and rank. The Memorial stands over the centre of the Turkish trenches and tunnels which were the scene of heavy fighting during the August offensive. Most cemeteries on Gallipoli contain relatively few marked graves, and the majority of Australians killed on Gallipoli are commemorated here.
Panel number, Roll of Honour, Australian War Memorial	71
Other details	War service: Egypt, Gallipoli Medals: 1914-15 Star, British War Medal, Victory Medal

FIGURE 5.5 Record from the 'The AIF Project' site. *Source:* http://www.aif.adfa.edu.au/showPerson?pid=48258.

In building the personal geography of warfare for John Henry Cartwright, these records, and other records, maps and imagery were used as resources to ascertain how best to build a geographic narrative to 'map' this personal geography.

Looking at samples of the online artefacts resourced, this would provide:

- His address when he volunteered – Korumburra, Victoria (Figures 5.1 and 5.6);
- His Battalion – 14th Battalion AIF (Figure 5.6);
- Departed Australia, 22 December 1914 – on the HMAT A38 'Ulysses' (Figure 5.8);
- Where he was wounded, 19 May 1915 – Dardanelles (Figure 5.7)
- Transfer to hospital, 20 June 1915 – Heliopolis (Figure 5.7);
- Transfer to Convalescent Camp, 22 June 1915 – Heliopolis (Figure 5.7);
- Embarkation on Scotia, 4 July 1915 – to Dardanelles (Figure 5.7);
- Reported missing, 8 August 1915 – Gallipoli Peninsula (Figure 5.7);
- Reported killed in action, 8 August 1915 – Gallipoli Peninsula (Figure 5.7);
- Memorial at Lone Pine Memorial, Gallipoli, Turkey (Figure 5.9);
- Correspondence regarding awarding of medals,13 December 1919 – Princes Street Korumburra and 4 May 1920, Errol Street North Melbourne (Figure 5.10)
- Application for 'New Anzac medal' by sister, Henrietta Barrot, 10 May 1967 – Elymminyt, Victoria (Figure 5.11).

These record examples provided both information and rich media artefacts. They were incorporated into a timeline, here constructed from sourced 'places'.

5.6 WEB 2.0 GEOLOCATED NARRATIVE PILOT PROJECT

This pilot project used geographically referenced interactive integrated media artefacts to tell this particular story – the personal geography about one individual who participated in the Gallipoli campaign – John Henry Cartwright. It was developed to address the methods needed to intertwine real geographies with personal geographies. Also, it explored how links could be made between georeferenced significant landscapes, places, and art that can be used to 'ground truth' the collection of personal geographies – building a narrative that is linked to space and place. This product was built using a contemporary Web 2.0 integrated media 'montage'-building tool. The tool can be employed to navigate through the narrative 'place' and to link between the personal and physical geographical spaces, providing insight into an individual's part in a campaign, where the wholesale weight of warfare and the war 'machine' can hide individual sacrifices. The prototype was developed using the *Timeline* package from VeriteCo (http://timeline.verite.co/).

The first attempt to build the *Timeline* was done using the spreadsheet template provided by the software provider. Whilst easy to use, the methods were far from satisfactory. The spreadsheet proforma, to which the developer replaces sample information with actual project information, is shown, below, in Figure 5.12.

This approach was abandoned and the user interface developed by Chris Marmo, from RMIT University, was employed. Chris has programmed an interactive interface that generates the timeline online. 'Places' located in the timeline thus far are shown in Figures 5.13–5.19.

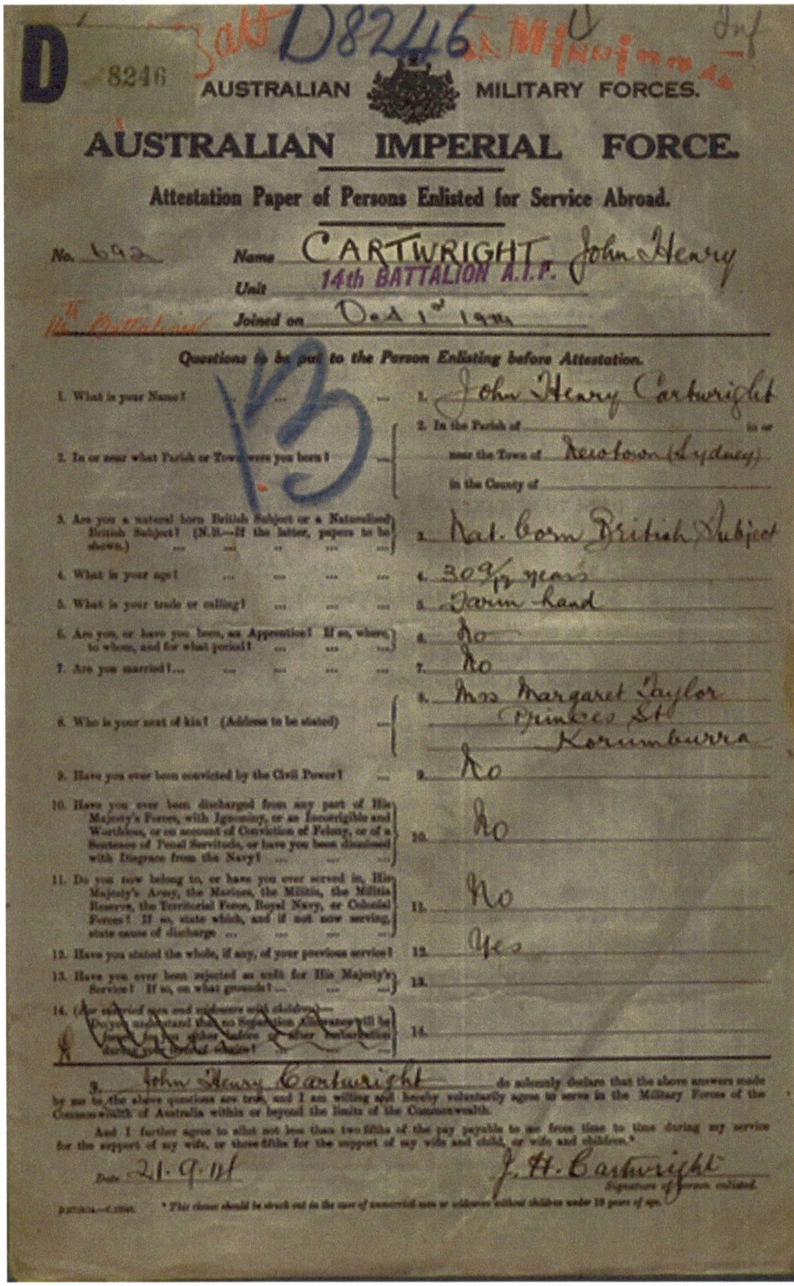

D 8246

D8246

AUSTRALIAN MILITARY FORCES.

AUSTRALIAN IMPERIAL FORCE.

Attestation Paper of Persons Enlisted for Service Abroad.

No. 692 Name CARTWRIGHT John Henry

Unit 14th BATTALION A.I.F.

Joined on Oct 1st 1914

Questions to be put to the Person Enlisting before Attestation.

1. What is your Name? ... 1. John Henry Cartwright
2. In or near what Parish or Town were you born? ... 2. In the Parish of ______ in or near the Town of Newtown (Sydney) in the County of ______
3. Are you a natural born British Subject or a Naturalized British Subject? (N.B.—If the latter, papers to be shown.) ... 3. Nat. born British Subject
4. What is your age? ... 4. 30 9/12 years
5. What is your trade or calling? ... 5. Farm hand
6. Are you, or have you been, an Apprentice? If so, where, to whom, and for what period? ... 6. No
7. Are you married? ... 7. No
8. Who is your next of kin? (Address to be stated) ... 8. Mrs Margaret Taylor, Princes St, Korumburra
9. Have you ever been convicted by the Civil Power? ... 9. No
10. Have you ever been discharged from any part of His Majesty's Forces, with Ignominy, or as Incorrigible and Worthless, or on account of Conviction of Felony, or of a Sentence of Penal Servitude, or have you been dismissed with Disgrace from the Navy? ... 10. No
11. Do you now belong to, or have you ever served in, His Majesty's Army, the Marines, the Militia, the Militia Reserve, the Territorial Force, Royal Navy, or Colonial Forces? If so, state which, and if not now serving, state cause of discharge ... 11. No
12. Have you stated the whole, if any, of your previous service? 12. Yes
13. Have you ever been rejected as unfit for His Majesty's Service? If so, on what grounds? ... 13.
14. ~~(For married men and widowers with children)—~~ 14.

I, John Henry Cartwright do solemnly declare that the above answers made by me to the above questions are true, and I am willing and hereby voluntarily agree to serve in the Military Forces of the Commonwealth of Australia within or beyond the limits of the Commonwealth.

And I further agree to allot not less than two-fifths of the pay payable to me from time to time during my service for the support of my wife, or three-fifths for the support of my wife and child, or wife and children.*

Date 21.9.14

J. H. Cartwright
Signature of person enlisted.

* This clause should be struck out in the case of unmarried men or widowers without children under 18 years of age.

FIGURE 5.6 Attestation paper. *Source: Mapping Our Anzacs. Record NAA: B2455, Cartwright J H. Page 1.* mappingouranzacs.naa.gov.au.

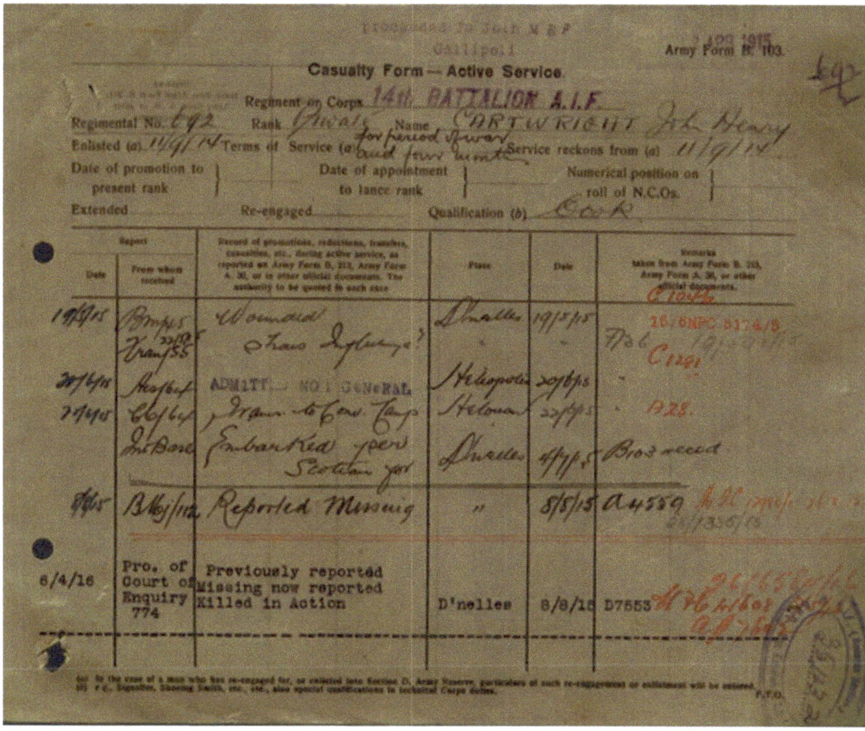

Army Form B. 103.

Casualty Form — Active Service.

Regiment or Corps 14th BATTALION A.I.F.

Regimental No. 692 Rank Private Name CARTWRIGHT John Henry

Enlisted (a) 11/9/14 Terms of Service (a) for period of war and four months Service reckons from (a) 11/9/14

Date of promotion to present rank | Date of appointment to lance rank | Numerical position on roll of N.C.Os.

Extended | Re-engaged | Qualification (b) Cook

Date	From whom received	Record of promotions, reductions, transfers, casualties, etc., during active service, as reported on Army Form B. 213, Army Form A. 36, or in other official documents. The authority to be quoted in each case	Place	Date	Remarks taken from Army Form B. 213, Army Form A. 36, or other official documents.
19/5/15		Wounded	D'nelles	19/5/15	
20/6/15		ADMITTED HOSP GENERAL	Heliopolis	20/6/15	
		Embarked per Scotian	D'nelles		
8/8/15		Reported Missing	"	8/8/15	
6/4/16	Pro. of Court of Enquiry 774	Previously reported Missing now reported Killed in Action	D'nelles	8/8/15	D7553

FIGURE 5.7 Casualty Form – Active Service. *Source: Mapping Our Anzacs. Record NAA: B2455, Cartwright J H. Page 12.* mappingouranzacs.naa.gov.au.

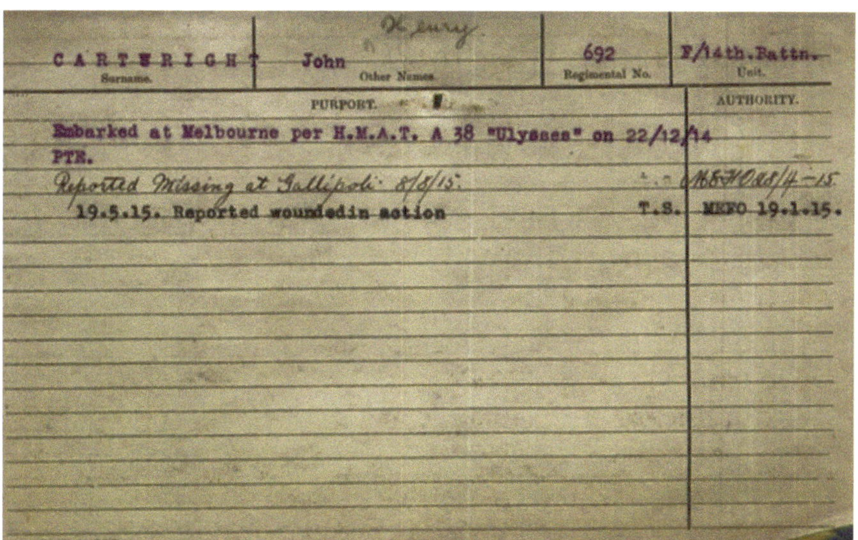

Surname	Other Names	Regimental No.	Unit
CARTWRIGHT	John Henry	692	F/14th.Battn.

PURPORT.	AUTHORITY.
Embarked at Melbourne per H.M.A.T. A 38 "Ulysses" on 22/12/14	
PTE.	
Reported Missing at Gallipoli 8/8/15.	
19.5.15. Reported woundedin action	T.S. MEFO 19.1.15.

FIGURE 5.8 Index Card. *Source: Mapping Our Anzacs. Record NAA: B2455, Cartwright J H. Page 10.* mappingouranzacs.naa.gov.au.

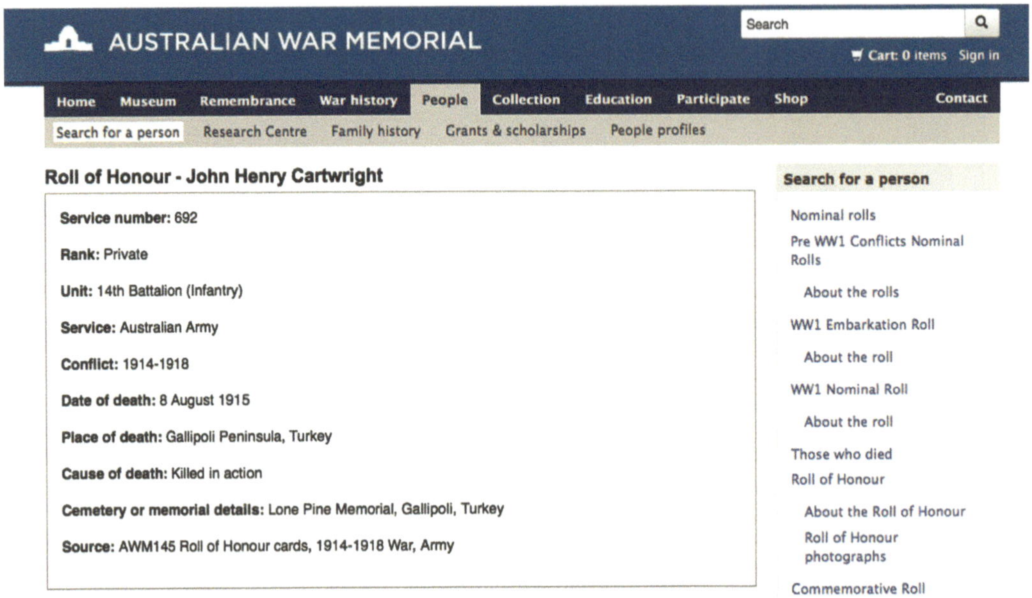

FIGURE 5.9 Record from the Australian War Memorial Web site. *Source:* http://www.awm.gov.au/research/people/roll_of_honour/person.asp?p=521496.

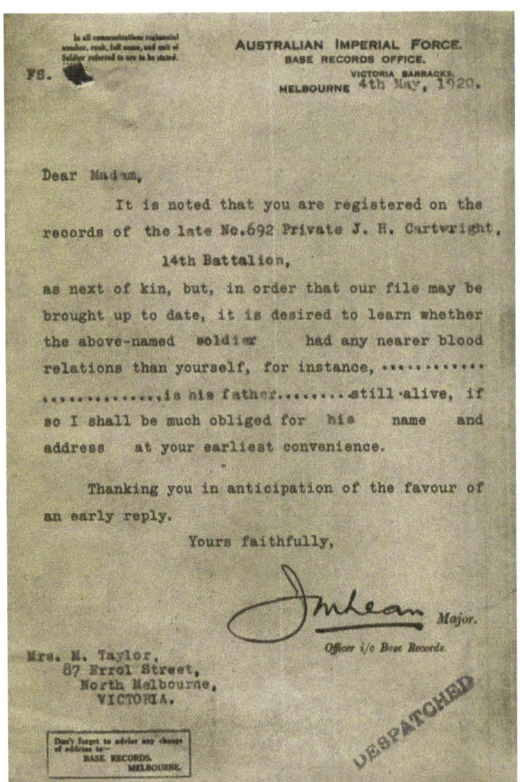
In all communications regimental number, rank, full name, and unit of Soldier referred to are to be stated.

FS.

AUSTRALIAN IMPERIAL FORCE.
BASE RECORDS OFFICE.
VICTORIA BARRACKS,
MELBOURNE 4th May, 1920.

Dear Madam,

It is noted that you are registered on the records of the late No.692 Private J. H. Cartwright, 14th Battalion, as next of kin, but, in order that our file may be brought up to date, it is desired to learn whether the above-named soldier had any nearer blood relations than yourself, for instance, ...is his father........still alive, if so I shall be much obliged for his name and address at your earliest convenience.

Thanking you in anticipation of the favour of an early reply.

Yours faithfully,

J. M. Lean Major.
Officer i/c Base Records.

Mrs. M. Taylor,
87 Errol Street,
North Melbourne,
VICTORIA.

DESPATCHED

Don't forget to advise any change of address to— BASE RECORDS, MELBOURNE.

G.4522.

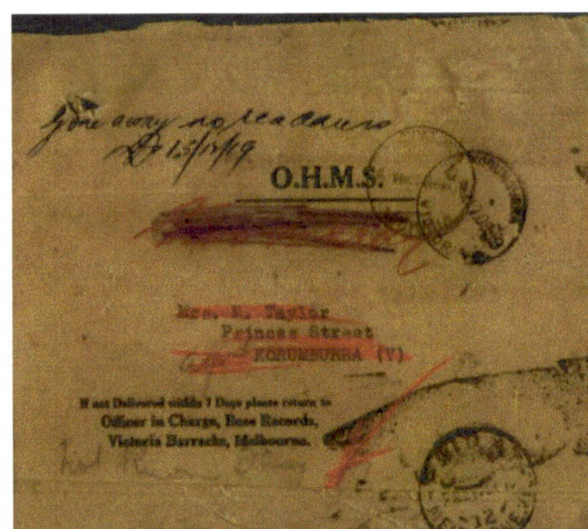

FIGURE 5.10 Left: Envelope: Letter to Mrs M. Taylor from Officer in Charge, Base Records, Victoria Barracks, Melbourne. *Source: Mapping Our Anzacs. Record NAA: B2455, Cartwright J H. Page 20. Right: Letter to Mrs M. Taylor from Officer in Charge, Base Records, Victoria Barracks, Melbourne Record NAA: B2455, Cartwright J H. Page 24* mappingouranzacs.naa.gov.au.

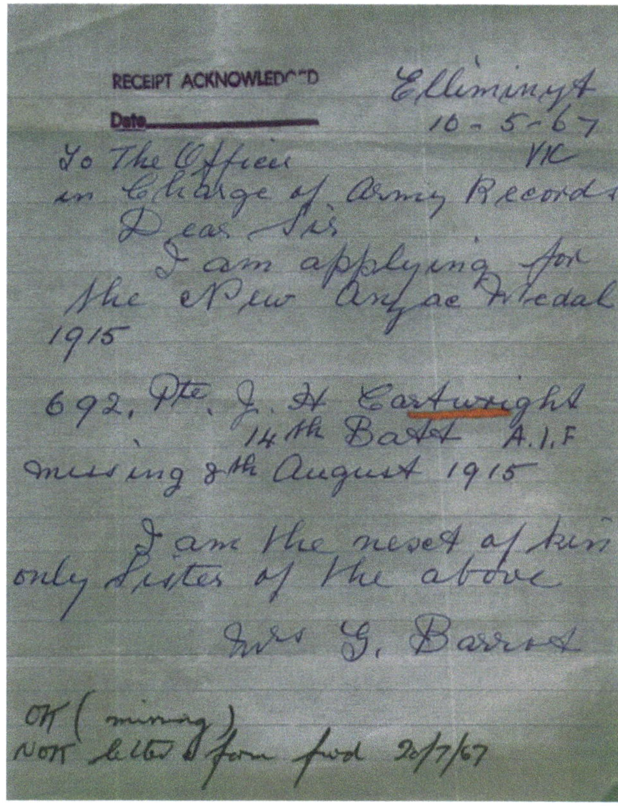
RECEIPT ACKNOWLEDGED
Date
Elliminyt
16-5-67
VIC
To The Officer
in Charge of Army Records
Dear Sir
I am applying for
the New Anzac Medal
1915

692, Pte. J. H. Cartwright
14th Batt A.I.F
missing 8th August 1915

I am the next of kin
only Sister of the above
Mrs G. Barrot

OK (missing)
NOK letter & form fwd 20/7/67

FIGURE 5.11 Letter from Mrs G. Barrot to Officer in Charge, Base Records, Victoria Barracks, Melbourne. *Source: Mapping Our Anzacs. Record NAA: B2455, Cartwright J H. Page 32.* mappingouranzacs.naa.gov.au.

What these example pages show is that a 'place' can be a town, a ship at sea, a battalion and a place of rest. The timeline links to images in the Australian War memorial Archive, to pictures in the German Military archive and to a rare digitized film of the initial landings at Anzac Cove, Gallipoli. The 'work in progress' can be viewed at: http://timeline.prosimian.com.au/index.php/timeline/view/22.

5.7 FURTHER DEVELOPMENT

The project continues to complete the prototype timeline, with the addition of additional artefacts and information related to John Henry Cartwright. As well, other personal geographies of individuals will be added. Research is being undertaken to identify other volunteer soldiers from the township of Korumburra who went to Gallipoli. The starting point for identifying these men is the inscriptions on the War memorial in the town of Kurumburra (Figure 5.20).

To automate the processes involved in producing the timeline work is underway to develop techniques to automatically extract 'place' information from archive artefacts. It is considered that things like addresses, location, the name of a transport ship, etc. could be automatically extracted from documents that have been scanned and made available as PDFs.

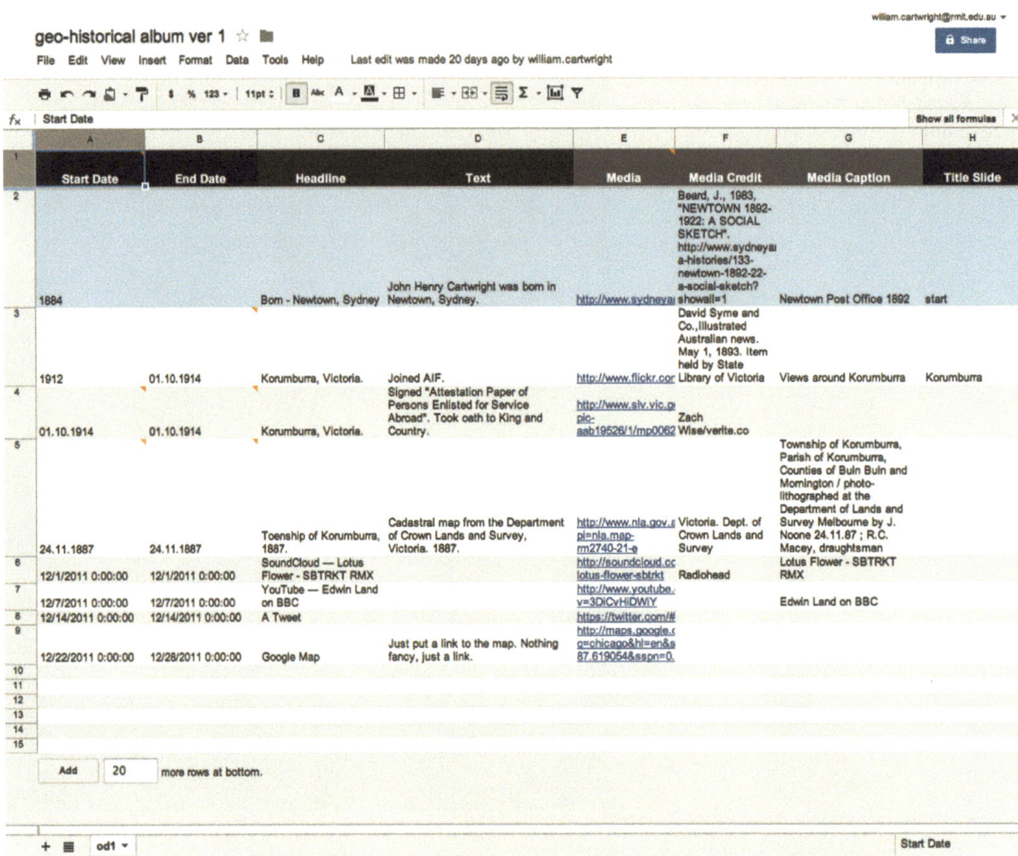

FIGURE 5.12 Timeline development spreadsheet.

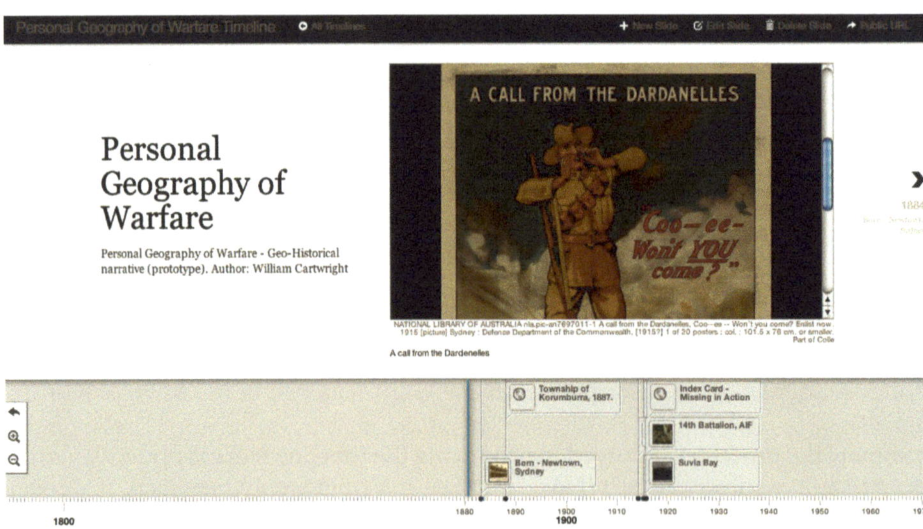

FIGURE 5.13 Personal Geography of Warfare Timeline – Title page.

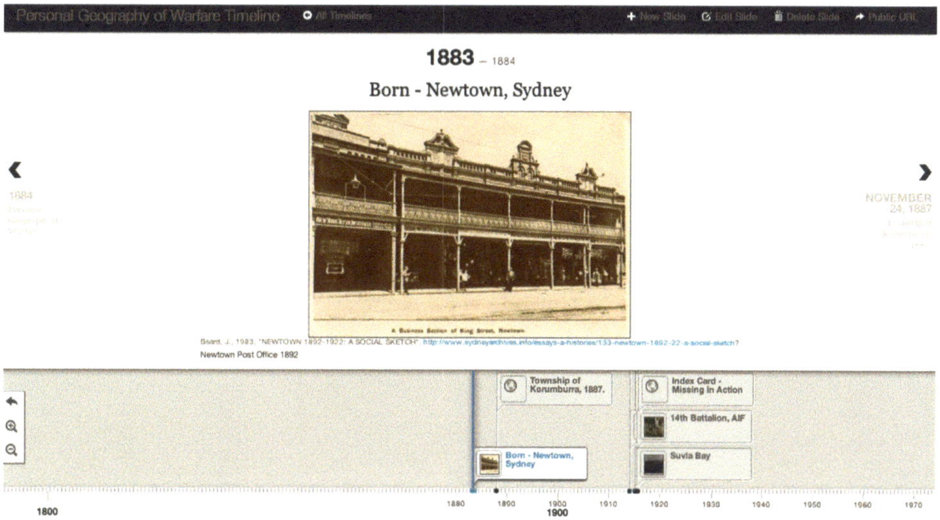

FIGURE 5.14 Personal Geography of Warfare Timeline – 1883.

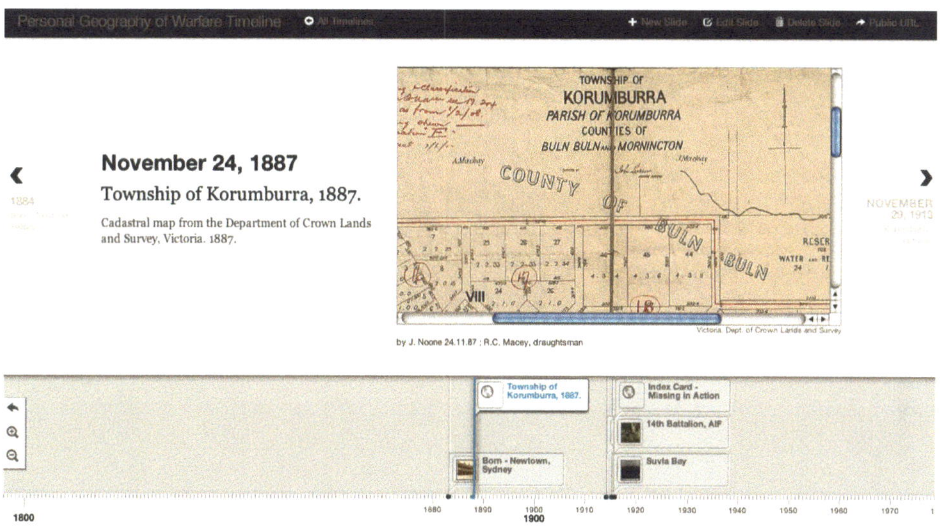

FIGURE 5.15 Personal Geography of Warfare Timeline – November 24, 1887.

5.8 POTENTIAL FOR BUILDING GEOLOCATED NARRATIVES USING WEB 2.0 AND PARTICIPATORY SOFTWARE

Rather than just develop the timeline personally and add additional nodes about other soldiers from Korumburra that would be sourced from archives a better methodology would be to invite families of these soldiers to contribute via a Web 2.0 resource. As noted

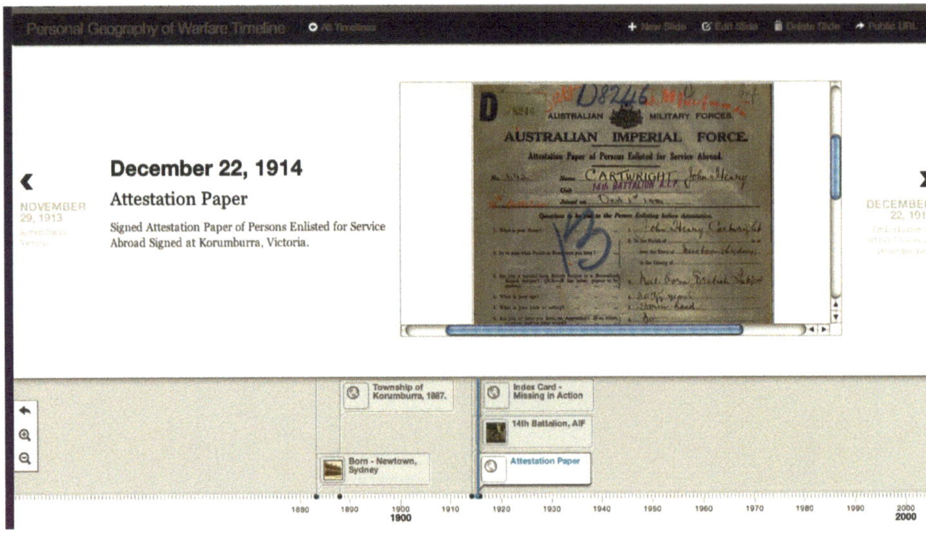

FIGURE 5.16 Personal Geography of Warfare Timeline – December 22, 1914.

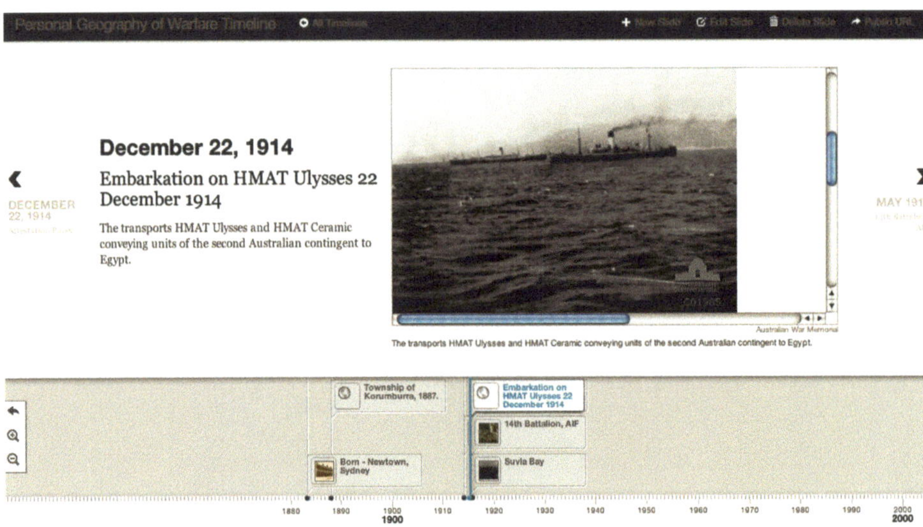

FIGURE 5.17 Personal Geography of Warfare Timeline – December 1914.

earlier in the chapter, a Timeline application was developed by Chris Marmo, which made building the Timeline much more efficient than using the spreadsheet technique recommended by the Verite company. Further development of the prototype will include participatory tools.

During the development of the prototype, it was obvious that archives, including war-related archives, are a source of not just information, but also of rich media artefacts. These

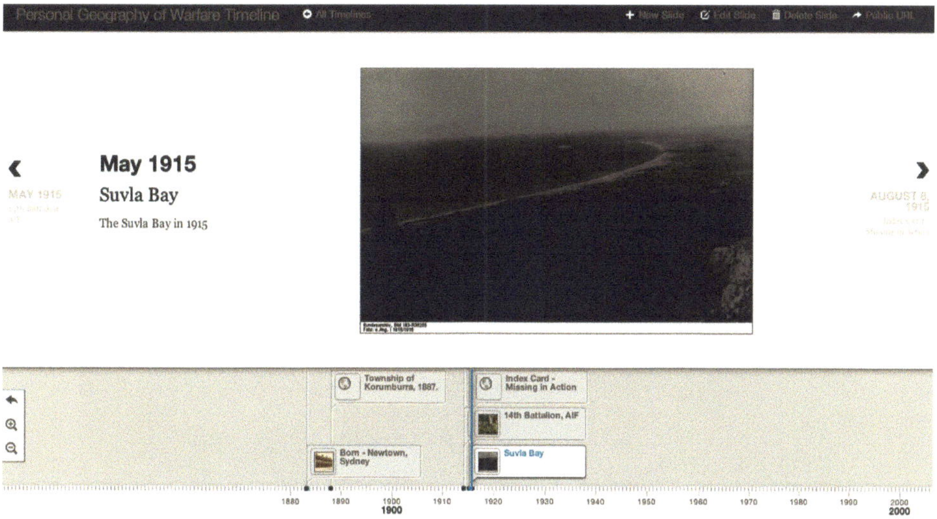

FIGURE 5.18 Personal Geography of Warfare Timeline – May 1915.

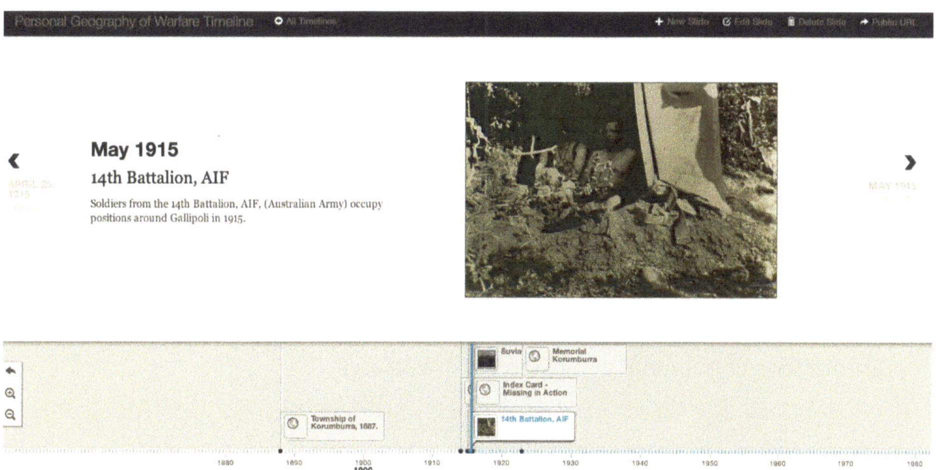

FIGURE 5.19 Personal Geography of Warfare Timeline – May 1915.

artefacts can be 'placed' – in three space + time – by extracting relevant information from the documents. This can be a street address, and army battalion, a battle site, a transport ship, the location and date of a parade, or addresses from where correspondence was sent and received. If the notion of 'place' is applied in the broadest sense, narratives can be geolocated and personal geographies depicted. Web 2.0, the development of appropriate software and the provisioning of tools that allow and encourage the collaborative building of digital, annotated narratives affords the development and provisioning of resources that can inform affectively.

FIGURE 5.20 War Memorial Korumburra, Victoria, Australia. *Source: Image: William Cartwright.*

5.9 CONCLUSION

What has been outlined here is a work in progress. It has described the concept of developing a timeline – depicting 'place' and time – to 'map' a personal geography of warfare. The prototype is developing a representation of John Henry Cartwright's personal geography of warfare, which began at Korumburra, Victoria and ended 'at the top of Australia Gully, near Hill 50', Gallipoli Peninsula (Australian Red Cross Society, 1916). Work is continuing on the prototype. It is hoped that the techniques developed might be applied to the development of other representations of personal geographies of warfare.

5.10 POSTSCRIPT

John Henry Cartwright's personal geography continued after his demise. Well after the War his family received his personal belongings from the Australian Military. His sister, Henrietta (my Great Aunt), applied for the 'Anzac Medal'. His name was included at the Lone Pine Memorial in Gallipoli, with other Australians who sought adventure in a far place. The story of his sacrifice in World War I began when he enlisted in December 1914, but it continued after his death in May 1915 for decades after. The 'placed' story really finished in 1967, when Henrietta Barrot applied for the 'New Anzac Medal 1915'. Her letter can be 'placed' using her address when sending the letter.

One interesting sideline to this research is that the beginning and end of this story has the same family name: 'Barrot'. It is the family name of the soldier who annotated the map found (by chance) in the archives of the Service Historique de la Defense, Department de l'Armee de

FIGURE 5.21 Barrot's signature on verso. *Source: Service historique de la Défense, Département de l'armée de la division des Terres, Château de Vincennes, Paris. Image: William Cartwright. 29 June 2009.*

Terre Division, Château de Vincennes, Paris (Figure 5.21) and also the married family name of John Henry Cartwright's sister, Henrietta, Mrs George Barrot (nee Cartwright) (Figure 5.11). This requires further investigation.

Acknowledgements

Research to build a bibliography and cartobibliography was funded by a grant from the School of Mathematical and Geospatial Sciences, RMIT.

Research at archives in London and Paris was supported by Research Sabbatical leave from RMIT.

Thank you to research assistants who sourced references and online map archives. My thanks to staff at the archives and map repositories that were visited during this research. Without this invaluable assistance this stage of the research could not have been completed.

Thank you to Chis Marmo, who kindly provided the interface to the Timeline package.

References

Australian Red Cross, 1916. Wounded and Missing Enquiry Bureau Files, 1914-1918 War: 1DRL/0428 – 692 Private John Henry Cartwright, 14th Battalion. Available at: https://www.awm.gov.au/collection/records/1drl0428/2/70/9/1drl-0428-2-70-9-4.pdf.

Caquard, S.S., Pyne, H., Igloliorte, K., Mierins, A., Hayes, A., Taylor, D.R.F., 2009. A 'living' atlas for geospatial storytelling: the cybercartographic atlas of indigenous perspectives and knowledge of the Great Lakes region. Cartographica 44 (2), 83–100.

Nicot, J., Schillinger, P., Ficat, J., 1972. Inventaire sommaire des archives de la guerre 1914–1918. Ministère d'état charge de la défense nationale.

Pyne, S., 2009. Spatializing history: the cybercartographic atlas of the Lake Huron Treaty process. In: Hele, K. (Ed.), This Is Indian Land: The Robinson Huron Treaties of 1850.

Taylor, D.R.F., 1997. Maps and mapping in the information era. In: Ottomson, L. (Ed.), Proceedings 18th International Cartographic Conference, Stockholm, vol. 1. Swedish Cartographic Society, Galve, pp. 1–10.

Taylor, D.R.F., 2003. The concept of cybercartography. In: Peterson, M. (Ed.), Maps and the Internet, Elsevier Science, Oxfod, pp. 405–420.

Taylor, D.R.F., Pyne, S., 2010. The history and development of the theory and practice of cybercartography. International Journal of Digital Earth 3 (1), 1–14.

CHAPTER

6

Tile-Based Mapping with Opacity

Andrew Clouston[1], Michael P. Peterson[2]

[1]Geography, Environment and Earth Sciences, University of Victoria, Wellington, New Zealand,
[2]Department of Geography/Geology, University of Nebraska at Omaha, Omaha, USA

OUTLINE

6.1 INTRODUCTION

Maps became a major component of the online information landscape in 1993 after graphical illustrations could be added to World Wide Web pages (Peterson, 2003). All types of maps became available, from subway maps of cities to maps of the moon. Map users could even choose between alternative representations of the same environment and determine the one

Developments in the Theory and Practice of Cybercartography, Second Edition, ISSN 1363-0814
http://dx.doi.org/10.1016/B978-0-444-62713-1.00006-4

that best suited their needs. Interactive sites made it possible to center the map on an area of interest and include features requested by the user.

The two decades since 1993 can be seen as a period of discovery as map providers experimented with the new medium for map distribution. Paper maps were initially scanned. Then, cartographers alternated between various vector and raster formats and forms of interactivity. A major goal was to increase the speed of map delivery. Map users did not want to wait for the map and a method was devised to appease the impatient online map user. This method involved dividing the map into pieces or tiles.

The tiling of maps, along with a more interactive form of communication, transformed the online mapping experience. Tiling is arguably the single major development in the first two decades of maps and the Internet. Soon after the method was introduced through Google Maps, all the other online map providers switched to the new form of map distribution. The introduction of Application Programming Interfaces (API) that allowed user information to be added to the maps solidified this form of online map presentation. Whether for good or bad, online mapping is currently in a tile-based mapping era and will likely be so for the foreseeable future (Peterson, 2011).

Map tiling addressed the continuing problem for the distribution of maps through the Internet: the size of the digital files (Sample and Elias, 2010). The slow dissemination of a map to a computer or mobile device is the single major limitation to their use. An additional concern is the usage impact on limited mobile data plans. Tile-based map delivery helps to speed the display of map through the Internet, reduce data transmission volumes, and to provide a number of other benefits when combined with transparent overlays.

This chapter explains tile-based mapping, the reasons for its development, current usage, and examines the use of an opacity control in a multilevel Google Maps environment.

6.2 DATA TRANSMISSION CONSIDERATIONS

Early Internet maps were simply scans of traditional paper maps. The size of the subsequent scanned digital map is a function of the original paper size, the resolution of the scanning, and the raster file format utilised for data storage.

Incorporating a colour further increases the size of files. Depending on the colour definition used, the size of the resultant file could be doubled or tripled. This is because only one byte can be used to display 256 shades of grey but multiple bytes are used to display shades of different colours. A two-byte colour image results in 'high colour' with 65,536 possible colours. 'True colour' with three bytes enables the storage of the red, green, and blue values for each pixel, thus providing nearly 17 million colours (256^3). Four bytes then enables transparency, by essentially adding an 'alpha' channel to store the percentage of translucency that will be applied to the red, green, and blue values (Wikipedia, 2012b). This then allows the background image to show through or blend with the foreground image, depending upon the compositing algorithm. Table 6.1 presents a summary of file sizes for a 600 dpi scan of A0 through A4 documents. Table 6.2 shows file sizes at three resolutions when scanned at 32-bit 'true colour' which is required to enable opacity.

TABLE 6.1 Files Sizes (mb) for a 600 dpi Scan of A0 through A4 Documents

Paper Size	NumPixels	1-bit (2 Colours)	8-bit (256 Colours)	16-bit (High Colour)	24-bit (True Colour)	32-bit
A0 1189 × 841	557972658	68.1	544.9	1089.8	1634.7	2179.6
A1 841 × 594	278751690	34.0	272.2	544.4	816.7	1088.9
A2 594 × 421	139375845	17.0	136.1	272.2	408.3	544.4
A3 421 × 297	69687922	8.5	68.1	136.1	204.2	272.2
A4 297 × 210	34843961	4.3	34.0	68.1	102.1	136.1

TABLE 6.2 File Sizes for a 32-bit 'True Colour' Scan for Various DPI Settings

Paper Size	300 dpi (mb)	600 dpi (mb)	1200 dpi (mb)
A0 1189 × 841	544.9	2179.6	8718.3
A1 841 × 594	272.2	1088.9	4355.5
A2 594 × 421.5	136.1	544.4	2177.7
A3 421 × 297	68.1	272.2	1088.9
A4 297 × 210.25	34.0	136.1	544.4

Transmitting files of this size through the Internet might take hours if not days depending upon the speed of transmission. As a result, four techniques have been applied to effectively distribute raster maps via the Internet:

1. Minimising the resolution;
2. Limiting the number of colours;
3. Compressing the image; and
4. Tiling.

Unfortunately, most static maps available through the Internet have focused on the first three options resulting in online maps with poor graphic quality. Often, image compression techniques render the map illegible.

6.2.1 Minimising Resolution

Not surprisingly, maps were often scanned at a low resolution, as halving the resolution yields an approximate 4-fold size reduction. Therefore, simply halving the resolution twice provides a near-95% size reduction. However, scanning at too low a resolution makes the map illegible, so a balance between file size and usability is required.

A common choice is to scan the map at the smallest resolution that still enables the map to be used. The resolution of choice was the screen resolution of the monitor – about 72 dpi. With the advent of laser printers, output resolutions increased to 300 dpi. Even when printer resolutions increased, scanning resolutions often didn't because computer monitor resolutions remained low, data transmission was still slow and disk storage was relatively expensive.

6.2.2 Limiting Colour

Similar to reducing resolution, the net effect of limiting colour is to reduce the number of bytes required to store and display the map. Because the human eye can only perceive a limited number of colours, scanning maps in 256 colour (8-bit) was common especially as existing paper maps tended to use limited colour palettes. To obtain greater clarity and definition, especially when using aerial photography as a backdrop, a larger colour palette associated with more bits was required. However, to utilise opacity, a high number of bits is required and therefore the option of limiting colour is now used less frequently.

6.2.3 Image Compression

Image compression applies algorithms to identify and remove information that is not critical to the perception of the image and then encode the remainder in a compact form. Developed primarily for pictures, there are two primary types: lossy and lossless. In general, lossy techniques are not advised for maps as, despite their excellent compression ratios, the quality of the image is irreparably damaged. The most common compression method, JPEG, tends to blur the appearance of lines – especially at low resolutions. In contrast, lossless image compression techniques retain the original quality but do not compress the image as efficiently. Examples include TIFF, GIF, and PNG. The 'multiresolution seamless image database' widely known as MrSID, a propriety format that is commonly used for geographic information systems imagery, is also considered a lossless format. Although enabling the map to be more efficiently distributed, all forms of image compression fail to meet the needs of those who require access to a digital map that retains similar accuracy to the original version.

6.2.4 Tiling

To address the data transmission problem, splitting the maps into very small digital 'seamless' tiles has become commonplace. Each tile is typically composed of 256 × 256 pixels for a total of 65,536 pixels. This technique is not entirely unlike traditional paper 'series' mapping, such as that commonly undertaken by government mapping agencies. For example, in topographic mapping a country is covered by adjoining individual maps at a single scale. In the New Zealand context, the current Topo50 and Topo250 map series are examples (see Figure 6.1).

Despite similarities to 'series' mapping, tiling an image for the computer has its roots in computer video games and was developed as a means to minimise the use of computer memory (Wikipedia, 2012a). Because a game may extend beyond the available space on the computer screen and memory was both scarce and expensive, only the area immediately required would be loaded into the computer's memory and displayed. To achieve this, the full extent of the game was divided into very small sections (tiles) with an index to adjacent tiles so that they could be quickly added as the game progressed.

The tiling technique applies easily to scanned maps and other raster images because these digital maps are usually rectangular and have a defined size and orientation, usually via a standard mapping projection. However, greater interoperability can be achieved by prespecifying a standard tile size and levels of detail (LOD) along with where tiles fit on the earth's

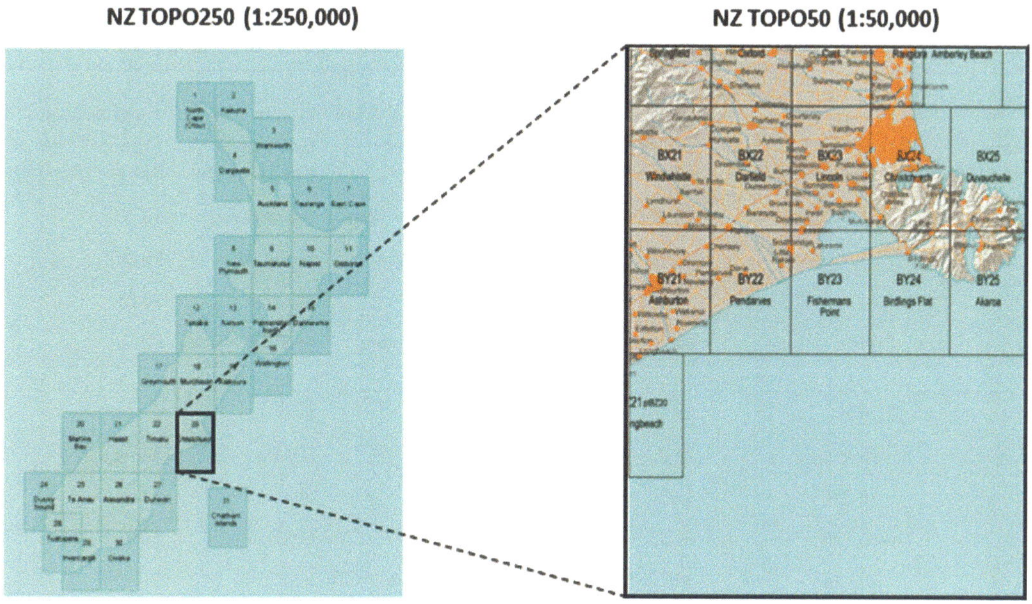

FIGURE 6.1 New Zealand Topographic Mapping.

surface. This means that any map can be cut to the same pattern. So, like the computer video game, as a user requires a specific area, the appropriate tile can be identified and displayed. This enables panning for every location for which a predefined tile exists.

By reducing resolution, tiles can also be created for multiple LOD to enable zooming – a primary benefit of this form of map delivery. Creating tiles at multiple scales actually increases the size of the tiled database because their total combined size is greater than that of a single scanned map. The cost of data storage can be excessive, especially at higher LOD. Table 6.3 shows the number of tiles, ground distance per pixel, and data storage requirements, and hard drive and RAM storage costs for a map of the entire world. While the use of tiles exacerbates the data storage problem, tiling does significantly improve the response rate for serving the map. This is because data volumes sent to the user are reduced as the users rarely examine the entire world at the highest level of detail. In addition, the user receives tiles at an appropriate resolution for the zoom level they are viewing.

6.3 TILING AND ONLINE MAPPING SERVICES

Google, OpenStreetMap, Nokia, Yahoo (now using Nokia), MapQuest and Microsoft Virtual Earth (now Bing) all implement tile mapping. The tile referencing system used by all of these services is based on a world map in the spherical Mercator projection. The initial tile and all subsequent tiles are divided into four tiles at each increase in zoom level (see Figure 6.2). This means that tiles designed for these systems are mutually compatible and can be interchanged as long as a conversion is undertaken to accommodate their slightly different tile indexing systems (Aitcheison, 2011; Google, 2012; MapTiler, 2008b; Microsoft, 2012; OSGeo Wiki, 2012).

TABLE 6.3 The Number of Tiles, Storage Requirements, and Storage Costs Used by a Tile-Based Online Mapping System to Represent the World at 20 Levels of Detail (LOD) or Zoom Levels (Peterson, 2011)

Level of Detail (LOD)	Number of Tiles	Ground Distance per Pixel in Metres	Storage Requirements at 15 Kilobytes per Tile	Disk Storage Costs at $100 per Terabyte	RAM Memory Storage Costs at $30 per Gigabyte
1	4	78,272	60 Kilobytes (kb)	$0.000006	$0.002
2	16	39,136	240 kB	$0.00002	$0.007
3	64	19,568	968 kB	$0.0001	$0.03
4	256	9784	3.75 Megabytes (MB)	$0.0004	$0.11
5	1024	4892	15 MB	$0.001	$0.44
6	4096	2446	60 MB	$0.006	$1.76
7	16,384	1223	240 MB	$0.02	$7.03
8	65,536	611.50	960 MB	$0.09	$28.13
9	262,144	305.75	3.75 Gigabytes (GB)	$0.37	$112.50
10	1,048,576	152.88	15 GB	$1.46	$450.00
11	4,194,304	76.44	60 GB	$5.86	$1800.00
12	16,777,216	38.22	240 GB	$23.44	$7200.00
13	67,108,864	19.11	968 GB	$93.75	$28,800.00
14	268,435,456	9.55	3.75 Terabytes (TB)	$375	$115,200.00
15	1,073,741,824	4.78	15 TB	$1,500	$460,800.00
16	4,294,967,296	2.39	60 TB	$6000	$1,843,200.00
17	17,179,869,184	1.19	240 TB	$24,000	$7,372,800.00
18	68,719,476,736	0.60	960 TB	$96,000	$29,491,200.00
19	274,877,906,944	0.30	3.75 Petabytes (PB)	$384,000	$117,964,800.00
20	1,099,511,627,776	0.15	15 PB	$1,536,000	$471,859,200.00
Total	1,466,015,503,700		20 Petabytes	$2,048,000	$629,145,600.00

6.3.1 Tile Overlay Limitations – Edge Effects

One major limitation of tiling is that adjacent images, such as those from series mapping or orthophotography, must be merged or mosaiced together prior to the image being tiled. This is because these images will rarely have their edges perfectly coincident with the tiles that are used by the various web mapping services. Without mosaicing the image prior to tiling, the tile displayed along the coincident edges will be missing part of one map (see Figure 6.3).

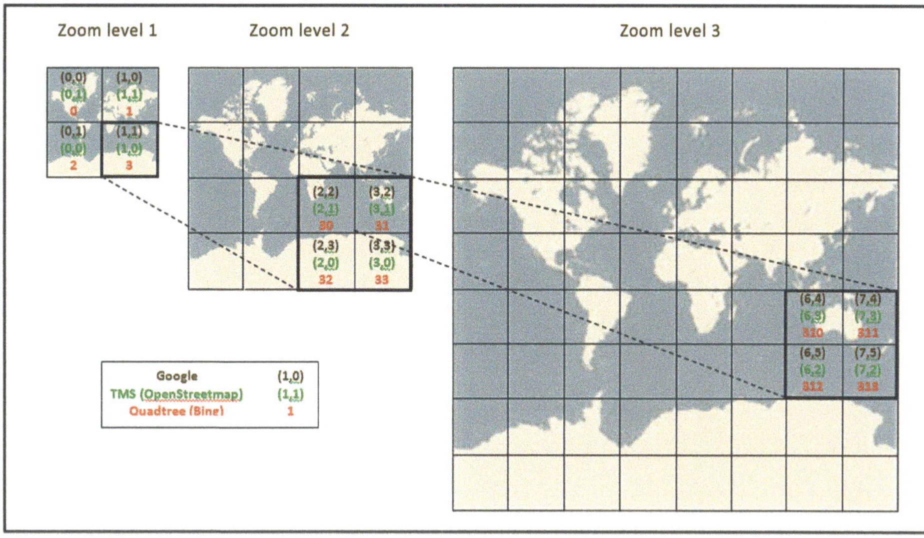

FIGURE 6.2 The method of tile indexing used by the major online map providers.

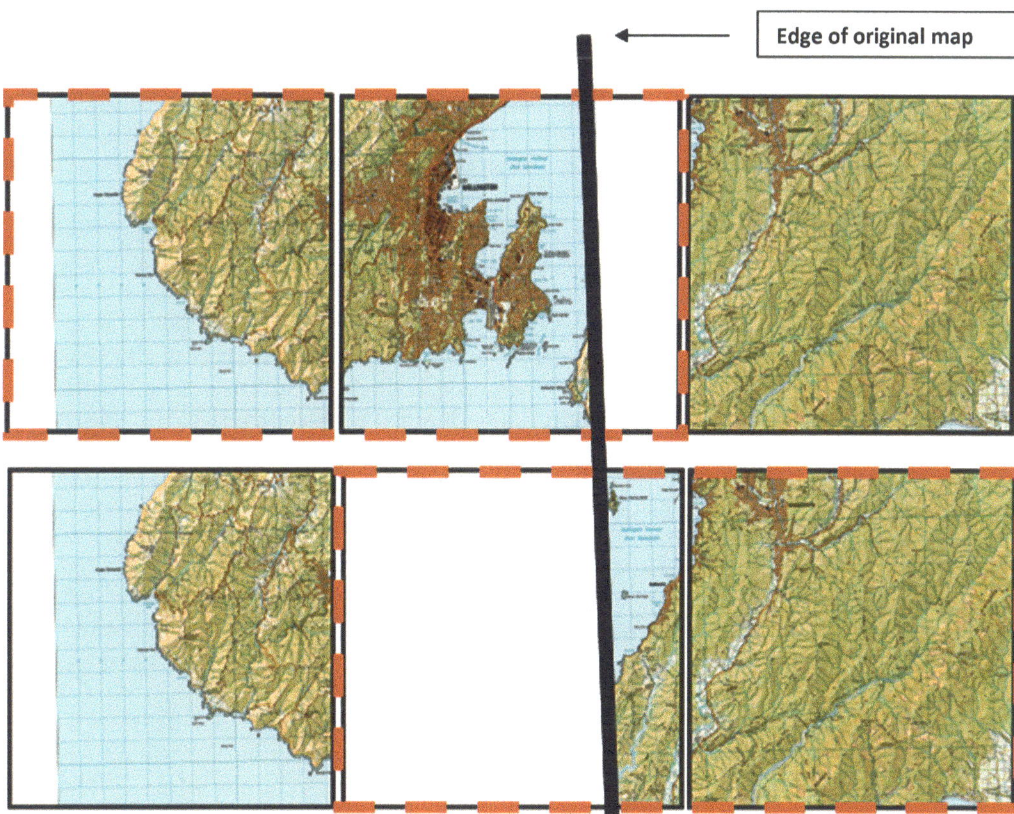

FIGURE 6.3 Tiles from two maps that have not been mosaiced prior to tiling.

Images can be stitched together in photo editing packages, such as Adobe Photoshop™, but this can be quite labour-intensive. There are also open source image processing tools that provide this capability.

6.3.2 Creating Tiles

Tiles can be generated at any time once a digital map is available, but creating tiles directly from vector data is very compute-intensive because the intersections of lines and tile boundaries must be determined. Google, with their extensive resources, have been able to move towards generating vector tiles on demand. However, the only practical solution for generating tiles for most users, including OpenStreetMap and all other online map providers besides Google, is to generate the tiles in advance as raster files. The simplest method to create tiles is to use readily available software and then transfer the tiles to the server (Přidal, P., and Žabička, P., 2008). Three options for creating tiles for subsequent upload are:

1. The *Geospatial Data Abstraction Library* (*GDAL*) (http://www.gdal.org) provides extensive capability and options to manipulate and process raster information. GDAL is provided by the Open Source Geospatial Foundation and is command-line driven. This combined with its many options and utilities can make it challenging for the less experienced. The output tiles are structured using the Tile Mapping Service (TMS) specification used by OpenStreetMap and other open-source systems.
2. *MapTiler* (http://www.maptiler.org) while built upon GDAL has the single purpose of creating tiles. Its simple intuitive interface steps the user through the process of registering the image, specifying the desired zoom levels and then generating and exporting the resultant tiles into a z, x, y file structure that mimics the actual tile pyramid (Maptiler, 2008a).
3. *TileMill* (http://mapbox.com/tilemill) also utilises GDAL and provides additional tools for combining multiple data sources. Its focus is enabling the creation of new maps and its tiles are in the MBtiles format developed by MapBox (2012). A significant advantage of the MBtiles format is that the tiles are structured and stored in a SQLlite database (a version of MySQL). The use of a database provides the potential to offer multiple layers or reduce the need for duplicate tiles. For example, a single ocean tile can be stored and used at high zoom levels instead of many 1000 identical blue-ocean tiles.

6.3.3 Displaying Tiles as a Google Mashup

To display the tiles with Goggle Maps or a similar API, the tiles simply need to be uploaded to a server while preserving the naming convention. Each individual tile is named with the Y-tile coordinate and sits within a directory named with the X-tile coordinate. This 'X' directory in turn lies within a directory that is named according to the zoom level (see Figure 6.4).

Maptiler and other GDAL-based tile utilities conform to the TMS specification. The primary difference between TMS and the Google-referencing system lies in the definition of the tile origin. With Google, the Y-tile coordinates run north to south while in TMS they are opposite, running south to north. The swap is achieved with the following code that eliminates the need to rename the tiles to the Google specification. (see Figures 6.5 and 6.6)

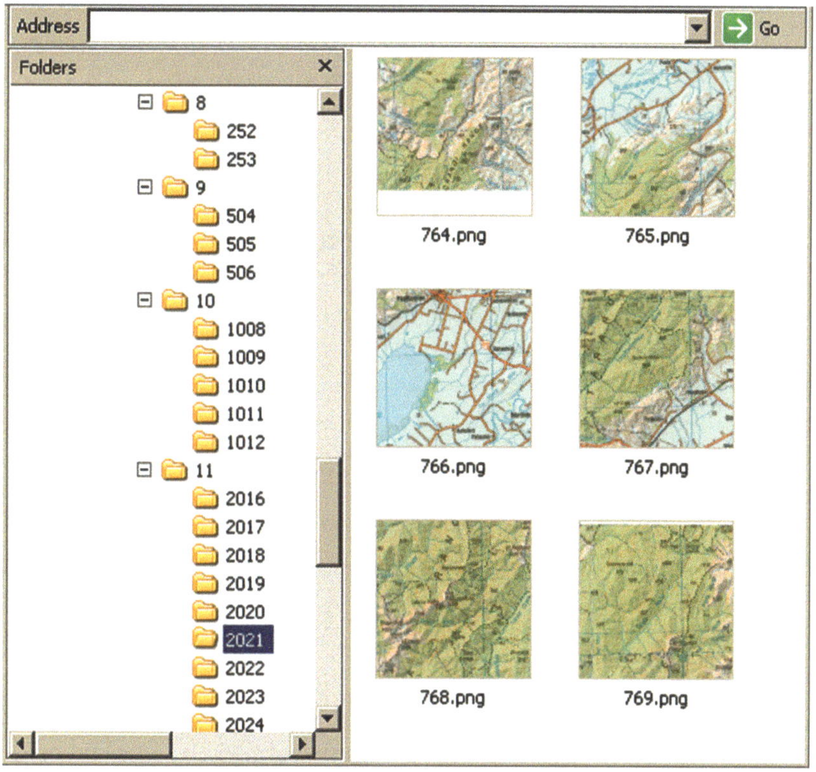

FIGURE 6.4 The hierarchical Maptiler file structure for the different zoom levels.

```
//Swap the origin of Ycoord
var tile_max1 = 1 << zoom;
var ymin = tile_max1 - tileCoordinate.y - 1;
var xmin = tileCoordinate.x;
```

FIGURE 6.5 Code for converting tile mapping service references to Google references.

Within the Google Maps API, the getTileUrl API call retrieves the appropriate tiles. The following example code uses the getTileUrl API call to obtain the tiles directly from the directory structure, where they fall within the current zoom level and are within the defined tile ranges.

Specifying tile ranges and placing the tile directories within a named directory makes for easier data management, especially when this code is extended to enable multiple levels.

6.3.4 Building a Multizoom Level (Hierarchical) Google Mashup

Tiles are commonly created for more LOD than are necessarily intended to be used. This is because only a few tiles are needed to cover the low levels and therefore creating them adds little processing or storage overhead.

```
// ***********************************************************************
//     Update with your variables for your own tile set
// ***********************************************************************
CustomTileOverlay.prototype.getTileUrl = function (coord, zoom) {

//Alter this to the name of the directory containing zoom layered directories of tiles
from maptiler
    var map_name = "BQ31 - Wellington" //Alter the "BQ31 - Wellington" to your directory
    var mapMinZoom = 9;                //Alter the "9" to your Minimum Zoom
    var mapMaxZoom = 14;               //Alter the "14" to your Maximum Zoom

    if (zoom >= mapMinZoom && zoom <= mapMaxZoom ) {
            return map_name  + "/" + zoom + "/" + coord.x + "/" + coord.y + ".png";
    }
}
// ***********************************************************************
```

FIGURE 6.6 Code for retrieving the tiles from the directory structure where map_name is the top level directory and the zoom levels are limited to between 9 and 14.

The tiles at low levels, however, may not *respect* the intent of the map. For example, a 1:50,000 map is unlikely to be intended for use at low zoom levels. This can be defined as an 'under-zoom'. A 1:50,000 map displayed at very low zoom levels will be unreadable due to the amount of detail. Using overview mapping scales such as a 1:1 million map may be more appropriate. Equally, the tiling may create tiles at an 'over-zoom' where excessive magnification enables the image to be examined and measured more closely than what would be possible with the original paper version. This can lead to meaningless measurements as the underlying data do not support that level of precision and the tiles may appear 'pixelated' if the scanning resolution was too low.

Altering the resolution of a single map to create a Google mashup that uses a wide range of zoom levels is not the best solution. A better option is to use a number of different maps with different scales and match them to the various the LOD that are provided. Care must be taken to respect the original map design limitations to ensure that they are not 'over-' or 'under-' zoomed so that the original intent of the map is retained.

The example in Figure 6.7 'replaces' the entire Google Maps for the Christchurch, New Zealand, area with two different topographic maps. A multilevel mashup of this type has the advantage of enabling the presentation of multiple maps through a single interface.

This multilevel mashup example is achieved by extending the tiled single-map code demonstrated earlier and specifying individual zoom levels for each map, along with its own section in the 'if statement'. Additional levels can be easily added by specifying a new map, zoom range and further extending the 'if' statement, as shown Figure 6.8.

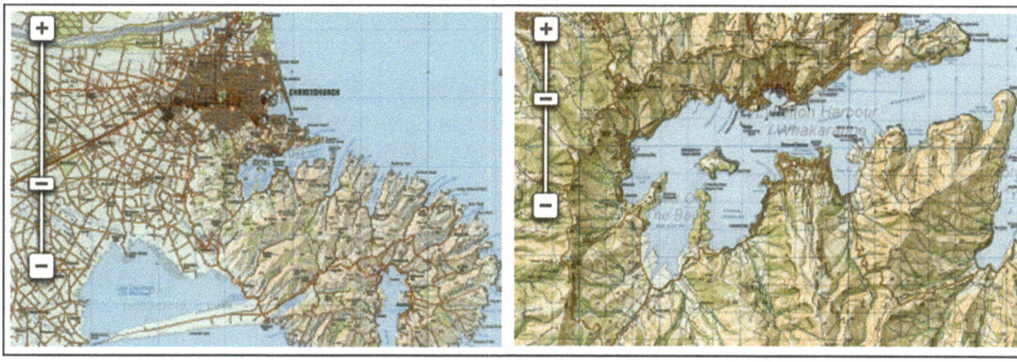

FIGURE 6.7 A portion of the New Zealand Topo250 map (1:250,000) is displayed in the left panel. The map at right is the New Zealand Topo50 map (1:50,000).

```
// ***************************************************************
//     Update with your variables for your own tile set
// ***************************************************************
CustomTileOverlay.prototype.getTileUrl = function (coord, zoom) {

//Alter this to the name of the directory containing zoom layered directories of tiles
from maptiler
        var map_name21 = "ChCh/Topo250/Map23_ChCh";
        var mapMinZoom2 = 9;
        var mapMaxZoom2 = 10;
        var map_name3 = "ChCh/Topo50/BX24";
        var mapMinZoom3 = 11;
        var mapMaxZoom3 = 12;

    if (zoom >= mapMinZoom1 && zoom <= mapMaxZoom1 ) {
            return map_name1  + "/" + zoom + "/" + coord.x + "/" + coord.y + ".png";
    }
    else if (zoom >= mapMinZoom2 && zoom <= mapMaxZoom2 ) {
            return map_name2  + "/" + zoom + "/" + coord.x + "/" + coord.y + ".png";
    }
}
// ***************************************************************
```

FIGURE 6.8 Code for creating a multilevel overlay consisting of two different topographic maps. Each map is associated with specific zoom levels in the Google Maps display.

6.4 OPACITY

A significant disadvantage of much of the scanned mapping available on the Internet is that they totally obscure the tiles from the underlying mapping service (see Figure 6.9). Occasionally, we wish to overlay raster maps or orthophotography and utilise the underlying information available in the Web mapping service of choice. This can be easily achieved by applying opacity, or making the overlay semitransparent. Google and other mapping services appear to use opacity in their tiles in order to add additional information to the map. An example is the traffic layer that displays the relative speed of traffic for many cities around the world. Most of the traffic tile is transparent, allowing the Google base map or image to remain visible. To achieve this, it also appears that Google may be dynamically generating these tiles in order obtain the advantages of both tiles and vector overlays.

Extending the use of opacity can be achieved by passing the control of amount of opacity to the user via a 'slider bar' – thereby making the use of opacity interactive. A 'slider bar' provides the map user full control to view both the overlay and the underlying map service, to transition between them, and where required, determine an optimal balance.

6.4.1 Opacity Control

The concept of the opacity slider bar for use with Google maps was developed by Petr Pridal at Klokan Technologies for the Google Maps version 2 API (see Figure 6.10) (www.maptiler.org/google-maps-overlay-opacity-control/). Its purpose is to provide an additional interactive control element, allowing the user to alter the opacity of an overlay image.

The user shifts the vertical bar along the slider, with the extreme left full opacity (i.e. the overlay image is completely transparent) to the extreme right where the overlay image has no opacity and the underlying Google Map tiles are totally obscured.

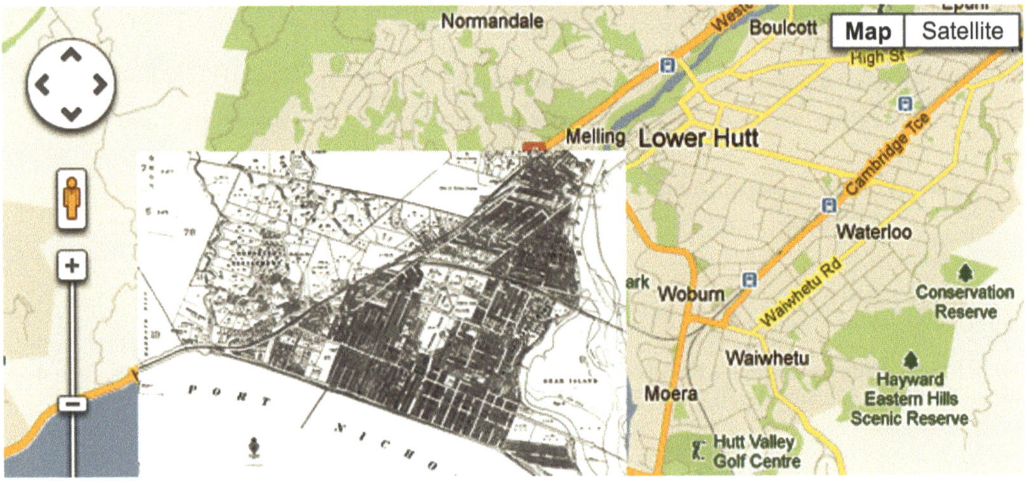

FIGURE 6.9 Totally obscuring the underlying map tiles.

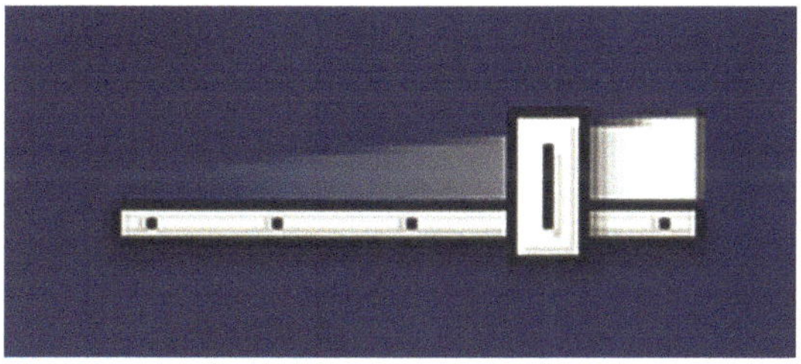

FIGURE 6.10 The Opacity Slider Bar.

Harris (2012) developed a version of the slider bar that works with the Google Maps version 3 API. Google has announced that version 2 of its API will no longer be supported. Thus, the version 2 slider will no longer be available. The remaining examples of user-controlled opacity are therefore based on the version 3 code.

6.4.2 Comparing Two Different Sets of Data

Traditional maps are not very good at demonstrating change. Even dynamic mapping available via computers does not easily provide the opportunity to 'review' the past because previous maps are often not stored or made available. The 'past' information, though, often still exists as a static map. The use of the opacity slider provides the user the ability to move between two versions of the map. The map user can then easily compare the two images and visually observe the differences by switching between the two views.

The following example (See Figure 6.11) shows the effects of the February 2011 Christchurch Earthquake on the Pyne Gould Corporation (PGC) building. The Google imagery is pre-earthquake and post-earthquake imagery is overlaid. The collapse of the PGC building is clearly visible.

FIGURE 6.11 Two images of the PGC building in downtown Christchurch, New Zealand, before and after collapse as the result of an earthquake.

6.4.3 Utilising Google Map Dynamic Data with Opacity Mapping

The map provided by Google Maps often does not provide enough detail or context. Replacing the Google Map with another map can provide additional detail; however, without an opacity-enabled map, the additional services that Google Maps provide – such as the traffic layer – would be unusable.

An advantage of adding the opacity slider is an ability to set an optimal balance between the two overlays. This is important because preferences vary for the amount of opacity that is required and the optimal balance is likely to be different for the various zoom levels. Different individuals may also have varying preferences.

Figure 6.12 shows an NZ topographic map overlaid onto a Google Map at two different scales. The middle map with the slider centered is a combination of the Google Map, the NZ topographic map, and Google's traffic layer (Table 6.4).

6.4.4 Applying Coordinate Bounding Limitations

The mapping services and base-maps provided by Google and others cover a significant part of the world. In exploring the map, the user might find it difficult to locate the overlay maps that may only be visible at higher zoom levels. By applying bounding information, tiles that do not overlay the custom map tiles can be blacked-out. Applying the opacity control enables the user to change this to a 'greyed-out' effect, or essentially remove the overlaid tiles entirely.

This masking capability provides a strong visual cue to help identify the area of interest and 'frame' the overlay map tiles. This has the additional advantage that portions of different custom overlays at lower levels are trimmed from the map. The example in Figure 6.13 has bounds set that relate to the extent of the Christchurch aerial photograph and each stepped image is an increment of two zoom levels. As a result, users are easily able to locate the single orthophoto. Figure 6.13 shows black tiles on the lower LOD that surround the area of interest.

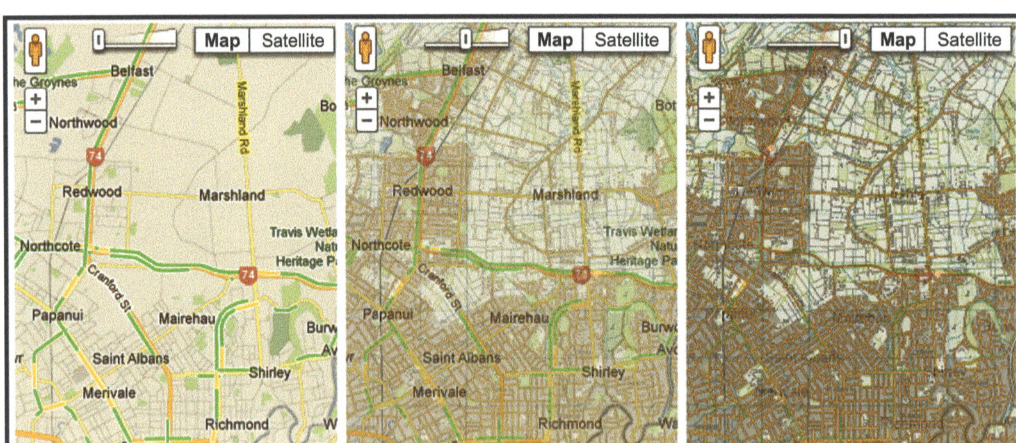

FIGURE 6.12 Merging using a partially transparent topographic overlay.

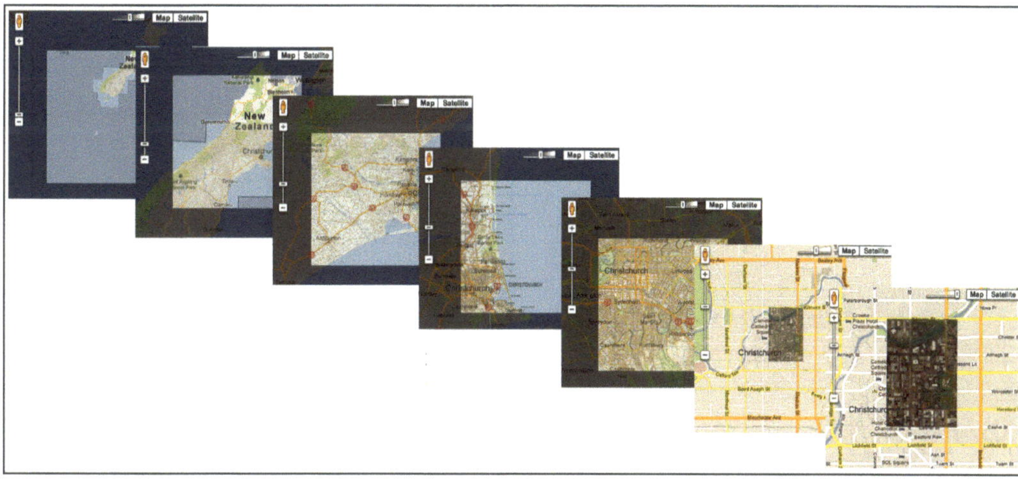

FIGURE 6.13 Zooming through an opacity enabled, tile limited multilayer mashup.

The process to create the mask is based on the multilevel code shown in Figure 6.8 except the tiles outside of the defined area of interest are replaced with a single 'black tile'. There are five primary steps:

1. Define the coordinate bounds;
2. Calculate the tile coordinates for these bounds;
3. Define zoom variables;
4. Insert a check for each zoom layer that tests whether the tile being returned is within the bounds; and
5. Return a 'black' tile instead of the map tile for all area outside of the coordinate bounds.

6.5 FUTURE WORK

The use of tiles best suits rectangular areas. However, most countries or administrative areas are irregular in shape. This means that additional tiles covering 'blank space' need to be generated. These additional tiles are commonly made transparent so as not to be visible. For example, New Zealand lies entirely within one tile at zoom level 2 but 65,536 tiles are needed for the same rectangular area at zoom level 10. But, the number of tiles required to cover just the New Zealand land areas including Stewart Island (i.e. excluding the ocean) is less than 500. This is approximately 0.7% of all level 10 tiles that are required for the area of a single level 2 tile, as shown in Table 6.4. This clearly highlights that many tiles are potentially being created over areas that may be managed differently. It would be of benefit to create a series of customised bounds that return generic tiles for these areas.

The database-driven MBtiles approach would be an excellent way for managing tiles and controlling masking. Spatial extensions are available for SQLite, so MBtiles may also provide

TABLE 6.4 Percentage Estimate of Effective Tiles for NZ

Zoom Level	Tiles Required to Cover the Area of a Single Tile at Zoom Level 2	Approximate Count of Tiles Actually Covering Mainland NZ	Difference (Percent)
2	1	1	100.0%
3	4	2	50.0%
4	16	2	12.5%
5	64	3	4.7%
6	256	8	3.1%
7	1024	17	1.7%
8	4096	46	1.1%
9	16,384	134	0.8%
10	65,536	438	0.7%

the potential to define 'exclusion' polygons and then discard unnecessary tiles through the use the Open Geospatial Consortium's precise spatial functions such as ST_Intersect.

6.6 SUMMARY

By removing the need to minimise image resolutions, limit colour depth, or use lossy image formats, tile mapping is an excellent means to preserve the quality of an original map, while enabling it to be transferred to users quickly and efficiently. It has the potential to be applied to any digital map whether a scanned historic one, an orthophoto or a modern computer-generated thematic map.

When 32-bit images are available, manipulating the opacity of images is possible. With the opacity slider, users are able to manage the look and feel of a map rather than accept a static view, and are in effect provided with a useful animation tool. The slider overlay can still utilise the underlying capabilities of the mapping service such as the imagery, terrain, and dynamic data services (such as the traffic layer) because they are not totally obscured.

Opacity, combined with tile masking, provides a new and subtle way to influence the maps users' behaviour by encouraging them to view specific areas of the map. The combination of tile mapping and opacity is a significant step forward in enabling user interaction with the map and provides a new tool for the modern cartographer.

References

Aitcheison, A., 2011. Converting TMS Tile Coordinates to Google/Bing/OSM Tile Coordinates. Available at: http://alastaira.wordpress.com/2011/07/06/converting-tms-tile-coordinates-to-googlebingosm-tile-coordinates.

Google, 2012. Google Maps JavaScript API v3. Available at: https://developers.google.com/maps/documentation/javascript/maptypes?hl=en#MapCoordinates.

Harriss, G., 2012. Opacity Control for Google Maps v3 Overlay. Available at: http://www.gavinharriss.com/code/opacity-control.

MapBox, 2012. Tile Mill Help Working with SQLite. Available at: http://mapbox.com/tilemill/docs/tutorials/sqlite-work.

MapTiler, 2008a. MapTiler – Map Tile Cutter. Available at: http://www.maptiler.org.

MapTiler, 2008b. Tiles à la Google Maps: Coordinates, Tile Bounds and Projection. Available at: http://www.maptiler.org/google-maps-coordinates-tile-bounds-projection.

Microsoft, 2012. Bing Maps Tile System. Available at: http://msdn.microsoft.com/en-us/library/bb259689.aspx.

OSGeo Wiki, 2012. Tile Map Service Specification. Available at: http://wiki.osgeo.org/wiki/Tile_Map_Service_Specification.

Peterson, M.P., 2011. In: Zentai, L., Nunez, J.R. (Eds.), The Tile-Based Mapping Transition in Cartography, Springer Verlag, Berlin.

Peterson, M.P., 2003. Maps and the Internet. Elsevier, Amsterdam.

Přidal, P., Žabička, P., 2008. Tiles as an approach to on-line publishing of scanned old maps, vedute and other historical documents. e-Perimetron 13 (1), 10–21, Available at: http://www.e-perimetron.org/Vol_3_1/Pridal_Zabicka.pdf.

Sample, J.T., Elias, I., 2010. Tile-Based Geospatial Information Systems: Principles and Practices. Springer, New York.

Wikipedia, 2012a. Tile Engine. Available at: http://en.wikipedia.org/wiki/Tile_engine.

Wikipedia, 2012b. RGBA Colour Space. Available at: http://en.wikipedia.org/wiki/RGBA_color_space.

CHAPTER

7

Advances in Location-Based Services

Georg Gartner

Vienna University of Technology, Vienna, Austria

OUTLINE

7.1 INTRODUCTION

Information is available anytime and anywhere nowadays. In its provision and delivery it is increasingly tailored to the user's context and needs. Location and maps play a key role in what information is provided and how it is presented. Location-based Services (LBS) are widespread and are being used with increasing frequency in a truly ubiquitous manner. Individuals often feel spatially blind without access to their LBS, which enables them to see who or what is near them and receive supporting information. They can also conduct searches on their current location, and collect data on their site in an accurate and timely manner. Mobile technologies have demonstrated their huge potential and change how we work, how we live and how we interact.

The paradigm of Cybercartography, introduced by Taylor (1997, 2005) and further developed in this volume, is a conceptual framework for these developments, as it clearly emphasizes the 'holistic approach where both mapping as a process and the map as a product are expanded' (Taylor, 1997). In this chapter, a number of major developments that are influencing the development of LBS, especially as they affect cartography, are discussed.

Developments in the Theory and Practice of Cybercartography, Second Edition, ISSN 1363-0814
http://dx.doi.org/10.1016/B978-0-444-62713-1.00007-6

7.2 DEVELOPMENT PHASES OF LBS

LBS are very much a development of the last decade but three phases in what is an increasingly rapid development can be determined. It should be noted that given the relatively short time-frame involved these phases are not distinct and developments have been more of a continuous process.

The early stages of LBS developments were basically driven by technology (Gartner, 2005). Whenever innovative positioning, telecommunication, or device technologies were introduced, new ideas, applications, and products in the domain of LBS were triggered. That period was very much characterized by efforts to deal with research questions such as 'How to locate or track a user?' as well as data modeling and data representation on small display devices. Significant progress was made in determining the positions of devices and users in both outdoor and indoor scenarios as well as in mixed environments and in 3D. The need for real-time tracking was tackled as well as attempts to combine direct and indirect positioning methods by using smart environment sensors. The initial period of LBS development was also strongly influenced by the rapid availability of an increasing number of new handheld devices. By creating more and better availability and accessibility to sensors, processing performance and communication channels, these handheld devices became quickly both smart and mass-market acceptable. A number of applications were developed influenced simply by the availability of a new smart device, including those that took the idea of LBS into new domains such as the entertainment and gaming, business, disaster management, and urban and regional planning, among others. Most of those applications were constrained by the limitations of the technologies available, forcing many users and use scenarios to adapt to the technical constraints rather than having systems available that had been designed with particular uses and users in mind.

The second phase was characterized by data-driven approaches. With the number of available technologies that have become available and the pace of technological innovations slowing, the high quality of mass-market technologies resulted in a shift of the focus to the availability of data. A major challenge was to overcome the obstacles between data acquisition and data consumption as well as to establish interoperability between the three main parallel worlds of authoritative, commercial and open data providers. The development of technical, legal, and economic standards and procedures was of great benefit to the LBS-environment as these helped to define accessibility and availability of data in an acceptable way. Additional data acquisition and usage from ubiquitous sensory sources as well as from user behaviour and use patterns was now not only possible technically but also was embedded in a legal and economic framework. Dealing with volunteered data posed new challenges, such as the necessity to learn about the motivation, the structure, and the representativeness of the volunteers as well as developing mechanism to determine the reliability of the available data. These and other aspects of volunteered geographical information are discussed in Chapter 4 of this volume. The huge quantities of data being generated created a requirement for enhanced data management systems. The need to address this problem drove one of the main trends – an increasing use of and reliance on Big Data technologies – technologies that enable the analysis of vast quantities of information within useable and practical time frames. Massively scalable, distributed systems for processing unstructured and semistructured data emerged as a result of this need, and became widely accepted and relied upon in the management and interpretation of geospatial information. Given the vast amount of data being generated, particularly through use of the Web, and the need to make sense of this data, the ability to link information on the Web became increasingly important. The network built on increasing numbers of sensors and the resulting increase in data, produced a hyperconnected environment or 'Internet of Things'. Location provides a vital link between the sensors that generate the Internet of Things and thus LBS making use of location are key elements to make use of these networks.

Finally, the third major and current phase of LBS developments can be characterized as focusing on the individual user and thus fitting the conceptual framework of Cybercartography, which is increasingly user-centric (Taylor, Chapter 1). Smart LBS support decision-making either on a collaborative or an individual level, help to act in space, support spatial enabled societies, enhance spatial awareness, or simply support daily routines. Everything happens somewhere, and thus location is a key selector for what we do and how we do it. In this phase, context-awareness is as interesting research and development topic as use and user modeling or developing adaptive algorithms. Cartography and data presentation remain the language through which the data explosion is spatially interpreted. The fact that increasing amounts of geospatial information are consumed and interpreted through mobile devices also led to the development of new visualization techniques including 3D and especially 4D – representations.

The successful development of LBS requires integrated interdisciplinary approaches from domains such as computer science, communication science, human–computer interaction, telecommunication sciences, cognitive sciences, law, economics, geospatial information management, and cartography to allow human-centred application developments by applying innovative engineering methods and tools in a highly volatile technological framework. In this respect, the interdisciplinary nature of LBS is very similar to that of cybercartography. The rapidly increasing use of geospatial information has led to a growing recognition amongst both governments and the private sector that an understanding of location and place is a vital component of effective decision-making. Citizens with no recognized expertise in geospatial information and who are unlikely to even be familiar with the term are also increasingly using and interacting with geospatial information and in some cases they are contributing to its collection as several chapters in this volume illustrate. A number of important technology-driven trends have had a major impact, creating previously unimaginable amounts of location-referenced information thus making LBS a central focus of research and development and an important component of cybercartography.

7.3 SOME EMERGING RESEARCH TOPICS

There is a number of important emerging research topics. Three of these will be considered here – indoor navigation, integrating social media and interface design.

7.3.1 Indoor Navigation

Outdoor navigation has been a research topic for LBS applications from the outset. However, indoor navigation is now emerging as a research and development focus.

Research has shown that people tend to lose orientation a lot easier within buildings than outdoor (Radoczky, 2003; Hohenschuh, 2004). Indoor navigation systems are designed to meet this need.

Navigation systems have three major components: positioning, route planning, and route communication/presentation (Huang and Gartner, 2010). In terms of positioning, outdoor navigation often employs global positioning system (GPS), which is not usually available for indoor navigation. Positioning in indoor navigation is often provided with the help of WiFi, Radio-Frequency Identification (RFID), Ultra-Wideband (UWB), Bluetooth, and dead reckoning. With the exception of dead reckoning, other positioning techniques require infrastructure that is not ubiquitously available. Due to the differences between outdoor and indoor data, route-planning techniques in outdoor navigation are not easily applied to an indoor navigation. Route planning in indoor environments is strongly influenced by the positioning of structures such as doors and corridors, which may change location frequently and need to be modeled in a real-time manner. In terms of route communication, different interface techniques such as mobile maps, verbal instructions, 3D, and augmented reality (AR) could all be useful. However, the contents or instructions must be adapted to the indoor environment, which is restricted, smaller, and has different landmarks. A similar comparison is given by Karimi (2011) that analyses the differences between outdoor and indoor navigations with respect to travel modes supported, map data, impacting factors, and routing criteria and concludes that these differences mainly stem from the differences in physical spaces in which navigation takes place.

The early research on indoor navigation mainly focused on how to obtain reliable and accurate positioning by using technologies and techniques like WiFi, RFID, and read reckoning (Gartner et al., 2004; Retscher, 2007; Zhang et al., 2009). While indoor positioning is still a research focus, recent research has started to explore the aspects of indoor data modeling, indoor route planning, and route communication.

In contrast to the outdoors, where data are often standardized and readily available from commercial data providers and mapping agencies, indoor data have not yet been standardized or widely collected. Afyouni et al. (2012) classify indoor spatial models into geometric and symbolic spatial models. They conclude that geometric models can efficiently integrate metric properties to provide highly accurate location and distance information, and that they require semantic annotations for different application needs. On the other hand, symbolic models semantically represent indoor space with human-recognizable objects (e.g. rooms) as well as their properties and relationships (e.g. connectedness), but lack geometric details on entities and places represented in space. Afyouni et al. (2012) propose that hybrid models of indoor space should be developed. Other similar classifications of indoor models and recommendations for the use of hybrid models can be found in Worboys (2011). Data about indoor spaces can be obtained from field surveying or extracted from Computer-Aided Design (CAD)s. Recently, following the success of mapping outdoors by using crowdsourcing approaches, the OpenStreetMap (OSM) community has begun to map indoor environment. Different data schema proposals have been developed, but these vary substantially from each other. One of the most popular schemes is IndoorOSM proposed by Goetz and Zipf (2011). However, it is important to note that there is currently no community-agreed standard schema (Goetz and Zipf, 2013) that hinders the mapping activities of indoor spaces. Another challenge of indoor data modeling and collecting is the consideration of real-time aspects, which has not been well addressed in literature.

Route planning is a basic element in navigation, and aims at computing an optimal route between an origin and a destination. Vanclooster and Maeyer (2012) provide a state of the art review of common route planners, such as Bing and Google Maps, and conclude that due to a lack of available indoor data, most existing route planners do not incorporate indoor route calculation. In the research community, Yuan and Schneider (2010) propose a model to construct an indoor navigation network from architectural plans, and use the network to support length-dependent optimal routing. Accessibility of network elements is also considered. Liu and Zlatanova (2011) propose a 'door-to-door' route-planning algorithm based on 2D floor plan. The algorithm first determines a coarse route between rooms (doors), and then computes a detailed route for each room. Dijkstra's algorithm is used to compute shortest or fastest routes (in normal situations) or safest routes (in emergencies). However, the paper does not indicate how to measure or calculate a safety coefficient. Schougaard et al. (2012) extract navigation graphs and visibility graphs from building information (e.g. in Industry Foundation Classes IFC standard), and these two graphs are then employed to calculate a shortest route. In contrast to the approaches using floor plans or building information, Goetz and Zipf (2013) process IndoorOSM data to automatically derive a Weighted Indoor-Routing Graph, and this graph is then used to compute a shortest path between an origin and a destination. It is important to note that current route-planning approaches mainly aim for shortest routes. This is probably due to the fact that objective measures such as distance and time are easier to identify for routing graphs. Since people usually do not follow the shortest routes when navigating, more work should be done on providing routing with other characteristics, such as attractiveness, simplicity, and safety.

In terms of route communication (i.e. conveying route instructions/guidance), Huang and Gartner (2010) provide a survey on the related issues, and find that compared to outdoor navigation, which mainly uses smartphones with mobile maps and verbal instructions, indoor navigation uses various client platforms with more presentation forms, e.g. digital signs on public displays, vibration on wrist devices, and AR with wearable computers. Butz et al. (2001) argue that route communication should deal with the inaccuracy of position and direction information. For example, when accurate positioning information is available, a simple navigation instruction such as an arrow can be given, but when position or direction information is uncertain, more information must be given in order for the user to understand the navigation instructions. The PerPosNav platform addresses this issue, and proposes four kinds of interfaces for route communication in indoor environments with various positioning accuracies: augmented signs (e.g. presenting route information on digital signs or public displays), mobile maps, verbal instructions, mobile AR (Schougaard et al., 2012). The PerPosNav platform also provides a tracking service to obtain information of whether the users are on route or not, and where they are on the route. The approach of augmented signs and mobile maps was evaluated in two real-world experiments, of which both received positive feedbacks from the users. It is important to note that the concept of landmark has not been well addressed in the existing research. However, as emphasized in the literature, landmarks play an essential role during navigation. In order to provide more effective navigation guidance, a landmark-based approach should be developed. This is still challenging as taxonomy for indoor landmarks and the way to derive them has not been well studied.

Considerable progress has been made for indoor navigation, but more research is required on indoor data model, and indoor routing and communication.

7.3.2 Integrating Social Media

Recently, with increasing ubiquity of GPS-enabled devices, location-based social network and media-sharing services are quickly growing in popularity, such as FourSquare, Facebook Place, and Eventful. In the meantime, many traditional social media (such as Flick and Twitter) are also allowing users to geotag their information. As a result, more and more georeferenced data are becoming available. On the one hand, LBS can be viewed as a tool for contributing georeferenced social media data. On the other hand, such data also enable a variety of novel LBS applications such as in the 'emotional mapping' project (Klettner and Schmidt, 2012), where subjective relations to places can be expressed in situ by using mobile devices. This allows the production of a rich database of subjective perceptions of places in an urban environment. This demonstrates the role of LBS maps and holistic instruments as conceived in the Cybercartography framework (Figures 7.1 and 7.2).

FIGURE 7.1 Emotional Mapping project of TU Vienna: Harvesting subjective perceptions of space by mobile phone apps and allowing adapted routing due to optimization based on emotional, subjective perceptions. *Source: Map data: OpenStreetMap and Contributors, CC-BY-SA.*

FIGURE 7.2 A screenshot of the AR-based interface, with a real-world camera view, route overlay, street names, and relevant landmarks. *Source: © Salzburg Research.*

Much research focuses on studying people's travel behaviours (Jankowski et al., 2010; Zheng et al., 2011). These approaches often start with building a database of travel trajectories, and apply some data mining techniques for identifying locations of attraction and frequent travel sequences that are very important for LBS applications. There is also research focusing on constructing travel itineraries based on geotagged photos on the Web (De Choudhury et al., 2010). Metadata (e.g. timestamp, tags, and GPS) of the photos are used to generate past tourists' travel trails, which are then combined to generate travel itineraries for future tourists.

Research attention is also being given to modeling users' preferences from their information on social media, and directly developing smart LBS based on the information extracted. For example, McKenzie and Raubal (2011) argue that the data contributed to online social networks not only offer insight to the reasons why people travel from location to location, but also offer a means of predicting future activities and travel. They study these data to extract social constraints for LBS, and thus provide socially relevant information to LBS users. Ye et al. (2011) provide Point of Interest (POI) recommendations by considering user preference, social influence and geographical influence, which are all derived from social media, such as FourSquare. With real datasets, they show that the proposed solution significantly outperforms a wide spectrum of alternative recommendation approaches. Levandoski et al. (2012) use location-aware ratings to recommend nearby features. The datasets used are social data from Foursquare and a part of MovieLens movie recommendation data. Waga et al. (2012) propose a context-aware personalized recommendation system on web and mobile devices to recommend services, photos, and GPS routes in a user's surrounding based on user-generated data collection. The relevance of items is determined by location, content, time, and network. Savage et al. (2012) learn user preferences by mining a person's social network profile (e.g. FourSquare). This information is then combined with physical constraints such as current location and transportation mode to provide LBS users with location recommendations.

The research discussed above mainly focuses on using information from a single social media source, which might only reflect parts of users' preferences and activities. Combining data from different social media will contribute to a better understanding of users, and based on this, more relevant services can be provided for LBS users. However, this issue is still very challenging due to the fact that social media data are often very noisy, unstructured, and contain heterogeneous and multilingual content.

7.3.3 Interface Design

While the early LBS tend to use mobile map and verbal/written descriptions for communicating relevant information to users, recent research attention has been also paid to explore the suitability of other interface technologies such as AR and haptic for LBS.

Tactile Wayfinder (Pielot and Boll, 2010) uses a belt with tactile output to convey navigation information. The direction of the next waypoint and the waypoint after that is given by vibration with different rhythm and intensity on the belt. Similarly, Jacob et al. (2012) provide the user with location, orientation, and distance information using varying vibration patterns. Evaluation shows that with the haptic solution, the user can point and query the availability of POI along the street, and follow the route without increased cognitive burden of interaction with the mobile interface. The authors conclude that Haptics is powerful for conveying quantitative information. AR has attracted much interest in the last few years. Wikitute's World Browser (http://www.wikitude.com/) is one of the most popular ARs, and it supports users' exploration of the real-world environment by adding virtual information to the camera views. Walther-Franks (2007) has developed two image-based AR prototypes for navigation purposes, implementing implicit instructions ('follow me') and explicit instructions ('go there'). The evaluation shows that test persons using the implicit route visualization of a virtual path diverged less from the route as turning instructions were grasped more clearly. However, in a field experiment, Rehrl et al. (2012) compare AR with voice and mobile maps, and find that compared to voice and mobile maps, AR leads to a higher cognition load and lower usability. This is probable due to the fact that AR is relatively new and still suffers from poor and unstable positioning and direction information.

Research has also addressed the issues of multimodal interaction in LBS. For example, the EU project HaptiMap (http://www.haptimap.org/) investigates how, and in what ways, multimodal feedback can both augment and replace visual feedback for diverse users in diverse situations. The project has developed tools that make it easier for developers to add adaptable multimodal components in LBS applications. Different applications have been developed in this project to illustrate multimodal interaction.

It is important to note that existing interface techniques are mainly designed to help LBS users solve their tasks (e.g. wayfinding) with less cognitive load. Other important aspects such as the side effects of these techniques, such as poor spatial knowledge acquisition as shown in Ishikawa and Montello (2006), Münzer et al. (2006) and Huang et al. (2012), have seldom been addressed. As people rely more and more on LBS, developing solutions to alleviate these kinds of side effects has become a challenge.

7.4 CONCLUSION

In this chapter, major aspects of the ongoing development of LBS are discussed. Map-based LBS include aspects of integrative positioning, context-adapted data modeling, and trends like indoor positioning, data mining from social media, and innovative presentation interfaces. The importance of LBS within the context of Cybercartography is clear as LBS address many of the same issues in both theory and practice.

References

Afyouni, I., Ray, C., Claramunt, C., 2012. Spatial models for context-aware indoor navigation systems: a survey. Journal of Spatial Information Science 4.

Butz, A., Baus, J., Krüger, A., Lohse, M., 2001. A hybrid indoor navigation system. In: Proceedings of the 6th International Conference on Intelligent User Interfaces, ACM, New York, NY, USA, pp. 25–32.

De Choudhury, M., Feldman, M., Amer-Yahia, S., Golbandi, N., Lempel, R., Yu, C., 2010. Automatic construction of travel itineraries using social breadcrumbs. In: Proceedings of the 21st ACM Conference on Hypertext and Hypermedia, ACM, New York, NY, USA, pp. 35–44.

Gartner, G., Frank, A., Retscher, G., 2004. Pedestrian navigation system - in mixed indoor/outdoor environment – the NAVIO project. In: Proceedings of the CORP 2004 and Geomultimedia 2004 Symposium, Vienna, pp. 165–171.

Gartner, G., 2005. TeleCartography: a new means of GeoCommunication. In: Taylor, D.R.F. (Ed.), Cybercartogaphy – Theory and Practice, Elsevier, pp. 373–387.

Goetz, M., Zipf, A., 2011. Extending OpenStreetMap to indoor environments: bringing volunteered geographic information to the next level. In: Urban and Regional Data Management: UDMS Annual, pp. 47–58.

Goetz, M., Zipf, A., 2013. Indoor route planning with volunteered geographic information on a (mobile) web-based platform. In: Krisp, J.M. (Ed.), Progress in Location-Based Services, Springer, Berlin Heidelberg, pp. 211–231.

Hohenschuh, F., 2004. Prototyp eines mobilen Navigationssystems fuer die Stadt Hamburg. Master Thesis. Hamburg University.

Huang, H., Gartner, G., 2010. A survey of mobile indoor navigation systems. In: Gartner, G., Ortag, F. (Eds.), Cartography in Central and Eastern Europe, Springer, Berlin/Heidelberg, pp. 305–319.

Huang, H., Schmidt, M., Gartner, G., 2012. Spatial knowledge acquisition with mobile maps, augmented reality and voice in the context of GPS-based pedestrian navigation: results from a field test. Cartography and Geographic Information Science 39 (2), 107–116.

Ishikawa, T., Montello, D.R., 2006. Spatial knowledge acquisition from direct experience in the environment: individual differences in the development of metric knowledge and the integration of separately learned places. Cognitive Psychology 52 (2), 93–129.

Jacob, R., Mooney, P., Winstanley, A., 2012. What's up that street? Exploring streets using a haptic GeoWand. In: Gartner, G., Ortag, F. (Eds.), Advances in Location-Based Services, Springer, Berlin Heidelberg, pp. 91–105.

Jankowski, P., Andrienko, N., Andrienko, G., Kisilevich, S., 2010. Discovering landmark preferences and movement patterns from photo postings. Transactions in GIS 14 (6), 833–852.

Karimi, H.A., 2011. Universal Navigation on Smart Phones. Springer.

Klettner, S., Schmidt, M., 2012. Visualisierung von Emotionen im Raum. In: Reder, C. (Ed.), Kartographisches Denken, Springer, Wien, pp. 398–399.

Levandoski, J.J., Sarwat, M., Eldawy, A., Mokbel, M.F., 2012. LARS: a location-aware recommender system. In: Proceedings of the 2012 IEEE 28th International Conference on Data Engineering, IEEE Computer Society, Washington, DC, USA, pp. 450–461.

Liu, L., Zlatanova, S., 2011. A "door-to-door" path-finding approach for indoor navigation. In: Proceedings Gi4DM 2011: GeoInformation for Disaster Management.

McKenzie, G., Raubal, M., 2011. Adding social constraints to location based services. In: Proceedings of the 8th International Symposium on Location-Based Services.

Münzer, S., Zimmer, H.D., Schwalm, M., Baus, J., Aslan, I., 2006. Computer-assisted navigation and the acquisition of route and survey knowledge. Journal of Environmental Psychology 26 (4), 300–308.

Pielot, M., Boll, S., 2010. Tactile wayfinder: comparison of tactile waypoint navigation with commercial pedestrian navigation systems. In: Floréen, P., Krüger, A., Spasojevic, M. (Eds.), Pervasive Computing, Springer, Berlin Heidelberg, pp. 76–93.

Radoczky, V., 2003. Kartographische Unterstuetzungsmoeglichkeiten zur Routenbeschreibung in Fussgaenger Navigationssystemen im In- und Outdoorbereich. Master Thesis. Vienna University of Technology.

Rehrl, K., Häusler, E., Steinmann, R., Leitinger, S., Bell, D., Weber, M., 2012. Pedestrian navigation with augmented reality, voice and digital map: results from a field study assessing performance and user experience. In: Gartner, G., Ortag, F. (Eds.), Advances in Location-Based Services, Springer, Berlin Heidelberg, pp. 3–20.

Retscher, G., 2007. Augmentation of indoor positioning systems with a barometric pressure sensor for direct altitude determination in a multi-storey building. Cartography and Geographic Information Science 34 (4), 305–310.

Savage, S., Baranski, M., Elva Chavez, N., Höllerer, T., 2012. I'm feeling LoCo: a location based context aware recommendation system. In: Gartner, G., Ortag, F. (Eds.), Advances in Location-Based Services, Springer, Berlin Heidelberg, pp. 37–54.

Schougaard, K.R., Grønbæk, K., Scharling, T., 2012. Indoor pedestrian navigation based on hybrid route planning and location modeling. In: Kay, J., Lukowicz, P., Tokuda, H., Olivier, P., Krüger, A. (Eds.), Pervasive Computing, Springer, Berlin Heidelberg, pp. 289–306.

Taylor, D.R.F., 1997. Maps and mapping in the information era. Keynote address to the 18th ICC, Stockholm, Sweden In: Ottoson, L. (Ed.), Proceedings, vol. 1. Swedish Cartographic Society, Gävle, pp. 1–10.

Taylor, D.R.F., 2005. Cybercartography: Theory and Practice. Elsevier, Amsterdam 574 pp.

Vanclooster, A., Maeyer, P.D., 2012. Combining indoor and outdoor navigation: the current approach of route planners. In: Gartner, G., Ortag, F. (Eds.), Advances in Location-Based Services, Springer, Berlin Heidelberg, pp. 283–303.

Waga, K., Tabarcea, A., Fränti, P., 2012. Recommendation of points of interest from user generated data collection. In: IEEE International Conference on Collaborative Computing: Networking, Applications and Worksharing.

Walther-Franks, B., 2007. Augmented Reality on Handhelds for Pedestrian Navigation. Master Thesis. University of Bremen.

Worboys, M., 2011. Modeling indoor space. In: Proceedings of the 3rd ACM SIGSPATIAL International Workshop on Indoor Spatial Awareness, ACM, New York, NY, USA, pp. 1–6.

Ye, M., Yin, P., Lee, W.-C., Lee, D.-L., 2011. Exploiting geographical influence for collaborative point-of-interest recommendation. Proceedings of the 34th International ACM SIGIR Conference on Research and Development in Information Retrieval, ACM, New York, NY, USA, pp. 325–334.

Yuan, W., Schneider, M., 2010. iNav: an indoor navigation model supporting length-dependent optimal routing. In: Painho, M., Santos, M.Y., Pundt, H. (Eds.), Geospatial Thinking, Springer, Berlin Heidelberg, pp. 299–313.

Zhang, K., Zhu, M., Retscher, G., Wu, F., Cartwright, W., 2009. Three-dimension indoor positioning algorithms using an integrated RFID/INS system in multi-storey buildings. In: Gartner, G., Rehrl, K. (Eds.), Location Based Services and TeleCartography II, Springer, Berlin Heidelberg, pp. 373–386.

Zheng, Y.-T., Li, Y., Zha, Z.-J., Chua, T.-S., 2011. Mining travel patterns from GPS-tagged photos. In: Lee, K.-T., Tsai, W.-H., Liao, H.-Y., Chen, T., Hsieh, J.-W., Tseng, C.-C. (Eds.), Advances in Multimedia Modeling, Springer, Berlin/Heidelberg, pp. 262–272.

CHAPTER

8

Inclusive Cartography: Theoretical and Applied Issues in Brazil

Regina Araujo de Almeida

Department of Geography, University of São Paulo, São Paulo, Brazil

OUTLINE

8.1 INTRODUCTION

Maps have been relevant to society for centuries, today however, they have become vital means to collect, process, and represent all sorts of data and geographical information, both in digital and conventional formats. Technology has brought great innovations, and people have adapted to these changes and have become accustomed to images and maps at work,

Developments in the Theory and Practice of Cybercartography, Second Edition, ISSN 1363-0814
http://dx.doi.org/10.1016/B978-0-444-62713-1.00008-8

school, in daily life, leisure, anytime and everywhere, using a mobile device. People have generally become map users and creators taking greater advantage of cartography in old and new ways.

Images and maps can now be made accessible to everyone and equal rights, respect and value to all social, ethnic or cultural differences, are now expected, including to people with physical, cognitive, or sensorial deficiencies. For these reasons, cartography needs to be inclusive and adopt multisensory and embrace multicultural perspectives in terms of representations. This inclusive cartography can serve those with different needs, such as the hearing and visually impaired and be culturally attuned towards the worldviews of indigenous people. Old questions most often discussed in cartographic communication processes such as *what, why, how, to whom, and with which results* gain new meanings in this new field of social, inclusive, and participatory cartography.

In the next sections, tactile maps for people who are visually impaired is discussed from a theoretical and applied perspective by examining tactile map design, creation, dissemination, and use. Inclusive cartography also means embracing a multicultural perspective therefore becoming accessible to indigenous groups. There have been many programmes developed in the Brazilian Amazonian region, involving participatory mapping and social cartography; most apply geographic information system (GIS) and digital techniques in conjunction with mental maps rendered by indigenous or minority groups in danger of assimilation or extinction. The chapter also discusses inclusive cartography in Brazil and highlights future challenges and offers different approaches along with new digital technologies.

8.2 INCLUSIVE CARTOGRAPHY: AN OVERVIEW

8.2.1 Concepts and Outcomes of Inclusion

The concept of integration and inclusion is premised on equal rights and that all forms of discrimination are unacceptable. The Universal Declaration of Human Rights – UDHR (United Nations, 1948) addressed this in 30 articles. Those related to accessible cartography are as follows: *...to seek, receive and impart information and ideas through any media and regardless of frontiers (article 19), everyone has the right to freely to participate in the cultural life of the community, to enjoy the arts and to share in scientific advancement and its benefits (article 27.1).* Many statements, declarations, and actions followed the original UDHRs by international organisms such as United Nations and UNESCO, or national bodies and government. Education was perhaps the field, which gained the most attention in the last decades, and the slogan *Education for All* encapsulates the issue of equality, and non discrimination, and also the idea of social, racial/ethnic, cultural, cognitive, physical differences, which may thwart learning opportunities and also to promote innovative approaches to improve it (See chapter 20 for more on educational approaches). The UNESCO Salamanca Declaration was adopted at the 1994 World Conference on Special Needs Education: Access and Quality, held in Spain, which endorsed inclusive schools by implementing practical and strategic changes, aiming at an 'Education-for-All' approach (UNESCO, 1994). Guiding principle three of that declaration states that schools 'should accommodate all children regardless of their physical, intellectual, social, emotional, linguistic, or other conditions. This should include disabled and gifted children, street and working children, children from remote or nomadic populations, children from linguistic, ethnic, or cultural minorities and children from other disadvantaged or marginalized areas or groups' (UNESCO, 1994).

There has been much research and many discussions that were later followed by actions on inclusion related to education and school, as well as for minority groups in society, including accessibility for people disabilities and cultural inclusiveness for indigenous people. The way these issues have been thought of has evolved, with disability now being understood as being both a human rights issue as well as a development issue. For example, people are disabled by social norm and not solely by their bodies and barriers can be overcome, if governments, nongovernmental organizations, professionals, and people with disabilities and their families work together (WHO, 2011). Disability is an umbrella term, covering impairment, activity limitation, and participation restrictions. Impairment is a problem of body function or structure (e.g. blindness); an activity limitation is a difficulty encountered by an individual in executing a task or action; while a participation restriction is a problem experienced by an individual involved in a variety of life situations (e.g. work). Disability is thus not just a health or a physiological problem. It is a complex phenomenon, reflecting the interaction between features of a person's body and the social aspects and physical infrastructure of the society within which he or she lives. Overcoming the difficulties faced by people with disabilities requires interventions to remove both environmental and social barriers (WHO Fact Sheet 352, September 2013, World report on disability, 2011 and the Convention on the Rights of Persons with Disabilities at http://www.who.int/). By the end of 2013, world population estimates predict that there will be 7 billion people, 15% of whom, or one billion people, will most likely be living with a disability. People are most often unaware of the great number of persons living with disabilities around the world and the challenges they face (WHO, 2013).

Plurality, inclusion, and accessibility are still challenges for the field of modern cartography, and the cybercartographic paradigm provides the means to addressing these challenges (Taylor, 2003, 2005). As discussed, every person is entitled to be a map user and a map creator, which implies providing equal opportunities to people, to enable all kinds of users to get information through maps and cartographic images, and also to facilitate the map use and mapmaking for minorities and special needs groups. This implies adapting the cartographic approaches by adopting tactile formats or technological innovations or even new methods of approaching cartography such as participatory mapping and a social cartography directed towards indigenous populations. Both of these cases will be discussed and examples presented in the next sections of this chapter.

8.2.2 Cartography for All: New Creators and Users

Who are inclusive map users and who are the creators of an inclusive cartography? How many can benefit from this kind of cartography and take advantage of it? As stated earlier, by the end of 2013, one billion people, about 15% of the world's population will have some sort of disability and they may require different map content, new mapping techniques or methods, or an expansion of cartographic language and design, to name a few issues. It is estimated that 285 million people will have some sort of visual impairment (WHO, 2012). On the topic of cultural and social inclusivity, there are approximately 370 million indigenous people in the world, who speak over five thousand languages in over 70 countries on 6 continents (UNESCO, Conference on Indigenous Education in a Changing World, June, 2012 see at http://portal.unesco.org/en/).

Physical impairment, learning difficulties, social and cultural minorities, children and the elderly are persons and groups that, for different reasons, will require different images, maps, and different types of geospatial data. Inclusivity means, digital resources, information and communication technology (ICT) and innovative methods specifically produced to meet special needs enabling persons with serious disabilities to participate in social life, work, or school. Even if the user does not belong to a special needs group, people have a variety of learning styles as discussed in chapter 20 and learn from information presented in a variety of formats and multisensory technologies. An inclusive cartography can be more user-centric and interactive and may be a way for people to overcome communication barriers while also promoting accessibility and equal rights. A multisensory, multimedia and multimodal cartography is also a multicultural cartography very much in keeping with the elements of cybercartography discussed in chapter 1.

New digital technologies are however not universally accessible and this is often the result of politics and economics, as well as a lack of an inclusive cartographic education, both for map creators and users. Cybercartography, with its seven key elements and six central concepts, renders cartography more inclusive as discussed by D. R. Fraser Taylor (2005) and in Chapter 1 of this book. Also, the author has for many years discussed issues about a 'cartography for and by all' (Vasconcellos, 1993, 1995, 1996, 2001; Almeida and Tsuji, 2005), for both the visually impaired and indigenous people. Throughout a need to define the creators and users and to discuss the role of perspectives and dimensions towards a new vision of cartography was discussed towards a new inclusive interactive cartography as seen in Figure 8.1 below.

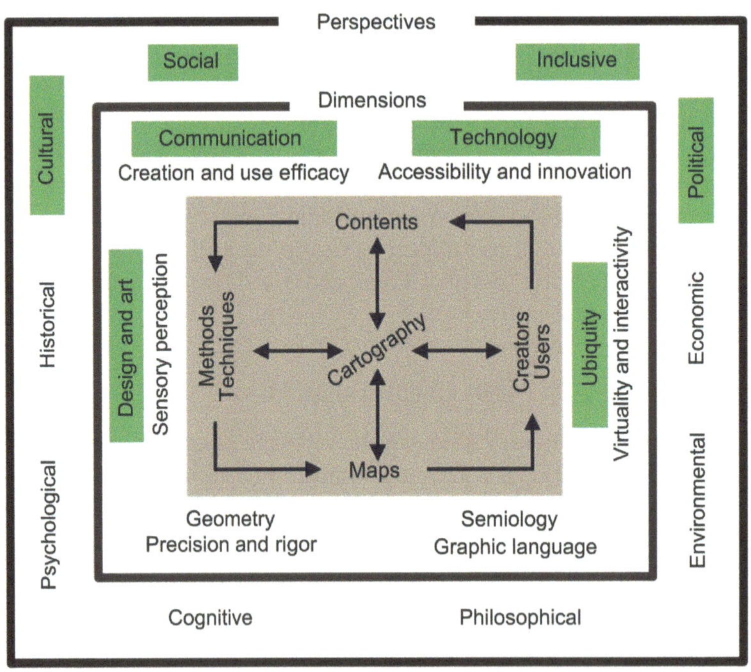

FIGURE 8.1 Theoretical framework of inclusive interactive cartography.

Figure 8.1 summarizes the theoretical framework for Inclusive Cartography, which has been refined recently (Vasconcellos, 1995; Almeida, 2011). This new framework differs from previous ones as it includes perspectives and dimensions directed to a cartography for everyone. For example, four perspectives are featured, namely, cultural, social, inclusive and political along with four dimensions, which are design and art, communication, technology and ubiquity, the latter four of which are considered vital to modern cartography. In addition, the same cartographic communication elements remain namely: by whom, what, why, how, through which signs, means and devices, to whom and with which results. For example, for visually impaired users the 'how' might be the most vital question since graphics, interfaces, modalities and cartographic considerations will play an important part in the process; while for indigenous people, questions such as what, by whom and why may be more relevant (as discussed in many chapters of this book). Even though technology enables maps to be created in new and interesting ways that can be used by people with different abilities than in the past, old challenges remain. For example, a lack of cartographic education and training remains as previously discussed.

8.2.3 The Relevance of Technology for Inclusive Cartography

As discussed, technology plays an important part in inclusive cartography and in the process of achieving accessibility, it can help people with different abilities to see, hear, or feel images, maps, and to navigate these in new ways. Assistive technology (AT) includes devices and software used by those with varying abilities, such as people with multiple disabilities, or people who are visually, hearing and/or mobility impaired. With AT distinctions are made between computer and no-computer aids and products, for instance, a person's visual impairment may require a sensory replacement, augmentation or substitution (see at http://www.perkins.org/). Almost all sensory substitute technologies are now computer-related, though several have their origins in predigital technology, such as tactile graphics, Braille displays, voice recorders, and video cameras. For an inclusive cartography and certainly for cybercartography, map creators and users must take advantage of this large variety of ATs.

The Global eSchools and Communities Initiative (started by the UN – ICT Task Force) does relevant work related to technology, innovation, and education. In their work, AT is organized in a system of low-tech, medium-tech and high-tech tools, and strategies matching a person's needs, abilities, and tasks (Hooker, 2007). This approach could be applied to inclusive cartography especially when considering that a large number of people with disabilities, approximately 800 million, are living in developing nations (WHO, 2012). Concurrently, many indigenous groups also live in the developing world. Neither group is benefitting from these innovative multisensory and multimodal technologies as these are rarely accessible in these countries.

Low- and medium-technology devices have been the norms for decades for tactile cartography, including some audio equipment used in combination with tactile graphics, like thermoform machines and Braille typewriters (see http://www.gesci.org/). Even these are not necessarily available in less developed countries. High-tech devices can make a real difference in tactile and inclusive cartography, for example, computer screens readers, voice synthesizers, a variety of software and hardware, adapted keyboards and mice, special glasses and cameras, global positioning system (GPS), mobile phones, and so on. These assistive computer technologies are vital for persons with visual impairment; they help them to 'see' images and graphics using tactile and sound resources. AT, directed to map design, creation, production, dissemination and use, will work towards an inclusive cartography for all, and not only for a special group of people.

8.3 TACTILE CARTOGRAPHY – FROM STONES IN THE SAND TO VIRTUAL MAPS IN THE CLOUD

8.3.1 Tactile Cartography in Brazil

People generally acquire and develop spatial understanding through vision, and visual perception is fundamental for understanding the environment. People deprived of vision, either totally or partially, depend on their other senses to gather information about the physical world around them and geography in general. Tactile perception works with movement (kinesthetic force) and touch, and haptic perception such as feeling raised surfaces with the skin allows visually impaired persons to explore the tangible world. Tangible and raised relief maps are currently applied examples of these. Tactility is however a coarse way to access spatial information but is useful to bring fragments of spatial information together to enable the formation of an image. Sight is more synthetic and is of a higher resolution (Vasconcellos, 2001). For visually impaired people to gather geoinformation, or to 'see' a landscape requires special cartographic materials and images and cybercartography brings new possibilities for these persons to use maps in an interactive and multisensory way.

8.3.1.1 Creation and Use of Tactile Maps

Tactile maps have been made for centuries as seen in stone tablets set in sand or raised lines in copper. By the mid-nineteenth century images in relief using Braille text began to be used by the visually impaired people. More recently, maps with raised symbols have been constructed using a wide variety of materials such as embossed paper and aluminum foil, moulded plastic, silk-screen printed fabric and microcapsule paper maps compatible with ink-jet printers and raised by heat, a technique also adapted to digital maps. There are two main groups of tactile maps, based on their purpose, scale, and features:

1. Orientation and mobility maps are directed only to blind and severe low-vision people who need to be aware of their physical environment in order to navigate in space. An orientation map gives an overview of the area. Mobility maps, as the name suggests, provide the kind of detailed information that is needed to enable a person to move safely through their environment. Mobility maps show obstacles, as well as elements to use such as sidewalks, or elevators. They can depict indoor and outdoor environments and both are always large-scale representations of buildings, routes, streets, transportation networks, neighbourhoods, and recreation areas. Verbal maps, also called sequential maps, can replace tactile mobility maps to explain routes to blind people.
2. Geographical maps represent larger areas in much smaller scale than mobility maps; they are either reference maps (political and physical tactile maps, ranging from cities, states, and countries to the whole globe) or thematic maps that portray qualitative and quantitative geographical information in tactile form. They show the locations, patterns,

and structures of spatial data on a wide variety of themes such as population, industry, climate, distribution of vegetation, at all scales from local and regional to global (Vasconcellos, 2001).

Many authors have contributed significantly to tactile cartography. Research on tactile map design, production, and use has been carried out by a large number of specialists and results have been widely published (e.g. Wiedel and Groves, 1972; Kidwell and Greer, 1973; James, 1982; Wiedel, 1983; Barth, 1987; Tatham and Dodds, 1988; Edman, 1992; Vasconcellos, 1996; Rowell and Ungar, 2003; Tatham, 2003; Almeida and Tsuji, 2005; Jacobson, 2007). Koch in his recent article (Koch, 2012) brings a very good overview of tactile cartography, covering design to new digital technologies and also provides a description of the chronological development of tactile maps.

Information about visual impairment and tactual perception are important to map and graphics creation and use. For example, distinguishing between an individual who acquired their visual impairment or were born blind is important, as is the type and degree of visual impairment as these affect the way maps and images can be read. Schiff and Foulke (1982) and Heller (1991) provide an excellent overview of these issues. Alternatively, Almeida and Tsuji (2005) provide a summary of these issues accompanied by an analysis of tactile map design, production and use, including a review of interactive multimedia maps. Research on tactile maps generally focuses on spatial knowledge and perception, orientation and mobility, navigation (wayfinding and locomotion), and much less so on thematic cartography, cartographic education, and the introduction to graphic language. As a contribution to these areas, the author has conducted studies on tactile map design, production and use, with an emphasis on education. Graphic semiology, as presented by Jacques Bertin in 1967, was applied to develop a tactile graphic and cartographic language and visual variables were translated to a tactual format and elevation was added to textures and colours, by either showing order or the differentiation of the data (Vasconcellos, 1993). Tactile maps can be designed for all users, combining visual and tactile symbols, preferably in colour. Findings highlighted clarity and simplicity as important qualities in tactile mapping, and map details have to match the resolution of touch, which is not as refined as it is for vision. Bertin's maps to 'see' are good for visually impaired people to see using touch, and also a collection of maps or a multilayered tactile map can help avoid complexity and excessive clutter, making tactual perception easier. Tactile map design, including the analysis of data, reduction, generalization, and choice of graphic language, is absolutely vital to achieve a useful map directed to people with any kind of visual impairment. It is also important to recall that impairment often leads to increased abilities in other areas; for example, people who are visually impaired have an extremely good ability to decode map keys because they are able to use both hands simultaneously, one on the legend and the other on the map, something sighted people have difficulty doing.

In general, tactile maps require more cartographic education and training, for both creators and users. For this reason, a set of activities was designed, as an introduction to cartography, that was grouped into six steps:

1. Developing a tactile graphic language (memory game with tactile graphic variables);
2. Understanding scale, reduction, and distance (activity using dolls, cars in several sizes, and a classroom map);

3. Presenting the concept of point of view (graphic representation of objects in vertical and horizontal views, avoiding perspective);
4. Seeing location and orientation (geographical battle game with coordinates and hemispheres);
5. Practicing map reading (the complete address of school, from classroom to the universe); and
6. Using map legends and combining the first five steps with the construction of a city (Vasconcellos 1993; Almeida, 2011).

The constructed materials and maps were tested with 200 visually impaired children and adults, and evaluated by 80 schoolteachers, and this informed the development of a set of guidelines directed to conventional tactile map creation and use (Vasconcellos, 1996).

8.3.1.2 *Tactile Cartography in Brazil*

Since 1989, the Laboratory of Geography Education and Didactic Materials (LEMADI), at the University of São Paulo (USP), Brazil has been working with tactile cartography and several research projects have been conducted over more than two decades (Vasconcellos, 1993; Almeida, 2011). LEMADI became a key organization in Brazil, not only because of the development of tactile maps and teaching materials, but also for the promotion of an inclusive geography and cartography, with an emphasis on education and teacher training. A very large number of courses and workshops about map creation and use, applying techniques and methods directed to students with special needs, were offered by the LEMADI team, under the coordination of the author. In addition, national and international conferences on these topics were organized.

There has been an ongoing involvement of graduate and undergraduate students conducting research on tactile cartography at USP. Some of these students continue to work in the field of tactile cartography up to the present. Carla Gimenes Sena, for example, received the International Cartographic Association (ICA) travel award in 2005 and has presented the results of her doctoral research (Sena, 2011). She developed 3D models with sound and presented the contribution of the use of technology to the teaching of geography. The city of São Paulo was her study area, and this included an emphasis on urban growth, seen at local and global scales. The teaching materials comprised five maps of the São Paulo municipality (from1881 to 2002), showing urban expansion in an interactive audio tactile scaled model of the city in the nineteenth century, along with graphic-tactile representations of historic buildings, a synthesis map of the urban area, and a topographical model of the city. This combination of tactile maps and images with sonorous scale models applied to Sao Paulo showed that a dynamic interdisciplinary approach to cartography can greatly improve geographic and cartographic learning. Waldirene Ribeiro do Carmo is working with teachers and provides training in tactile cartography and geography. She has also published and presented her work at conferences (Sena and Carmo, 2012).

The Brazilian team participated in the Tactile Atlas coordinated by D. R. Fraser Taylor (2001) with other Latin American countries. Since then, a research project funded by the Pan-American Institute of Geography and History (PAIGH) and the Organization of the American States (OAS) has been ongoing coordinated by Alejandra Coll at the Metropolitan Technological University (UTEM), in Santiago, Chile, which involves several Latin America countries, including Brazil. The international team meets yearly, organizes workshops and courses (in Chile, Argentina, and Mexico), and has demonstrated excellent results over the last decade. This tactile cartography initiative has enabled an interdisciplinary and international team to be consolidated, facilitating access by visually impaired people access to geographic knowledge and maps. Maps at different scales were developed and distributed to participating countries, accompanied by a book of geographic concepts that has been created using maps and models, to present rivers, lakes, hills, among others, with images and text in Braille (Coll and Pino, 2007). Between 2009 and 2012, the PAIGH-OAS Research Project *Tactile Mapping and Learning Materials for the Understanding of Global Warming* was also coordinated by UTEM with the participation of Argentina, Brazil, Chile and Peru, with Carmo representing Brazil. She is also the vice-chair of the ICA Commission on *Maps and Graphics for Blind and Partially Sighted People*, with Alejandra Coll as the chair. Carmo is responsible for the project on School Tactile Cartography, particularly tactile map production and use, and its relevance in the teaching of geography and cartography. Several examples of tactile materials can be visualized at this website http://lemadidg.wix.com/cartografia. In the last 6 years, LEMADI organized and participated in 8 international courses with 20 in Brazil, and trained over 1200 teachers and specialists. An important contribution to the field, from Brazil, is the publication of a book on tactile cartography, with articles from several researchers (Freitas and Ventorini, 2011). Several other initiatives in Brazil are taking place in universities, public institutions and NGOs. For instance, there is a commission studying tactile map standards for public areas in Brazil with the Association of Brazilian Technical Norms association and participation members of the LEMADI team. In addition to USP and University State of São Paulo (UNESP), other cartographers and geographers are working with tactile cartography in Brazil. At the Federal University of Santa Catarina, Ruth Nogueira created a Laboratory (see www.labtate.ufsc.br/) and has conducted studies on tactile maps, the standardization of symbols, in addition to the dissemination of maps through their website (Nogueira, 2009).

8.3.2 Audio-Tactile Visual Map Devices: New Tactile Cartographies

New technologies have introduced great changes and improvements in the area of tactile map creation, dissemination, and use. In the last 30 years, many products have been developed to support people with visual impairment enabling them to 'see' maps through touch and sounds. In the late 1980s, a tactile-audio system called NOMAD was developed, combining touch and sound so that a tactile map, when touched, spoke the information that was being touched. NOMAD was named after the guide dog of a blind Philosophy student, and this system was invented by Professor Don Parkes, the Director of the Institute of Behavioral Sciences at the University of Newcastle, in Australia, in 1986. The stimulus for this invention was the sudden onset of blindness of a friend and colleague at the University of California, Santa Barbara, Professor R. Golledge, with whom he continued to undertake collaborative work on audio-tactile systems for many years. NOMAD and all the tactile graphic design (TGD) components followed the lead of this first version of this audio-tactile device (Parkes, 1988). It included functions allowing for location, navigation, distance, scale, and a variety of different features to be shown in a tactile format, with graphics and maps combined with sounds. In 1988, NOMAD was tested by the Sensory Aids Foundation who comprehensively evaluated it and published very good reviews by blind users. Also, Parkes' invention and devices were presented and tested in several countries, including in Brazil during the 4th International Symposium on Maps and Graphics for Visually Handicapped People (Parkes, 1994).

NOMAD became an established multisensory device showing great potential. A detailed map of the entire London Underground system had been developed at King's College London with Dr. Andrew Tatham (Chair of the Tactile Mapping Commission of the International Cartographic Association) with Parkes when he was a Visiting Senior Research Fellow, in 1991. At the time the audio-tactile system had interactive capabilities and up to three layers of information could be embedded into every information point. Effectively an entire encyclopedia could have been embedded into NOMAD, but computer storage and retrieval capacities at the time were a limiting factor. The system was presented in Europe, North and South America, and in many Asian countries; and in April 1993, NOMAD was awarded the First Prize at the Autonomic Expo in Paris. Maps, spatial guidance systems, a capacity to paint in sound with frequencies changing at user-defined boundaries of any shape made a virtual 3D audio-tactile surface possible. It also read Braille and interacted with voice recognition to answer user questions. NOMAD was translated into Portuguese, Spanish, Japanese, French, Italian, German, Swedish, Malay, and Dutch. From the original NOMAD system, the TGD components of AudioPIX, Audio CAD, and Audio TRIP were developed and presented in journals and scientific conferences (Parkes, 1995, 1998). In 2004, the system comprised a number of components under the general name TGD, using an off-the-shelf touch pad rather than the original NOMAD tablet, and it was handed over to Duxbury Systems who applied many software systems to their established market for Braille translation systems and embossed dot graphics.

A few years ago, a Talking Tactile Tablet – TTT3 – with the same functionality as NOMAD, manufactured by Touch Graphics, came onto the market (see at www.touchgraphics.com/). The company also produced a talking pen, and a Power Chord Keyboard used to type Braille into a computer. There are new interesting devises being developed, mainly at universities and institutions, one of them being the camera for visually impaired and a voice technology now available worldwide (see www.seeingwithsound.com). The camera scans a visual field and converts images to computer-generated sound. For navigation, Braille Note GPS and various Adapted GPS devices make it possible for visually impaired people to travel independently. It is essential that a screen reader that shows images on the monitor through sounds and speech also has a Braille display (see http://www.humanware.com). Also now, text is easily translated to sound using any voice synthesizer software combined with the screen reader. In addition, Braille Touch allows people to use touch screen mobile phones by typing Braille (http://brailletouchapp.com). The American Foundation for the Blind (http://www.afb.org/) provides an extensive list of products and software, including all sorts of AT that help visually impaired persons to access computers, multimedia interactive maps, and images. The force feedback joystick and mouse is very interesting; however in terms of products and software availability, there are problems with discontinuity, as was the case with the Logitech's iFeel mouse, which is no longer available. This combined a tactile feedback with optical technology to give an interactive computing experience. Other research projects suffered similar fates such as the *Blind Audio Tactile Mapping System.*

Several specialists have dedicated time and effort to develop and study audio-tactile and visual-mapping devices and software, both proprietary and open source, directed to people with visual impairment. Golledge had made a very special contribution to the field of audio-tactile cartography, being himself a geographer and a visually impaired person and his work was considered relevant and original from the outset (Golledge et al., 1996, 2005). Relevant work was also done by Rowell (2007) and Jacobson (2007) and their research was mainly related to navigation and mapping for visually impaired people, focusing on new digital technologies and their possible outcomes. Habel and Graf (2008) discussed the application of You-are-here (YAH) maps as navigation aids for visually impaired people, redesigning them to enable tactile perception towards the creation of audio-tactile maps providing verbal descriptions in a multimodal format. Rowell (2007) conducted an interesting survey that compared conventional tactile maps with new wayfinding geotechnology, and informed his work by interviewing 30 visually impaired persons, asking them to evaluate methods (tactile maps, technologies, audio descriptions and fixed-electronic navigation aids). The results rated maps and GPS devices, including the advantages and problems and discussed the potential of these new technologies. Paladugu et al. (2010) wrote a very good article highlighting human–computer interaction issues related to audio-tactile maps.

Overall, multisensory interactive maps do work and can change the way visually impaired persons relate to cartography (Almeida and Tsuji, 2005). There are many technology and haptic studies and publications proposing new methodologies, innovative devices, and digital resources applied to audio-tactile cartography, and a few examples are provided here. Many types of ATs can help the visually impaired take advantage of ICTs and cybercartography, especially in a time dominated by mobile devices, smartphones, and tablets. Unfortunately, ubiquitous and telecartography as discussed by Gartner et al. (2007) are not everywhere and these technologies are not available to everyone. It is an exciting new field of research in cartography, but it mainly depends on ICT and related infrastructure. It is estimated that around 4.6 billion do not have access to the Internet (see http://www.internetworldstats.com/stats.htm). In Africa, only 15.6% of the population is Internet users (2012). Although, things are changing rapidly, as in the last 12 years, Internet users went from 361 million to 2.4 billion, a significant number of people still have difficulties in accessing the Internet, mobile devices, GPS-receiving satellite or radio signals, modern computers and proper software to take advantage of new technologies and digital cartography. In fact, ICT is not yet for all, but it could make a huge difference in an inclusive cartography. In places like public schools in less developed countries, a tactile map collage made by a geography teacher may still be the only way for visually impaired children to 'see' images.

8.4 CARTOGRAPHY FOR AND BY INDIGENOUS PEOPLE

Indigenous cartography can be considered as a form of inclusive cartography. This type of inclusive cartography may use some of the technologies discussed above, but the motivation would be access and cultural or ethnic representations of geospatial information. While this type of inclusive cartography differs from the type discussed above, socioeconomic and political issues remain, as do ICT limitations in developing countries. Furthermore, identity loss due to diminishing territorial integrity, discrimination, and lack of respect for ancestral history are issues making access to cartography a human rights and development issue (UNESCO, 2008).

Chapin (2006) defines indigenous mapping as mapping done by and for indigenous peoples to achieve political objectives. It was first carried out in Canada and Alaska in the 1960s and 1970s, primarily to claim and defend land and natural resources and also to strengthen political organization, record traditional history and culture, develop education programmes on a variety of topics (such as the environment), and plan for economic development. A variety of methodologies have been applied, ranging from highly participatory approaches involving village sketch maps to more technical efforts with GIS and remote sensing. In general, indigenous mapping has shown itself to be a powerful tool and it has spread rapidly throughout the world (Chapin et al., 2005). The International Fund for Agricultural Development (IFAD, 2009, 2010) discussed participatory mapping and outlined concepts and actions required to implement participatory mapping processes. It also presented several case studies. There were very interesting initiatives related to the use of ICTs applied to the reinforcement of culture and knowledge in Bolivia, such as the *Project Multimedia Systems* for Ethnographic Materials and the *Thailand Indigenous Knowledge Initiatives* (Resta, 2011). The Geomatics and Cartographic Research Center at Carleton University, Canada, has also achieved impressive results in mapping traditional knowledge such as the Nunaliit Cybercartographic Atlas Framework (Taylor, 2009), among several other actions and publications including several chapters in this book. These chapters discuss indigenous mapping with Inuit and First Nations in Canada. This chapter deals with a very different cultural context although many of the central purposes of the mapping are similar.

Three Brazilian initiatives involving digital technologies and mapping with indigenous communities were chosen as example of inclusive cartography. In addition to the case studies presented below, there are a large number of individual researchers, theses, and publications in many parts of Brazil, inside universities as well as in Brazilian NGOs.

8.4.1 Participatory Mapping in the Brazilian Amazon

Two successful projects that applied participatory mapping methodologies are discussed here. The first is the Instituto Socio Ambiental (ISA), established in 1994 with the objectives of defending the social good and rights, both collective and diffuse, relating to the environment, cultural heritage, human rights, and the peoples (see http://www.socioambiental.org/en/). From the outset, its attention to cartography has been remarkable and from its six permanent activities – administration, communication, institution development, documentation, information technology and geoprocessing – three are related to cartography. All ISA projects deal with mapping, such as the one on conservation units in Brazil (see http://uc.socioambiental.org/en), and their studies on indigenous and traditional peoples collect important information. Cartography is involved in most of the ISA's work and publications, including participatory mapping. In the Amazonian River Negro, which is one of the areas where they work, indigenous sacred places and cultural heritage are being mapped and several maps have been made available (see http://raisg.socioambiental.org/mapa#english). The *Mapeo* bi-national project of Cultural Cartography is mapping sacred sites in the northwestern Amazon, and it works with indigenous people who inhabit the border region and documents their cultural and natural (Brazil and Colombia). In Brazil, *indigenous people* are the native people and they differ from *traditional people*, such as *seringueiros* (rubber-tappers) and *castanheiros* (Brazil nut collectors) from the Amazon region. These groups are also distinguished by Brazilian laws, and the ISA defines these concepts by providing examples of other traditional groups such mussel collectors or fisherman in the south of Brazil (see http://uc.socioambiental.org/en). ISA also coordinates the Amazonian Network of Georeferenced, Socio-Environmental Information (RAISG) (raisg.socioambiental.org), which was formed by countries under the Amazonian forest. RAISG also provides cartographic materials using web map services (ISA, 2012).

The New Social Cartography of the Amazon Project is another Brazilian project, coordinated by Professor Alfredo Wagner Berno de Almeida, at the Federal University of the Amazon. His outstanding work was given the *Visionaries Awards* by the Ford Foundation in 2011 (see http://www.fordfoundation.org/newsroom/news-from-ford/480).

The Project New Social Cartography of Amazon (PNSCA) objective enables traditional peoples and communities in the Amazon region to map themselves. Cartography in this instance is a tool for social self-affirmation. Upon the request by an indigenous or traditional group to PNSCA, a mapping workshop is organized with the participation of about 30 local social agents and a few in cooperation with some of the 70 researchers available. During the workshop, researchers teach GPS and mapping techniques, talk to the social agents, and collect testimonies about the social history and problems of the community. The social agents produce layouts depicting their region, and specify which elements are relevant for map rendering. The social agents alone mark down the GPS points that they consider meaningful. The PNSCA receives these GPS points and georeferences them on a cartographic base, inserting the illustrations produced during the initial layout. These illustrations include drawings, sketches and the reproduction of symbols and objects (e.g. paddles, houses, boats, work instruments, animals, plants, etc.), which are transformed into icons that compose the map. Also excerpts of testimonies selected and transcribed into text, which the community reviews as a preliminary version of a brochure, as seen in Figure 8.2(a) and (b).

Each brochure is the result of a combination of collaborative efforts and relationships between a community or traditional group and the research team, and all brochures have pages of text and a map (A3 paper size). One thousand copies of each brochure are printed, most of them remaining with the local group for their use. Figure 8.2 shows two examples of this publication. Between 2005 and 2012, over 120 brochures, organized in six series, were produced, together with maps, books, and videos. During map workshops and activities, local people select the information they want to put on the map, expressing their perception and highlighting problems and conflicts. While doing so, they learn how to use new technologies. The Project expanded its mapping workshops to include areas outside of the Amazon in *Series 2 – New Social Cartography Project of Traditional Peoples and Communities of Brazil*. The PNSCA publications and more information can be downloaded from their website (http://www.novacartografiasocial.com/).

8.4.2 Ethnocartography and GIS Initiatives in the State of Acre, Amazon

The second Brazilian example is a local experience coordinated by the NGO Comissão Pró-Índio (CPI) (see www.cpiacre.org.br) in the western Amazon state of Acre, which has 15 native nations, with a population of 16.5 thousand native people living in 186 indigenous areas, most of them legalized and recognized by the federal government. CPI has coordinated projects, activities, and publications related to indigenous people and cartography for the last 34 years. In 1996, it initiated an action plan to educate and train Indigenous Agroforestry

(a)

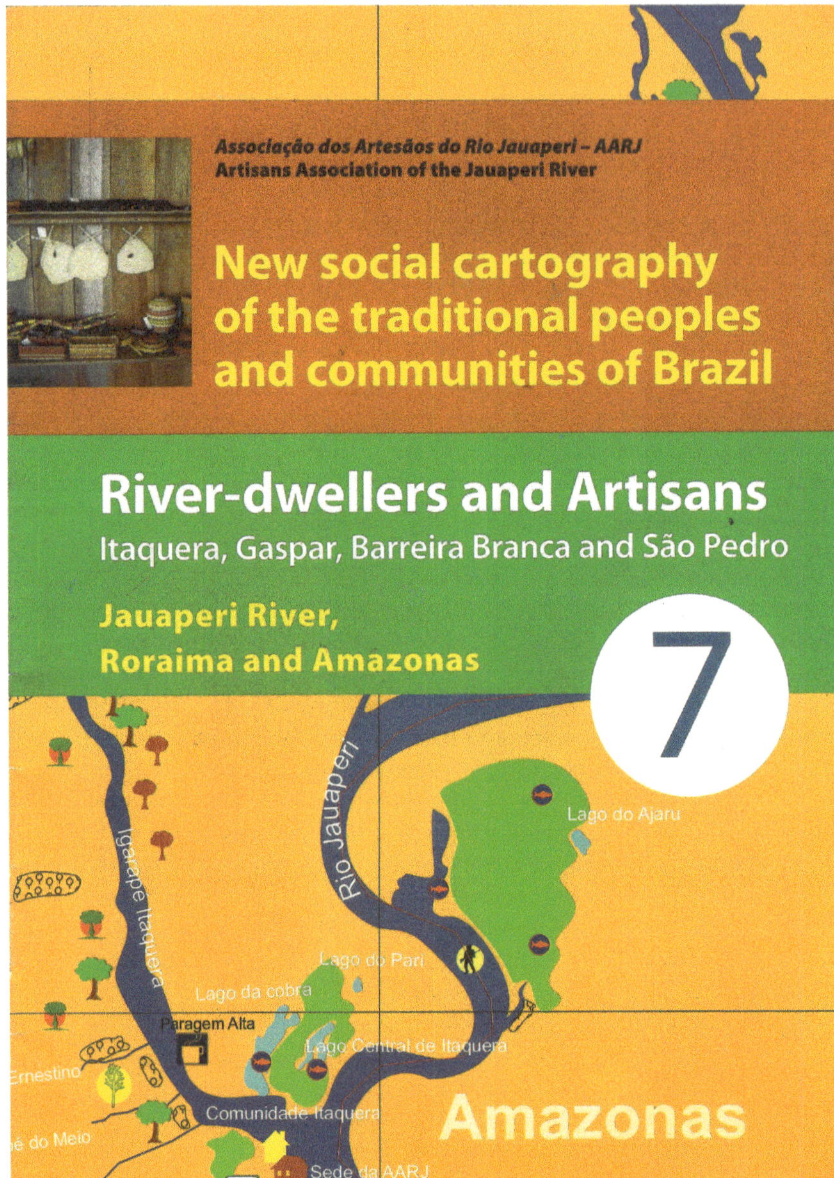

FIGURE 8.2 Examples of New Social Cartography publications (brochures 7 in (a) and 31 in (b)).

Agents (IAFAs) emphasizing environmental land management. This was done under the coordination of Renato Gavazzi, a geographer who has been involved with indigenous cartography for over 25 years. His latest work (Gavazzi, 2012) discusses two key aspects: agroforestry and indigenous mapping. This publication is in Portuguese, but many indigenous maps, photographs, and images can be seen. As part of the research process, agroforestry

(b)

FIGURE 8.2 Cont'd

is documented by the IAFAs through in diary notes, providing an ethnographic record of their reality with the use of written language, figurative drawings, and mapping. The IAFA's agroforestry practices have contributed to building new models and new productive spaces adapted to the ecological conditions of the rainforest, with the contribution of traditional, scientific, academic, and local knowledge and has effectively included the participation of communities in managing their lands. Indigenous cartography has a key role in this process, with

(a)

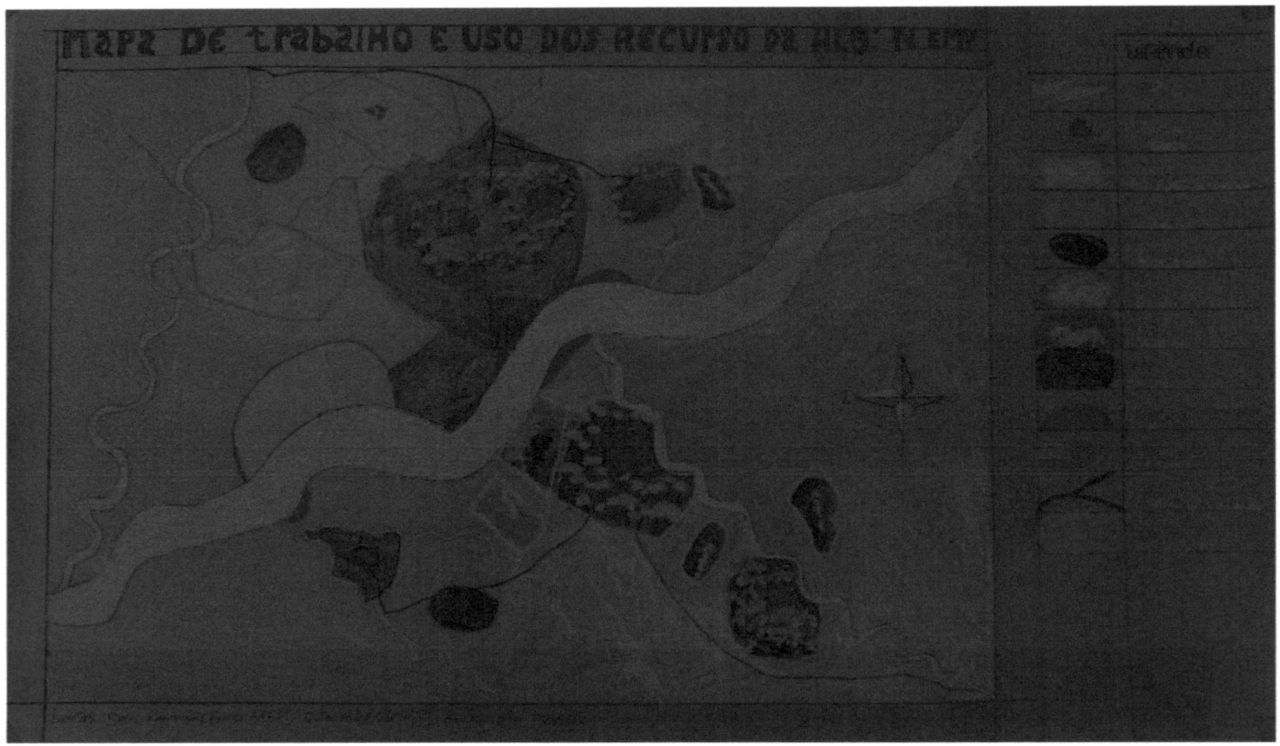

FIGURE 8.3 Natural resources and land use maps – Kaxinawa indigenous territory in Acre, Brazil ((a) mental map done by native people and (b) digital georeferenced map).

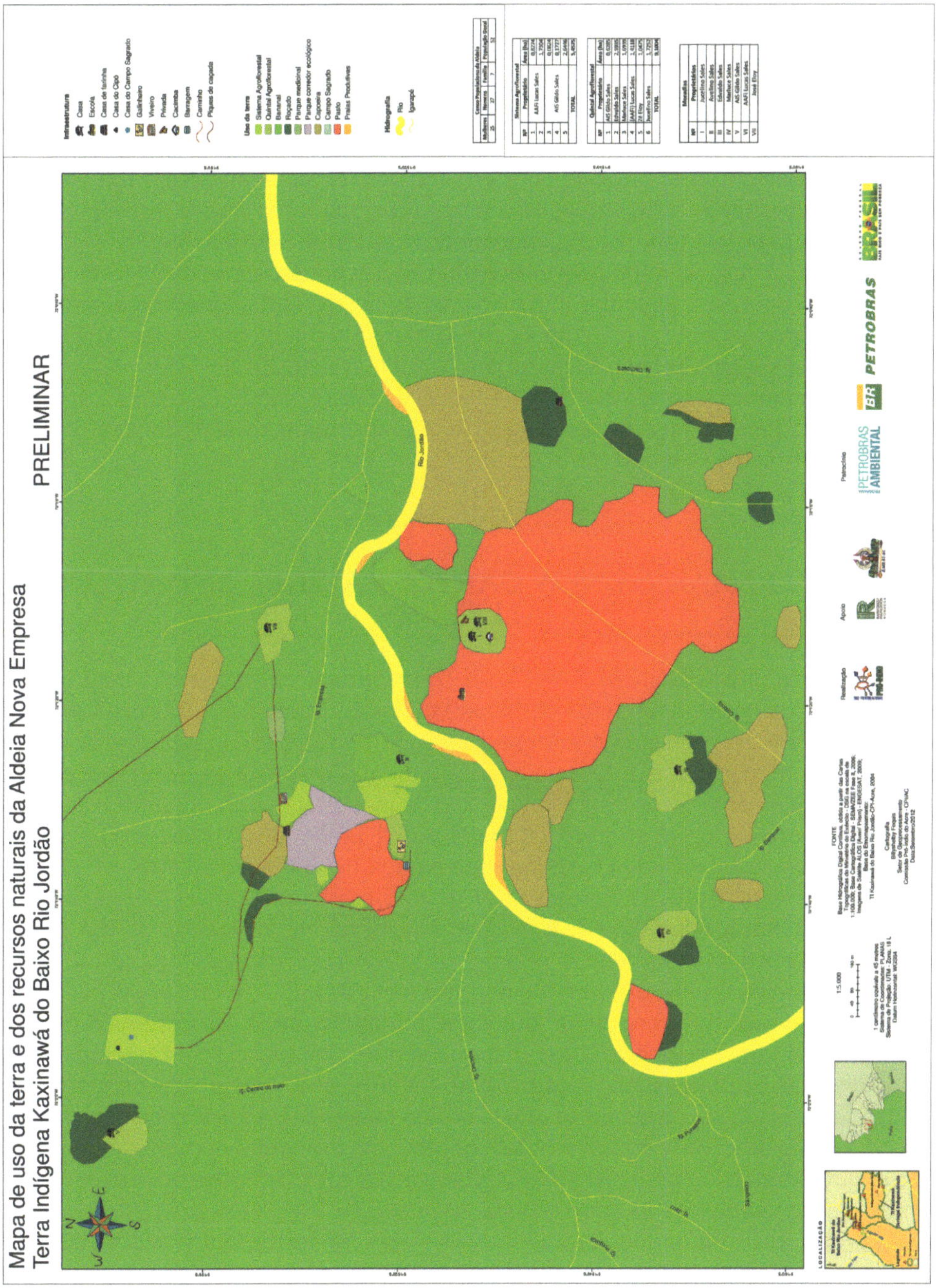

(b)

FIGURE 8.3 Cont'd

both mental and georeferenced digital maps using GIS. The work done in the last 10 years highlights the importance of indigenous traditional knowledge in the construction of individual and collective maps and plans, as tools targeted for biodiversity conservation, protection, and territorial environmental management of indigenous lands in Acre.

The methodology was applied in more than 50 courses organized at the Education Center for the Forest Peoples at CPI, in Rio Branco city, and also in the 11 native lands in Acre state. This Center has a Cartography laboratory, which is also used for indigenous peoples to write their books and use computers during courses, and where they can access the support of researchers. Figure 8.3(a) and (b) shows two maps from the same native territory, one was drawn by an IAFA to be used for the digital georeferenced map creation; both have the rivers in yellow, which is a cartographic convention that indigenous people decided to use long ago, during one of the first CPI courses. In previous work, the author presented results from several years of experience in giving courses to IAFAs and indigenous teachers at CPI, about geography and mapping (Almeida, 2001; Almeida and Gavazzi, 2011).

Mapping for and by Amazonian natives makes Brian Harley's and Woodward's arguments a reality (Harley, 1989; Woodward and Lewis, 1998), and more so as ICTs give to maps a future by representing the past and transforming the present. Indigenous groups in Brazil are witnesses to that.

8.5 FINAL REMARKS: NEW CARTOGRAPHIES AND FUTURE CHALLENGES

Inclusive cartography will always deal with conventional old-fashioned map in paper, fabric, or sand, at the same time, with multimedia, new technologies, and maps on mobile devices. There has always been many cartographies and the examples presented in this chapter demonstrates that all sorts of maps, from a simple mental map full of cultural expressions made by Amazonian native people or a collage tactile map made with different textures will survive, alongside new digital and virtual maps saved in the cloud. Neither can be disregarded because they fulfill different functions and have different 'raison d'être'. Inclusive cartography means creating maps applying a multisensory and multicultural graphic language, with 'images' that can reach different users, while also taking advantage of the available ITC. New formats such as those seen in telecartography, cybercartography, and the Internet demonstrate that maps can be used by all, and the right technology can assist people with any special need. Much has been done, but there remains a need for improved hardware and software development, such as those that enable visually and hearing impaired people to access cartographic information, using sound, tactility, and images. Vision and hearing are dominant senses when compared to touch, which means that able-bodied people do not often consider using other senses such as touch or olfaction to acquire or produce information about the environment or to navigate space. Cybercartography provides opportunities to do so by exploring new pathways to map creation and use.

In this chapter, the author highlighted the relevance of cartography and participatory mapping, to promote the empowerment of communities to deal with political issues and to seek social, environmental, and economic sustainability. In the case of regions such as the Brazilian Amazon, maps can help people to be the protagonists of their future towards greater and more inclusive community-led development. There have been many social cartography and participatory mapping projects all over the world; however, they need to be discussed and disseminated more broadly. Native people can share valuable lessons about the future while not forgetting the past, and on how apply innovative technology to draw maps of their knowledge. Besides innovation, plurality and accessibility are critical issues in modern cartography.

There is a responsibility regarding cartographic education at all levels and cartographers need to prepare new generations to be map users and creators, and this includes minority groups and persons with any kind of disability. As geographers and cartographers, we must develop new approaches, as well as new digital technologies. Maps give voice and empower minorities as has been demonstrated in several chapters in this book, they can be art, cultural images, planning tools, or teaching strategies. Maps can be translated to sound and be heard by visually impaired persons. We should aim at a *cybercartography for all*, based on interactive multisensory, multicultural, and multimodal maps. It is important to fulfill the social role of cartography in all its potential and to think about new cartographies for a future where telecommunication and computer technologies will bring unexpected outcomes.

References

Almeida (Vasconcellos), R., 2001. Cartography and indigenous populations: a case study with Brazilian Indians from the Amazon region. Proceedings of the 20th International Cartographic Conference Beijing, China.

Almeida (Vasconcellos), R., 2011. Bertin graphic semiology and its relevant contributions for research and teaching in Brazil. Proceedings, 25th International Cartographic Conference Paris, France.

Almeida (Vasconcellos), R., Tsuji, B., 2005. Interactive mapping for people who are blind or visually impaired. In: Taylor, D.R.F. (Ed.), Cybercartography: Theory and Practice, Elsevier, pp. 411–431.

Almeida, R., Gavazzi, R., 2011. Ethnocartography and native people in Acre, Brazil – experiences with participatory mapping. Proceedings of the 25th International Cartographic Conference Paris, France.

Barth, J.L., 1987. Tactile Graphics Guidebook. American Printing House for the Blind, Louisville, KY, USA.

Chapin, M., 2006. Mapping Indigenous Lands: Issues & Considerations. Center for the Support of Native Lands. Available at: http://globetrotter.berkeley.edu/bwep/colloquium/papers/Chapin2006.pdf.

Chapin, M., Lamband, Z., Threlkeld, B., 2005. Mapping indigenous lands. Annual Review of Anthropology 34, 619–639.

Edman, P.K., 1992. Tactile Graphics. American Foundation for the Blind, New York, NY.

Coll, A., Pino, F., 2007. Impact of Tactile Cartography on the Teaching of Geography in Latin America. Proceedings of 23th International Cartographic Conference. Moscow, Russia.

Freitas, M.I.C., Ventorini, S.E., 2011. Cartografia Tátil: Orientação e Mobilidade às Pessoas com Deficiência Visual. Paco Editorial, SP.

Gartner, G., Bennett, D.A., Morita, T., 2007. Towards ubiquitous cartography 34 (4), 247–257. Available at: http://personalpages.manchester.ac.uk/staff/m.dodge/teaching/envisioning/lecture10/Gartner_Towards_Ubiqutous_Cartography.pdf.

Gavazzi, R., 2012. Agroforestry and Indigenous Cartography: Territorial and Environmental Management in the Hands of Agroforestry Native Agents of Acre State (In Portuguese, with Maps and Images). University of São Paulo, SP, BrazilAvailable at: http://www.teses.usp.br/teses/disponiveis/8/8135/tde-25102012-121716/.

Golledge, R.G., Riceand, M., Jacobson, R.D., 2005. A commentary on the use of touch for accessing on-screen spatial representations: the process of experiencing haptic maps and graphics. The Professional Geographer 57 (3), 339–349.

Golledge, R.D., Loomis, J.M., Klatzky, R.L., 1996. Auditory maps as alternatives to tactual maps. Proceedings, GeoDigital '96 Symposium, Department of Geography, FFLCH, University of São Paulo, 25–28 November, pp. 129–136.

Habel, C., Graf, C., 2008. You-are-here maps as navigation aids for visually impaired people. In: Klippel, A., Hirtle, S. (Eds.), Workshop Proceedings You-are-here Maps, Spatial Cognition 2008, Freiburg/Breisgau, Germany, pp. 1–10.

Harley, J.B., 1989. Deconstructing the map. Cartographica 26 (2), 1–20.

Heller, M.A., 1991. Haptic perception in blind people. In: Heller, M.A., Schiff, W. (Eds.), The Psychology of Touch, Erlbaum Associates, Lawrence, Hillsdale, NJ, pp. 239–261.

Hooker, M., 2007. Concept Note: Developing a Model for Inclusive Education and Assistive Technology Appropriate for Teaching and Learning Contexts in Developing Countries. Available from GESCI at: http://www.gesci.org/old/files/docman/model_ie_at.pdf.

ISA – Instituto Socio Ambiental, 2012. Amazon 2012, publication also in pdf at: http://raisg.socioambiental.org/system/files/AMAZON2012_english.pdf or see the maps on line at: http://raisg.socioambiental.org/mapa-online/index.html.

IFAD, 2010. The IFAD Adaptive Approach to Participatory Mapping: Design and Delivery of Participatory Mapping Projects. IFAD, Rome.

IFAD, 2009. Good Practices in Participatory Mapping: A Review Prepared for the International Fund for Agricultural Development (IFAD). IFAD, Rome. Available at: http://www.ifad.org.

Jacobson, D., 2007. The future of tactile cartography: from static raised lines to multimodal dynamic portable computer interfaces. Abstracts of Papers. XXIII International Cartographic Conference, 4–10 August Moscow, Russia.

James, G.A., 1982. Mobility maps. In: Schiff, W., Foulke, E. (Eds.), Tactual Perception: A Source Book, Cambridge University Press, New York, NY.

Kidwell, A.M., Greer, P.S., 1973. Sight, Perception and the Nonvisual Experience: Designing and Manufacturing Mobility Maps. American Foundation for the Blind, New York.

Koch, W.G., 2012. State of the art of tactile maps for visually impaired people. In: Buchroithner, M. (Ed.), True-3D in Cartography – Autostereoscopic and Solid Visualization of Geodata, Springer-Verlag, Germany, pp. 137–151.

Nogueira, R.E., 2009. Standardization of tactile maps in Brazil. Proceedings, XXIV International Cartographic Conference Santiago, Chile.

Paladugu, D.A., Wang, Z., Li, B., 2010. On Presenting Audio-Tactile Maps to Visually Impaired Users for Getting Directions. CHI 2010, April 10–15, 2010, Atlanta, Georgia, USA.ACM 978-1-60558-930-5/10/04 at: http://dmrussell.net/CHI2010/docs/p3955.pdf.

Parkes, D.N., January 1988. Conference of the Australian and New Zealand Association of Educators of the Visually Handicapped, Audio-Tactile Mapping for the Visually Handicapped, Ormond College, University of Melbourne, Melbourne, Australia, pp. 18–22.

Parkes, D.N., 1994. Multi-media audio-tactile maps and plans: a sound space for blind users with the Touchblaster' Nomad system. Proceedings of the 4th International Symposium on Maps and Graphics for the Visually Handicapped People, University of São Paulo, Brazil, February, pp. 20–26.

Parkes, D.N., 1995. Access to complex environments for blind people: multimedia maps, plans and virtual travel. Proceedings of the 17th International Conference, vol. 2. International Cartographic Association, Barcelona, Spain, pp. 2449–2460.

Parkes, D.N., 1998. Tactile audio tools for graphicacy and mobility: a circle is either a circle or it is not a circle. The British Journal of Visual Impairment 16 (3), 99–104.

Resta, P., 2011. ICTs and Indigenous People. UNESCO at: http://iite.unesco.org/pics/publications/en/files/3214689.pdf.

Rowell, J., Ungar, S., 2003. Feeling your way: a tactile map user survey. Proceedings, XXI International Cartographic Conference – ICA, Durban, SA, August 2003, pp. 652–659, Available at: http://icaci.org/files/documents/ICC_proceedings/ICC2003/Papers/078.pdf.

Rowell, J., 2007. The end of tactile mapping or a new beginning: LBS for visually impaired people. Proceedings, XXIII International Cartographic Conference, 4–10 August 2007 Moscow, Russia.

Schiff, W., Foulke, E. (Eds.), 1982. Tactual Perception: A Source Book, Cambridge University Press, New York, NY.

Sena, C.C.R.G., 2011. Use of didactical resources for people with visual impairment in an application of tactile cartography for geography teaching. Proceedings, XXV International Cartographic Conference Paris.

Sena, C.C.R.G., Carmo, W.R., 2012. Tactile cartography and geography teaching: LEMADI's contributions. first ed. In: Zentai, L., Nunez, J.R. (Eds.), Maps for the Future, vol. 5. Springer, pp. 305–316.

Tatham, A.F., 2003. Tactile mapping: yesterday, today and tomorrow. The Cartographic Journal 40 (3), 258.

Tatham, A.F., Dodds, A.G., 1988. Proceedings of the Second International Symposium on Maps and Graphics for Visually Handicapped People. University of Nottingham, King's College, London, UK.

Taylor, D.R.F., 2001. Tactile Atlas of Latin America. Carleton University, Ottawa, Ontario, Canada.

Taylor, D.R.F., 2003. The concept of cybercartography. In: Peterson, M. (Ed.), Maps and the Internet, Elsevier, Amsterdam, NL.
Taylor, D.R.F. (Ed.), 2005. Cybercartography: Theory and Practice, Elsevier.
Taylor, D.R.F., 2009. Some new applications in the theory and practice of cybercartography: mapping with indigenous people in Canada north. Proceedings, XXIIII International Cartographic Conference Santiago, Chile.
UNESCO, 2008. UN Declaration on the Rights of Indigenous Peoples. Available at: http://www.un.org/esa/socdev/unpfii/documents/DRIPS_en.pdf.
UNESCO, 1994. Salamanca Declaration. Available at: http://www.unesco.org/education/pdf/SALAMA_E.PDF.
United Nations, 1948. The Universal Declaration of Human Rights at: http://www.un.org/en/documents/udhr/.
United Nations, 2009. Convention on the Rights of Persons with Disabilities. http://www.un.org/disabilities/documents/convention/convoptprot-e.pdf.
Vasconcellos, R., 2001. Tactile maps in geography. In: Hanson, S., Weinert, F.E. (Eds.), International Encyclopedia of Social and Behavioral Sciences, Elsevier, Boston, MA.
Vasconcellos, R., 1996. Tactile mapping design and the visually impaired user. In: Wood, C., Keller, P. (Eds.), Cartographic Design: Theoretical and Practical Perspectives, John Wiley & Sons, London, UK.
Vasconcellos, R., 1995. Tactile mapping for the visually impaired children. Proceedings 17th International Conference, vol. 2. ICA, Barcelona, Spain, pp. 1755–1764.
Vasconcellos, R., 1993. Representing the geographical space for visually handicapped students: a case study on map use. Proceedings, XVI International Cartographic Conference, vol. 2. ICA, Cologne, Germany, pp. 993–1004.
Wiedel, J.W., Groves, P., 1972. Tactual Mapping: Design, Reproduction, Reading and Interpretation. University of Maryland, College Park, MD.
Wiedel, J.W. (Ed.), 1983. Proceedings of the First International Symposium on Maps and Graphs for the Visually Handicapped, Association of American Geographers, Washington, DC.
Woodward, D., Lewis, G.M., 1998. The History of Cartography. Book 2/3. University of Chicago Press, Chicago and London.
WHO, 2011. World Report on Disability by WHO and World Bank. http://whqlibdoc.who.int/publications/2011/9789240685215_eng.pdf.
WHO, 2012. Global Data on Visual Impairment. World Health Organization. http://www.who.int/blindness/GLOBALDATAFINALforweb.pdf.
WHO, World Health Organization, 2013. Disability and health, Fact sheet N°352 Sept, 2013 at: http://www.who.int/mediacentre/factsheets/fs352/en/.

CHAPTER

9

The Nunaliit Cybercartographic Atlas Framework

Amos Hayes[1], Peter L. Pulsifer[2], J.P. Fiset[3]

[1]Geomatics and Cartographic Research Centre (GCRC), Carleton University, Ottawa, Ontario, Canada, [2]National Snow and Ice Data Centre (NSIDC), University of Colorado, Boulder, USA, [3]Class One Technologies Inc., Ottawa, Ontario, Canada

OUTLINE

9.1 INTRODUCTION

The Nunaliit Cybercartographic Atlas Framework (http://nunaliit.org) is an interactive data management platform for collecting, relating, presenting, and preserving information and its context, with a particular focus on using maps as a unifying framework. Nunaliit is a central part of cybercartographic technology. It began as an experiment in telling compelling stories using interactive maps and has evolved into a highly flexible data management system with the ability to connect information and present narratives that put the information into context (Taylor, 2005; Taylor and Pyne, 2010; Pulsifer et al., 2011; Brauen et al., 2011). It is capable of

Developments in the Theory and Practice of Cybercartography, Second Edition, ISSN 1363-0814
http://dx.doi.org/10.1016/B978-0-444-62713-1.00009-X

providing interactions with information through maps, graphs, timelines, and other expressions while integrating text, audio, video, and photo information. At the Geomatics and Cartographic Research Centre (GCRC), these are typically referred to as atlas deployments.

Uses of the Nunaliit platform to date have usually focused on data sets that have not normally been mapped for broad use before. Examples include spoken maps of the North in the native languages of Inuit peoples (Chapters 15, 16), interactive representations of the risk of homelessness in Canadian municipalities (Chapter 12), trade relationship visualizations combining mapping elements with graphing to show trade proximity rather than physical proximity (Caquard, S., Fiset, J.P., 2008), and many other unique examples of story-telling on top of a fully fledged information system.

Nunaliit development has been driven by the needs of users in a collaborative and iterative development process that has combined software design and development with user capacity-building. It is very much a 'bottom up' technology as outlined by Taylor in Chapter 1.

A number of unique features and capabilities have evolved to meet the needs of knowledge holders and researchers working in realms where existing tools and off-the-shelf software do not adequately meet user requirements or where the types of data have not been represented before. Nunaliit has also been developed with a strong vision to have the interactive elements be driven by data without any need to reformat or relocate those data elements. Using open standards, it is designed to work with data stored in different locations while still allowing new connections to be made and new stories to be told. This fundamental principle distinguishes Nunaliit from many other interactive story driven maps that typically require the data to be embedded in the application logic, making them difficult and costly to update with new information.

In the same vein, data that are received by Nunaliit are stored in a platform neutral way. This allows the use of other tools to access and manipulate these data without having to go through Nunaliit or to have any knowledge of Nunaliit's data structures. This is a significant difference from other popular open-source (OS) content management systems such as Drupal, Mediawiki, or Dspace that expect to be the centre of the system and broker all requests for information through their own programming interfaces. One of the development principles for Nunaliit is to avoid depending on rigid proprietary data structures and encourage interactions with other systems.

People deploying Nunaliit may decide to enable uploads and edits by users, with those users being either authenticated, moderated, or anonymous. Nunaliit is a flexible framework whose pieces can generally operate independently and be integrated into bespoke applications. The platform is developed and made available under an OS license that permits commercial and noncommercial reuse (i.e. New BSD Licence). It makes use of a number of other well-supported OS technologies, and puts a heavy emphasis on supporting open standards for data use and data sharing.

While the field is relatively small, other research and development efforts identify the challenges and opportunities related to supporting applications that require the use of narrative, story-telling approaches and where 'relationality'[1] is important (Christie, 2004; Dyson and Leggett, 2006; Stevens, 2008; van der Velden, 2010; Pulsifer et al., 2011). Nunaliit and the applications that it supports are making a significant contribution to providing an integrated,

[1] Relationality can be defined as an ontology and epistemology, which focuses on relationships between things (e.g. ourselves, others, environment, spirit, ideas) rather than the things themselves (Wilson, 2007).

flexible platform that is especially useful for rendering the types of data and information in the cybercartographic atlases described in various chapters in the book.

9.2 THE EVOLUTION OF NUNALIIT

Nunaliit has seen numerous significant changes to its software code over its 10 years of development. Early versions of Nunaliit used eXtensible Markup Language (XML) to define data sources and the interactions available between them in a flexible way that was independent of the way it would appear. It supported static built-in data in addition to data loaded in via other systems on the web using standards-based interfaces such as Web Feature Service (WFS). Nunaliit used custom, internal map rendering methods as the software libraries available in 2004 were inadequate to meeting the challenges encountered at the time. The *Cybercartographic Atlas of Antarctica* (Pulsifer et al., 2005), the *Atlas of Canada's Trade with the World* (Eddy and Taylor, 2005), and the more recent *Atlas of the Risk of Homelessness* (Chapter 12) were built on this generation of Nunaliit. The XML description and relatively accessible syntax made it possible for less technical users to build and package some complex renderings, such as the GraphoMap discussed in Chapter 12 and described by Caquard and Fiset (2008). This worked very well for less technical atlas builders provided they were prepared to use what Nunaliit offered out of the box. Adding additional functionality required designing the XML syntax to invoke it, writing code to compile it (written in Java), and writing the code that would ultimately be run by the browser (JavaScript + HTML) or server (Java). This process proved too demanding for web developers who might typically only know HTML and JavaScript. The XML-based Nunaliit had an advantage in that it could store the data sources and interactions, and render them out to diverse formats (such as print, web, DVD, CDROM, etc.). However, in practice, web and CDROM were usually the only platforms used and the limitations on development and adoption created larger problems.

To respond to these challenges, Nunaliit evolved. First, the customized browser-rendering engine was replaced with OpenLayers, an OS JavaScript library that renders maps and allows interactions on multiple platforms. The XML schema and compiler were set aside. Nunaliit became strictly a framework for telling stories with maps over the web. Also added was the ability for users to add and connect content via the web interface. This new functionality required the ability to edit and add to data repositories used by an atlas in addition to reading from them. Implementations of the transactional additions to the WFS standard, dubbed WFS-T, were not mature enough to support this work at the time. To meet these needs, Nunaliit was enhanced to allow integration with Postgres/PostGIS directly. Postgres/PostGIS is an OS relational database with geospatial extensions. This allowed for the building of the *Kitikmeot Place Name Atlas* (Chapter 15), the *Atlas of Arctic Bay* (Chapter 20), the *Lake Huron Treaty Process Atlas* (Chapter 17), and finally the *Siku (sea ice) Atlas* (Chapter 14).

9.3 SOME IMPORTANT CHALLENGES

Numerous lessons have been learned from the development, deployment, training with, and use of these Nunaliit-based atlases. Other chapters in this book have highlighted many of those including the importance of culturally appropriate representations such as spoken place names in Kitikmeot (Chapter 15), the link between data authenticity and contributor in a volunteered information system as shown in the *Siku (sea ice) Atlas* (Chapter 14), and the importance of local possession of knowledge as reflected in numerous projects and explicitly discussed in Chapters 1, 4, 18, and 19. The experience of creating all of these atlases revealed some major challenges that Nunaliit had to be able to meet. Three in particular required significant developments to solve.

9.3.1 Different Projects Have Different Data Needs

This may seem obvious at first glance. But what is interesting is not simply that the data are different across projects, but the relationships between data, the interaction model, and data representations are also different. Every atlas required significant work to map-out relationships between data and develop custom code to handle differences in how the information was to be displayed or interacted with. The *Siku (sea ice) Atlas* (Chapter 14) has a very large amount of data for this type of application. And although researchers made great efforts to organize it in a very well thought-out way in a relational database, it still required significant code to be written to have Nunaliit operate on top of that unique data model. Possibly more significant was that much of that development could not happen until the data model was finalized, which delayed the iterations of the *Atlas* and any feedback or changes that would be based on end-user interactions with each iteration.

9.3.2 Enabling Local and Internet Availability and Preservation Requires Replication

Two years after launching the *Kitikmeot Place Names Atlas* (Chapter 15), the GCRC received a request to deploy a kiosk version of the *Atlas* in the community to make the *Atlas* more accessible to all community members as well as to alleviate local Internet connectivity concerns. At the same time, the *Atlas of Arctic Bay* was slated to see classroom use (Chapter 20) and there was some interest in having a locally hosted copy of it. However, due to northern network bandwidth limitations, it was still necessary to host the atlases at the GCRC to allow other northern communities and the broader public to access and participate in the atlases. Additionally, the technical capacity and infrastructure necessary to safeguard data was very limited or absent within the communities. Any solution would need to be able to replicate data in both directions between multiple instances of the atlas. This would also help to address a broader concern of information and heritage data being displaced and the importance of local possession. This is clearly reflected in the Ownership Control Access and Possession guidelines offered by the First Nations Information Governance Centre in Canada (2010). These and related ethical issues are fully discussed in Chapters 18 and 19. Even outside indigenous communities, the need to be able to possess and control a data set that one owns while simultaneously sharing it to 3rd parties is a common issue and the foundation for the web-services standards such as those developed by the Open Geospatial Consortium. Two additional challenges in this general context were encountered. The system had to be able to accept contributions and edits of all supported formats in any instance of the information; and, connectivity between instances could be expected to be slow and prone to prolonged outages.

9.3.3 Plethora of Devices, Formats, and Tools for Contributing Content

Sharing content via online services has become easier over the last few years. Websites like YouTube can accept almost any type of video and perform conversions. Many social media sites now offer mobile applications. Combined with smart phones containing text, photo, video, and audio documenting capabilities, these mobile applications make it much easier to share data. That said, there are some significant drawbacks. Users typically have to be 'online' for these applications to work fully. The information lives in a system that users do not control and is presented in a context and format set by the service provider. Services tend to focus on certain types of media and rarely provide anything more advanced than a playlist or slideshow to permit users to try to link content or tell stories with it. Users may turn to yet another service such as a blog to add context and embed this content, but that requires another site with another set of tools.

In the context of community projects where a database of local knowledge is being created through collaboration and field work carried out by Elders and youth, data are typically collected on a mix of several different digital cameras, audio recorders, video cameras, handheld Global Positioning System (GPS) units, and personal computers. A typical computer lab in a remote community classroom is not uniform in operating system, configuration, or driver support. This creates a major challenge for students, teachers, and Elders who cannot count upon systems to behave the same way and cannot maintain documentation for every possible outcome or workflow when a device is plugged in.

What was needed was a mix of these approaches. A cheap commodity device with full multimedia and GPS support, combined with an application that supported multiple formats and offline use, would allow for relationships between documents to be created and transferred to a community-operated online atlas. That system could then render and share that information in a fashion devised by the community to tell their stories.

9.4 RECENT DEVELOPMENTS

In the creation of the *Siku (sea ice) Atlas* (Chapter 14), it was realized that a system that was truly flexible in the organization of information was required. The notion of a sparse relations table, where rows of any table could be linked to rows of any other table, provided a starting point. Code could be written that made use of this relations table and some other hints in the database to work with the data without requiring code that depended on understanding the organization of the database. Adding a new table into the data model required simply adding another column to the relations table rather than creating whole new key tables for each relationship. Although it was much easier to add new tables and connect these, knowledge of the database was still required as was administrative access to it. Individual tables were still constrained to the fields specified during creation, and that meant users and nontechnical atlas creators could not change the nature of what was stored in the database without help.

At the same time, the replication challenge surfaced. Common relational database implementations are not well suited to replication, particularly in ways that allow any of the instances to modify data. This is called 'multi master replication' and for systems that support it at all, it is complex to set up and administer, requires excellent connectivity between nodes, and typically requires administrator intervention if any of the nodes lose connectivity with the cluster of database servers. Some experiments with this technology clearly indicated this would not suit the needs of remote communities using Nunaliit.

The work started with a clean slate and the team began to investigate other database technologies. Document-oriented databases were emerging and one in particular named CouchDB (http://couchdb.apache.org) stood out for the emphasis developers placed on data replication and 'eventual consistency', meaning that the connectivity between database nodes could be slow or intermittent but that the databases would eventually regain synchronicity. A document-oriented database does not have the same structure as a relational database. There are no tables, rows, or columns. Each document can have an entirely unique collection of attributes. In CouchDB, documents are stored in JavaScript Object Notation (JSON) format. JSON is widely used and understood by web developers and permits strings, numbers, arrays, and objects to be the values associated with any key. This allows a rich nesting of attribute information rather than the two-dimensional table construct used in relational databases and spreadsheets. Documents must have a unique identifier attribute and CouchDB will also create a unique attribute for every revision of a document to allow it to make decisions about that document to keep when replicating. Documents can have attachments that are binary objects that have been specially encoded and stored in a reserved attribute within a document.

Interacting with CouchDB is done purely through an HTTP interface, so it is well suited to being accessed directly by web browsers. In fact, entire dynamic HTML applications can be stored as attachments inside the database, served out to web browsers directly, and replicated along with the data.

In CouchDB, it is easy to create new documents taking any form and add them to the database. Likewise, any existing document can be modified to include new attributes, ingest new attachments, or restructure existing attributes. This flexibility would be the key to solving the challenge outlined in Section 9.3.1, allowing for very different data needs in the system without having to re-engineer the application code that retrieves and stores information in the database. Nunaliit could look for documents possessing keys that it understood, and ignore those that didn't have what it was looking for. As the content stabilized and atlas creators settled on what keys and values they would use for their various data, Nunaliit could be given hints about those keys and values and how to display or interact with them. Somewhat ironically, these hints have become known as Nunaliit schemas even though the approach is schema-less.

Nunaliit schemas are documents with a structure that Nunaliit understands, whose purpose are to define a class of documents and indicate what attributes one might expect to find in documents declaring themselves to be of that class, and how to display those attributes in various circumstances. For example, a schema for a class of documents containing uploaded photos and videos might indicate that Nunaliit should offer a file upload button (and perform all the conversion and thumbnail tasks associated with that), ask for a caption and credit line, provide an interface for linking the media document to any place name document in the database, and always show the caption and credit under the photo when it is displayed – with different fonts and sizes to distinguish them from each other.

Documents are never restricted to what is specified in the schema. Nunaliit provides an advanced editor that allows for any attributes to be added, removed, or edited on any document by a user authorized to do so. The Nunaliit schema merely allows Nunaliit to treat specified attributes in a defined way when interacting with them for displaying or editing.

One benefit of this approach is that other tools or applications deployed by an atlas host may also retrieve and manipulate the documents directly in CouchDB without any understanding of Nunaliit's implementation. This provides a great deal of flexibility for interoperability with other systems as well as avoiding being locked in to one approach for managing the data.

Another critical distinction is that since schema documents are documents too, it is possible for them to be edited through the web interface by nontechnical people. This also makes it possible for the nature of the data and the interactions with them to be defined by atlas creators as they proceed, without having to preconceive everything before the site can be built. This provides for more frequent iterations of an atlas and therefore more engagement by atlas creators and users, more opportunities for feedback, and ultimately a better result.

The ability to replicate between instances of CouchDB is also a strong reason for its adoption in Nunaliit, particularly as our northern and remote indigenous community partners look to have local versions of their atlases. Because Nunaliit creates a new document with a universally unique identifier for every user contribution to the atlas, and because information linking one document to another is stored within the documents themselves, there is virtually no shared state information to get out of synchronization. Each document (including its attachments, if any) replicates in its entirety if there is connectivity, or will try again eventually when connectivity is re-established. Content that has not yet been synchronized simply does not appear and Nunaliit does not act on links to documents not yet present. On the rare occasion that documents have been modified on multiple nodes simultaneously, CouchDB nodes handle this collision by choosing the latest generation of the document and saving the losing version in case a user wants to revisit at a later time.

This allows for the kiosk scenario where users in the community are able to add and comment on, for example, video and photos at local network speeds. Nunaliit can then use CouchDB replication to make that content available on another version of the same atlas with better internet connectivity. An internet user could then ask questions (yet more new documents in the database) that make their way to the community kiosk.

The nature of the replication allows for multiple nodes to share the content of a Nunaliit atlas. Some nodes could be run by communities. Some could be run by organizations with a mandate to combine and/or preserve these digital holdings on behalf of communities. Communities could also share some documents with other communities. An example would be shared travel routes, cultural heritage documents and locations, hunting areas, or environmental data. This data replication has been tested between a few sites and a network of existing Nunaliit nodes and the current list of possible nodes is included as Figure 9.1.

Importantly, for the third challenge outlined in Section 9.3.3, replication can also serve the need for truly flexible mobile data.

To address the issue of field collection of local and traditional knowledge information, a prototype Nunaliit application for use on Apple's second generation iPad device was developed. The application runs its own copy of CouchDB and a simplified Nunaliit interface. Users are able to take notes, photos, video, audio, and position (GPS) recordings, even while

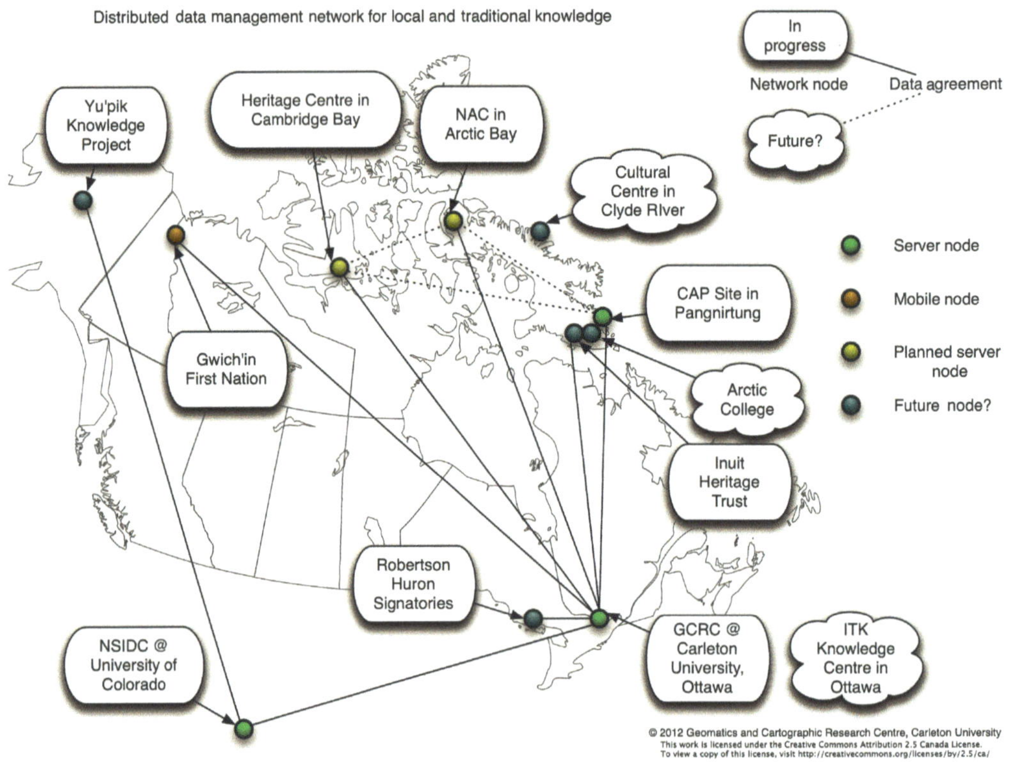

FIGURE 9.1 Map of the distributed data management network for local and traditional knowledge.

offline, and make the important connections between them (Figure 9.2). Because core Nunaliit functions are present in the application, along with CouchDB, users can design new Nunaliit schemas and modify existing ones if new needs are discovered, even while out in the field without connectivity. An advanced editor is also available to allow advanced editing of documents if required. Once returning to a location with wireless network access, the tablet application uses its replication functionality to exchange new and modified documents and media attachments with a server-based instance of Nunaliit, either in the community or hosted elsewhere on the Internet. With one button press, all the field information is made available in the atlas. Any relevant new data or schemas on the server side are also copied over to the tablet.

This ability to make changes to what is being collected and how it is related, while offline in the field, and yet still have near-immediate online access in a user-friendly atlas format, without time-consuming and error-prone processing and cataloguing upon return, greatly simplifies the process.

The Gwich'in Social and Cultural Institute was the first to test the new application in the field in 2012 and reported great success (Chapter 16). Ingrid Kritsch spent three weeks offline in remote communities recording over 800 spoken place names, photos, and videos with Elders on a single iPad (Figure 16.6). Upon her return to Yellowknife, much of the data were

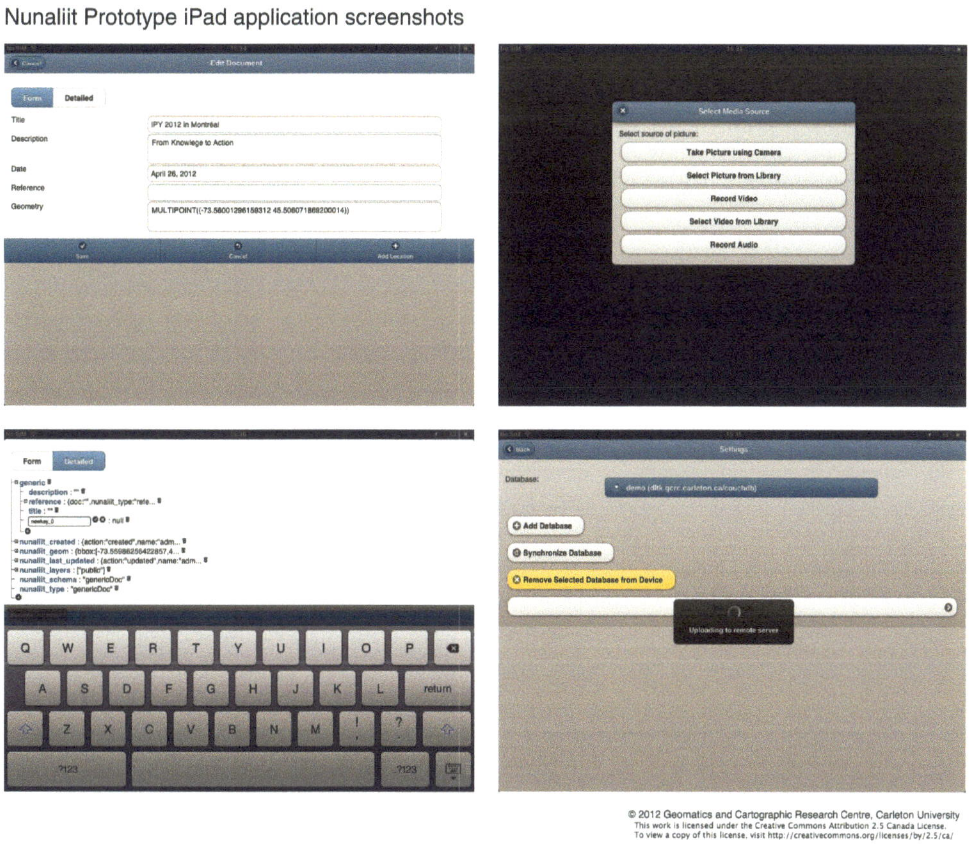

FIGURE 9.2 Nunaliit Prototype iPad application screenshots.

replicated to Ottawa in a matter of hours (Figure 9.3). Although an iPad system update initiated during the replication caused some unforeseen problems, the ease and speed with which all the information was made available to users of the *Atlas* was a remarkable improvement. The previous process would have taken months of manual effort to name recording files, match them with written or typed notes, locations, photos and videos, and upload them to an online system.

At the time of writing, the Nunaliit iPad application is still in prototype form and not yet available through the iPad App Store. Based on user feedback, hardware costs and specifications, and restrictions in Apple's iOS, the next iteration of the application will be implemented for Android devices. This will allow more flexibility in the design of the application. It will also permit users to acquire less expensive hardware for Nunaliit use. Android tablet devices with GPS are approximately one third the price of similar iPad devices with GPS. New hardware with additional capabilities such as fully fledged cameras running Android are envisioned for future *Views from the North Atlas* work (Chapter 13). Unlike iOS devices, Android devices also typically have the ability to write data to memory cards via a designated interface or via a USB connection. Being able to use a memory card with a device offers

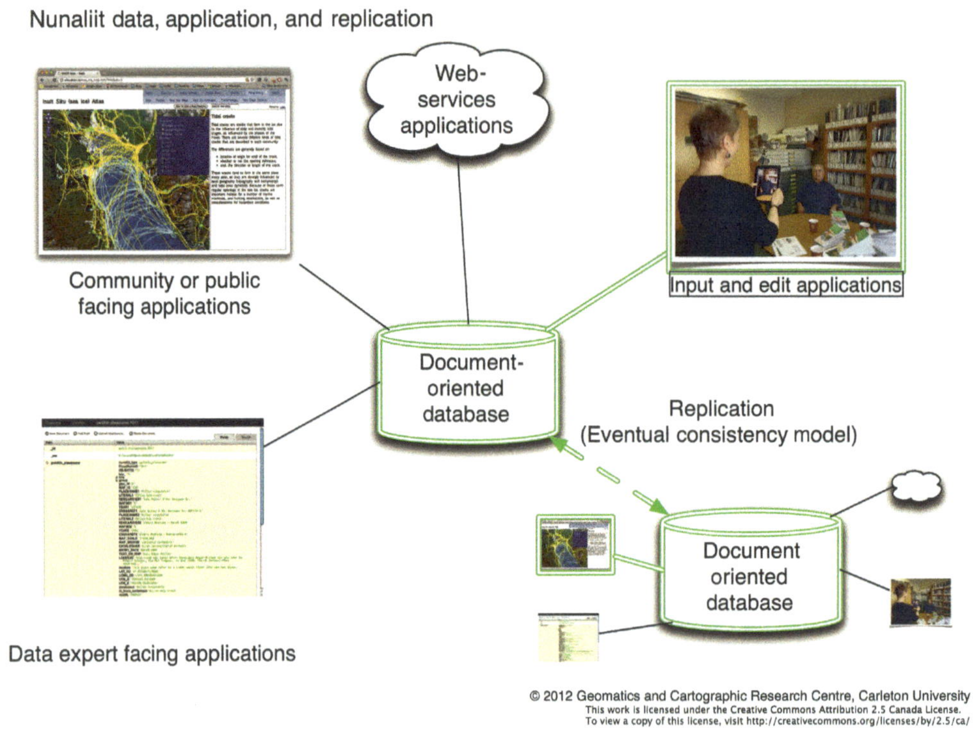

FIGURE 9.3 Database may be accessed and edited by Nunaliit web and/or mobile applications in addition to other applications and tools. Replication between one or more other instances allows synchronization between mobile and server nodes or between multiple server nodes. Green double-outline elements show path from Gwich'in iPad to web application hosted at GCRC.

the possibility of making backups that can be stored in small waterproof housings while out in the field for long periods.

9.5 FUTURE DEVELOPMENT OF NUNALIIT

Nunaliit has been built incrementally by pooling numerous sources of research, heritage, and community funding and focusing on the needs of specific projects. As functionality is added to the system, there is always significant time spent thinking about how to best implement it to be useful to other existing and future projects.

Future work with GCRC partners will address important issues such as capturing the motivation and intent of a contributor along with their consent to document their knowledge (Figure 9.4). This will help Nunaliit to provide respectful and informed decisions about how to display, store, and share information (see Chapters 18 and 19, which discuss ethics, consent, and legal issues regarding data collected from indigenous communities).

Joint development of the software, along with education and training materials and community programs are already pulling together networks of organizations that share the values upon which Nunaliit was built.

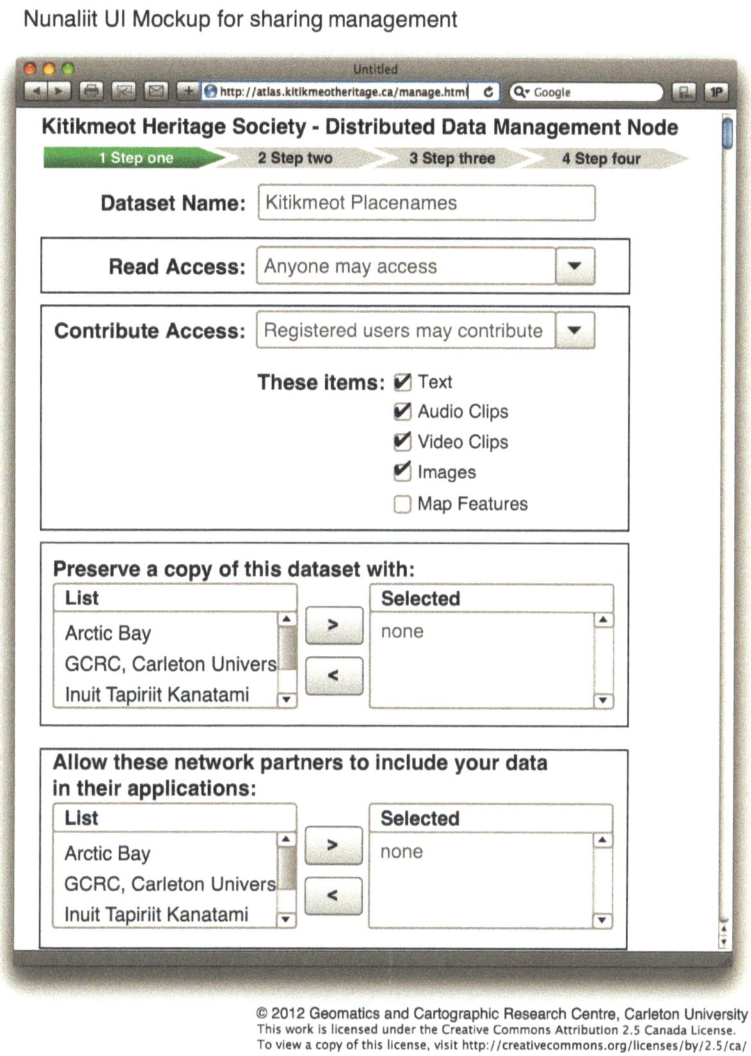

FIGURE 9.4 Mockup of potential interface for specifying allowable sharing for community content.

Development will continue to focus on making it easier to tell better stories with rich information both new and old, show the relationships to place and space, and do it all with data management concepts built in to the system.

The process of building an atlas, working through all the decisions about what information is important, and how it should interconnect, and then storing it in an accessible and flexible system are all important steps to ensure that the knowledge remains useable and workable for new stories and applications in the future as well. The more usable and immediately useful to the broadest audience Nunaliit can be, the more likely it is that digital artefacts will avoid the fate of being stored in a box on the shelf to be dealt with later. Instead, they become active resources available to be used and referenced in new and exciting ways.

References

Brauen, H.G., Pyne, S., Hayes, A., Fiset, J.-P., Taylor, D.R.F., 2011. Encouraging transdisciplinary participation using and open source cybercartographic toolkit: the atlas of the Lake Huron Treaty relationship process. Geomatica 65 (1), 27–45.

Caquard, S., Fiset, J.P., 2008. Un cyber cartogramme source ouverte et interopérable pour l'analyse de données spatiotemporelles complexes et diverses. Revue des Nouvelles Technologies de l'Information (RNTI) E13, Sélection des meilleurs articles de SAGEO 2007.

Christie, M., 2004. Computer databases and aboriginal knowledge. learning communities. International Journal of Learning in Social Contexts 1, 4–12.

Dyson, L.E., Leggett, M., 2006. Towards a metadesign approach for building indigenous multimedia cultural archives. In: Attwater, R., Merson, J. (Eds.), Proceedings of the 12th ANZSYS Conference Sustaining Our Social and Natural Capital, 3–6 December, Katoomba, NSW Australia, ISCE Publishing, Mansfield, MA.

Eddy, B., Taylor, D.R.F., 2005. Applying a cybercartographic-human interface (CHI) model to create a cybercartographic atlas of Canada's trade with the world. In: Taylor, D.R.F. (Ed.), Cybercartography: Theory and Practice, Elsevier, Amsterdam.

First Nations Information Governance Centre, 2010. First Nations Principles of OCAP. Available at: http://www.rhs-ers.ca/node/2 September, 2013.

Pulsifer, P.L., Parush, A., Lindgaard, G., Taylor, D.R.F., 2005. The development of the cybercartographic atlas of Antarctica. In: Taylor, D.R.F (Ed.), Cybercartography: Theory and Practice, Elsevier, Amsterdam.

Pulsifer, P.L., Laidler, G.J., Taylor, D.R.F., Hayes, A., 2011. Towards an Indigenist data management program: reflections on experiences developing an atlas of sea ice knowledge and use. The Canadian Geographer 55 (1), 108–124.

Stevens, A., 2008. A different way of knowing: tools and strategies for managing indigenous knowledge. Libri 58, 25–33.

Taylor, D.R.F., 2005. Cybertography: Theory and Practice. Elsevier, Amsterdam.

Taylor, D.R.F., Pyne, S., 2010. The history and development of the theory and practice of cybercartography. International Journal of Digital Earth 3, 11–14.

van der Velden, M., 2010. Design for the contact zone. In: Sudweeks, F., Hrachovec, H., Ess, C. (Eds.), Proceedings of the Seventh International Conference on Cultural Attitudes towards Communications and Technology, 15–18 June, Murdoch University, Vancouver/Murdoch, pp. 1–18.

Wilson, S., 2007. Guest editorial: what is an Indigenist research paradigm? Canadian Journal of Native Education 30 (2), 193–195.

CHAPTER

10

Interactive Audiovisual Design for Cartography: Survey, Prospects, and Example

Glenn Brauen

Geomatics and Cartographic Research Centre (GCRC), Carleton University, Ottawa, Canada

OUTLINE

10.1 INTRODUCTION

This chapter discusses prospects for designing maps as distributed, network-enabled audiovisual media to express spatial data through the inclusion of digital visual and acoustic elements. The perspective used throughout this chapter was developed while working on cybercartographic atlases or related stand-alone projects and was, in part, motivated by Taylor's (2003; see also Taylor and Pyne 2010) call for greater research into the use of multimodal interfaces. This perspective would be compatible with other conceptions of cartography including *multimedia cartography* (Cartwright and Peterson, 2007), *web mapping* (Kraak and Brown, 2000), and *maps for the Internet* (Peterson, 2008).

As will be discussed, although it is possible to find map examples online with names suggesting the use of sound as part of the map, the most common design paradigm currently

Developments in the Theory and Practice of Cybercartography, Second Edition, ISSN 1363-0814
http://dx.doi.org/10.1016/B978-0-444-62713-1.00010-6

used results in maps that provide access to acoustic media organized by location tags, playable through audio interfaces that are separated to greater or lesser degrees from the map itself. The prospects for cartographies that use sounds as map design elements, as will be discussed here, haven't been extensively researched, warrant continued attention, and provide scope for productive research into alternative models for the construction and uses of digital map designs.

When considering sound in relation to map making, map distribution, and map use, sound or noise (i.e. unwanted sound) could be a geographical topic, or part thereof, to be studied and mapped, often but not always resulting in a strictly visual depiction of the acoustic phenomena (Arana et al., 2010; Dannhorn, n.d.; MarSensing Lda, 2012; Müller and Scharlach, 2001; Porteous and Mastin, 1985; Schafer, 1977; Servigne et al., 1999; Steel et al., 2009). The generation of sounds and analysis of acoustic properties from reflected sounds could be the basis of methodologies for studying other geographical phenomena (Aprea et al., 2009; Barbagli et al., 2010; Chesterman, 1968; Komatsu et al., 2003; Roberts et al., 2005; Rona et al., 1997). This chapter instead examines possibilities for sound as a design element, along with visual map elements, for the creation and use of interactive digital maps. The research described here has focused on the use of maps on the World Wide Web and a distributed network such as that is assumed as the context in which digital acoustic and visual information is being distributed for the composite map to be used. This focus has been adopted simply because the web is now the main distribution channel of maps for many purposes (Peterson, 2007) and it is a context within which many forms of media including sound are now widely used. Therefore, it has been a context in which theoretical perspectives on the use of sound as part of media can be studied and applied, design possibilities and examples can be developed and tested, and in which the technical difficulties encountered in attempting to use sound and visual elements together in the creation of interactive audiovisual maps can be understood.

10.2 SURVEY

Experiments incorporating sound with maps are certainly not new but the literature on the cartographic uses of sound is sparse and there is as yet little guidance concerning how to use sound effectively. Discussions of methods that would usefully include sound in an interactive map design have also been rare (cf. Brauen, 2011b; Brauen and Taylor, 2008; Krygier, 1994). Examples of maps incorporating sound in conjunction with visual maps have been developed over at least the last two decades as an outgrowth of research into *animated* and *interactive* cartography and geographic information system. In these examples, sound has been used to provide narration explaining the function of a map or a multimedia interface (Harrower, 2003; Krygier, 1994; Monmonier, 1992); to draw a user's attention to a visual map component (Harrower, 2003); as an abstract language providing a set of variables (pitch, tempo, timbre, etc.) adding information to a thematic map without overloading the visual display (Fisher, 1994; Krygier, 1994; Müller and Scharlach, 2001; Servigne et al., 1999); and as media to be accessed through the map, alone or as part of a video or an animation (Hu, 2003).

Multimedia cartography (Cartwright et al., 2007) incorporates audio and video clips, still images, animation, and text into dynamic cartographic products, often using the map as an interface organizing and providing access to the multimedia content. Examples include

Wula Na Lnuwe'katiyek, an atlas of Mi'kmaq places and stories (Francis, 1996); *Introduction to Iqaluit*, a set of maps combining imagery and place name pronunciations (Mouafo and Müller, 2002); and *Gwich'in Place Name Map*, a set of maps organizing access to written stories, videos, and place name pronunciations concerning Gwich'in use of their territory (Gwich'in Social and Cultural Institute, 2003).

Location-aware navigation systems incorporating audio cues have become common in vehicles and on mobile devices (Gartner, 2004) and experiments to replace visual navigation cues with acoustic beacons have been conducted (Walker and Lindsay, 2006). Artists and digital media designers have experimented with sound and maps to address a range of themes.[1] Caquard et al., (2008) used sound within a cybercartographic atlas project to enhance and highlight the narrative character of an atlas and the potential of that narrative to encompass multiple perspectives.

Although these examples show that interest exists in the use of sound as part of map-based artefacts, the number of maps designed using sound is tiny in comparison to silent maps available in books or on the World Wide Web. Furthermore, a review of maps available on the World Wide Web that incorporate sound, initially conducted during June 2010 and updated in April 2013,[2] showed that most included sound in such a way that it is possible to argue that map designers predominantly still think of these 'sound maps' as silent visual organizers for acoustic media. The sounds still seem to be logically separated from the map in most cases and as a result most of these maps cannot be called audiovisual devices. Tables 10.1 and 10.2 present results from the survey.

Table 10.1 lists web-mapping applications that incorporate sound as part of an audiovisual map interface. Although using diverse designs, maps have been included in this category if sounds play while a user interacts with the visual map interface without the use of intermediate visual devices such as graphical audio controllers[3] that obscure, hide, or in another way withdraw a user's attention from the map while sound is playing or in order for a user to initiate audio playback. These maps seem to have been conceptualized as audiovisual devices, the map itself using visual and auditory modes of signification. Audiovisual map interfaces present information on themes including soundscape ecology (e.g. *Gordon Soundscape*, *Soundtracks*, *Forest Hills Sound Map*, *SoundMap*), musical events as part of place identity (e.g. *Manchester: Peripheral*, *Cinco Cidades*), narrations concerning particular

[1] Levine et al., (2004) created a web application as part of *Baghdad San Francisco*, an installation incorporating maps and geocaches geographically shifted from Baghdad to San Francisco documenting the bombing of Iraq by US armed forces during the initial invasion of Baghdad in March 2003, in which an audio recording evokes the bombing to underscore the political message of the piece. Levine (2006) created an installation documenting the impact of the 1906 San Francisco Bay area earthquake on the city of Santa Rosa using audio derived from seismic recordings of the earthquake to accompany visual distortions of a map of the city. Thirion (2007) created a map-based data visualization of Madrid automobile traffic projecting an explicit authorial perspective onto the underlying traffic flow data through the use of a *traffic as noise* metaphor.

[2] These web map applications were found by: (1) initially querying the Google and Google Scholar search engines using the terms 'sound map', 'acoustic map', and 'audiovisual map'; and (2) by following URLs to 'similar' or 'highlighted' web applications listed on sites identified in the step-one searches. Web applications created by projects with which the author was associated have been omitted from the review.

[3] Examples of graphical audio controllers can be seen in, for example, *Montreal Sound Map* (Stein, 2008) and *SoundMap* (Minard, 2010).

TABLE 10.1 Web-Map Applications with Sound Embedded in the Map. Results of Review of Audiovisual Web Maps, June 2010 and April 2013. Application URLs are Listed in Reference

			Status*	
#	**Map Title**	**Attribution**	**2010**	**2013**
1	The Austin Zoo sound map	Kite's in the Tree Media (2012)	∄	
2	BadiaFonia	Cordobés and Picanyol (2008)		X
3	Cinco Cidades	Dant et al. (2007a)		
4	Echo location: a sound map for Bedford UK	Hinde and Holroyd (2011)		
5	Favourite sounds	Cusack (2012)	∄	
6	Folk songs for the five points	Dant et al. (2006)		X
7	Forest Hills sound map	Kiser (2012)	∄	
8	Gordon soundscape	Stollery (2005)	∄	
9	Isle of dog music	Chapman (2008)	∄	
10	Manchester: Peripheral	Dant et al. (2007b)		
11	Sound map: The Caldeonian Road	Dein and Panetta (2010)		
12	SoundMap	Minard (2010)	∄	
13	Soundtracks	Kinayoglu (2010)		
14	Soundwalks	Soundwalks.org (2010)		∄
15	Tactical sound gardens	Shepard et al. (2011)	∄	

* *Inventory of websites conducted in June 2010 and reassessed in April 2013. Sites found in June 2010 continue to exist and work if not marked. Alternatively, they may no longer exist (∄) or may exist but no longer work (X). Other sites have been added by the April 2013 survey and are marked as not having existed or not having been found in June 2010 (∄).*

places (e.g. *The Austin Zoo Sound Map*, *Sound Map: The Caledonian Road*), or as part of hybrid situated/online interactive acoustic installations (e.g. *Tactical Sound Gardens*). Note that the classification of maps as integrating sound into the design of the map itself is nuanced, especially considering the diversity of designs, and the mere presence of a graphical audio controller, for example, has not in all cases prevented a map from being included in this group. For example, *SoundMap* (Minard, 2010) allows a user to initiate simultaneous playback of up to four sound loops by selecting locations on the map. Each sound loop is associated with an audio controller overlaid on the map that allows a map user to adjust the loop's amplification or to stop its playback. While each audio loop is playing, the location marker for the place associated with the sound is continuously visually animated to create a connection between the visible and acoustic design components of the map. Therefore, the map behaves as and was clearly conceived as an integrated audiovisual design.

By contrast, the web-mapping applications listed in Table 10.2 use a map to organize sounds geographically but separate playback of those sounds from the visual map through the use of an intermediate visual device, often a pop-up speech bubble overlaid on the map

TABLE 10.2 Web-Map Applications Organizing Sounds within a Map. Results of Review of Audiovisual Web Maps, June 2010 and April 2013. Application URLs are Listed in References

#	Map Title	Attribution	Status*	
			2010	2013
1	12 Gates to the city: acoustic map	Prior (2009)		
2	Andalucía soundscape	Cantizzani et al. (2006)		X
3	audioBus	Callard et al. (2006)		
4	BBC save our sounds	BBC (2009)		∄
5	Belfast sound map	Rebelo et al. (2012)	∄	
6	Berlin wall of sound	Netaudio Festival Berlin (2009)	∄	
7	Chicago sound map	Chicago Public Radio (2007)		∄
8	Columbia River sound map	Ecklund (n.d.)	∄	
9	Common senses sound map	Museum of Modern Art (2013)	∄	
10	Dubai sound map	Simon Charles Audio (2012)	∄	
11	Escoitar	Molina et al. (2006)		
12	European soundscape map	European Acoustic Heritage (2011)	∄	
13	Favorite Chicago sounds	Seay et al. (2009)		X
14	Firenze sound map	Radicchi (2009)	∄	
15	Fleurieu sound map	Louth-Robins (2011)	∄	
16	Gwich'in place name map	Gwich'in Social and Cultural Institute (2003)		
17	Inukjuak sound map	Yoganathan et al. (2010)	∄	
18	Knox College sound map	Hope et al. (2011)	∄	X
19	Listen to Africa	Williams and Sumner (2009)		
20	Listening to the deep ocean environment	Laboratory of Applied Bioacoustics, Technical University of Catalonia (2010)	∄	
21	Listening to nature: a sound walk across California	California Library of Natural Sounds and Oakland Museum of California (2003)		
22	Locus Sonus	locusonus.org (2006)		X
23	London sound survey	Rawes (2010)		
24	Madrid soundscape map	Madrid Soundscape.org (2008)		
25	Map of geotagged sounds	Freesound Project (2005)		
26	A map Of our own: Kwun Tong culture and histories	Mak et al. (2009)	∄	

Continued

TABLE 10.2 Web-Map Applications Organizing Sounds within a Map. Results of Review of Audiovisual Web Maps, June 2010 and April 2013. Application URLs are Listed in References—(*cont'd*)

#	Map Title	Attribution	Status*	
			2010	**2013**
27	Memoryscape audio walks	Butler and Whiting (2005)		
28	Mississauga sound map	Centeno and Sinclair (2008)		
29	Montreal sound map	Stein (2008)		
30	Murmur	murmurtoronto.ca (2003)		
31	Open sound new Orleans	Booth and Brancasi (2008)		
32	Puget soundscape	Becker (2005)		X
33	Radio Aporee	Noll (2006)		
34	The smalls street sounds	The Smalls Limited (2010)		∄
35	SeoulSoundMap	Sound@Media (2010)	∄	
36	Soinu Mapa	Erkizia et al. (2009)		
37	Sons de Barcelona	Freesound Project (2010)		
38	Sound and story project of the Hudson Valley	Metzner and McAdam (2010)		∄
39	Sound around you	Mydlarz et al. (2009)		
40	Sound map of the Paiva River	Binaural/Nodar (2011)	∄	
41	Sound maps	British Library (2010–2011)	∄	
42	Sound seeker	Poll and Feng (2009)		
43	Sound tourism: a travel guide to Sonic Wonders	Cox (2010)		
44	Sound transit	Holzer et al. (2004)		
45	Soundlandscapes	Coulam (2008)	∄	X
46	Soundcities	Stanza (2010)		
47	Sounds like Staten Island	Monda (2009)		
48	Soundscape of China	Eason and Lewis (2007)		
49	Stockholm sound map	Lidbo et al. (2009)	∄	
50	Toronto sound ecology	Ritts et al. (2010)		∄
51	Toronto sound map	Russo et al. (2012)	∄	
52	Ursinus sound map	Hovick et al. (2012)	∄	
53	Wild sanctuary	Wild Sanctuary Inc. (2009)		∄

** See note for status columns of Table 10.1.*

but also sometimes a separate HTML page, providing playback controls and information about the audio. The result is the creation of a procedural and perceptual distance between the visual and auditory signification of the map application. Despite the description of these maps using terms such as 'sound map' or 'acoustic map', the sounds are organized *by* the map but are not part *of* the map.

Almost four-fifths (79%) of the maps examined during this review (53 of 68) are designed such that playback of acoustic media is separated from the map. Of those using such a design, the majority offer access to recorded audio and other media such as images or videos through the use of a pop-up bubble that appears on the map when a map feature (usually a point) is clicked (30 of 46 examples working in April 2013), obscuring or hiding at least a portion of the map. The next most common design (11 of 46) provides access to the recorded audio and other media through a separate HTML page reached by clicking on mapped entities or, less directly, hyperlinks within a pop-up displayed after clicking on the map. Despite being described as a sound map, the map itself has disappeared before a user has access to audio. The remainder (5 of 46) use a hybrid design that provides a visual representation of an audio player or the sound (e.g. a sound wave amplitude graph) either overlaid on the map or adjacent to the map. Again, sounds are presented using a common metaphor for audio media that, if not obscuring or hiding the map, at least draws attention and responsibility for the sound away from the map. The map designs included in this category are designed such that, most commonly, sounds are only initiated through a graphical or visible device accessed from the map and sounds play only while that device remains open and visible. Less commonly one or more audio players are created adjacent to or overlaid on the map (e.g. *London Sound Survey* and *Columbia River Sound Map*), each sound is dependent on and controlled by an audio player, and the perceptual connection between the sounds and areas or locations on the visual map are weak or nonexistent. These maps seem to have been designed to organize sounds without being audiovisual devices themselves.

Tables 10.1 and 10.2, in addition to categorizing the integration of sound into the map interfaces examined, also note those cases in which maps inventoried in 2010 no longer exist online, in which new maps have been launched or found since the 2010 survey, or cases in which maps still exist but no longer function properly. Although a few maps have been found by the 2013 survey that apparently would have existed at the time of the 2010 survey and could have been found, the most usual case is new maps that have been developed and launched since 2010. For both of the categories listed in the tables, there were more maps available online during the later survey. The number of maps examined and classified as entailing integrated audiovisual design, listed in Table 10.1, approximately doubled between 2010 and 2013. Although there has been greater absolute growth in the number of maps listed in Table 10.2 as organizing but not integrating sounds, the more recent survey indicates that growth in this subset has been relatively slower although still very significant.

In those cases where maps no longer function, the most common problem was that the sounds were no longer playable through the map as interface once locations were selected. If the map interface itself was no longer visible as part of the website or application, it would have been coded in the tables as no longer existing. The phenomenon of broken or lost maps is not unique to audiovisual maps and has been noted as a problem with the maintenance and archiving of digital interactive or animated maps in general. Although the absolute number of maps marked as no longer working in the most recent survey is not great, as a percentage of the number of maps found during the surveys these are significant. For example, two (2) of the eight (8) maps found and categorized as integrating sound into the map during the 2010 survey are still available but the audio no longer works properly (*BadiaFonia* and *Folk Songs for the Five Points*). That the maps no longer work as originally intended could be the result of updated software components in the data flow of the application (e.g. web browser security features preventing access requests to sound files located on web servers other than the one hosting the map application itself); could, possibly in conjunction with changes in the world wide web infrastructure, indicate a lack of maintenance by the author(s); or could indicate changes in priorities and/or the perceived importance and usefulness of the map interface in particular projects in comparison to the collection and posting of the sounds themselves.[4] From the perspective of archiving and preservation of data, a theme to be discussed in more detail in Chapter 21 of this volume, the proportion of maps in the survey that no longer work as originally intended so quickly after being deployed is cause for concern. Although the World Wide Web is used increasingly for the distribution and streaming of audio content, the technological basis of these capabilities within network environments (composition, encoding, compression, transmission) continues to be the subject of ongoing experimentation and development. Technological choices in the creation of an audiovisual map to support application longevity or to simplify forward-conversion to new technologies for long-term maintenance or archiving are still uncertain. If audiovisual maps and their components were originally developed under research-based funding grants, as at least some of those listed in the tables appear to have been, funds to assist in maintenance or forward-conversion are often limited or nonexistent.

The classification of online maps designed to include sounds presented here involves some interpretation and other researchers could assign a small number of the maps differently between the categories than I have. The broad groupings would, nonetheless, remain very similar. This classification is not intended as criticism of those designs that do not integrate sounds into the map interface, but simply as a means of describing the existing situation. The main conclusion drawn from this review is that design possibilities for utilizing sound and for creating complementary visual and auditory designs for web-mapping applications are not yet nearly exhausted. There is still much scope for research on these and related topics. It is perhaps not surprising that this review has shown that a visual bias remains in map design, even when acoustic multimedia is to be included.

[4] The map interface for *Soundlandscapes* attempts to access sound files stored on a cloud-based audio service, resulting in web browser errors complaining that such an access would be a potential security vulnerability and have been blocked. Prior (2013) decided to archive his *12 Gates to the City* sound map project in favour of plans to produce and publish a field recording at least weekly along with contextual information that he found was missing, for his purposes, from the sound map interface. Although *Soundlandscapes* was not found during the 2010 survey, the sounds included in the map presumably worked at some time and then were moved to the cloud-based audio server, possibly to improve the work flow associated with collecting and posting field recordings despite the impact of causing the sound map to stop working. Among the acoustic ecology projects included in the surveys, a large subset of the full set of projects, one or two others now also collect their audio recordings on cloud-based servers with similar results.

10.3 PROSPECTS

Designing, distributing, and using audiovisual maps over the World Wide Web highlights processual aspects of the visual and auditory elements of the map and relations between them, and demonstrates many of the characteristics of digital media discussed by Manovich (2001). Whether implemented using raster or vector graphics, visual map elements use modular, digital media technologies based on numerical representations, and are amenable to variability in display and automation. Similarly, acoustic elements of the map are implemented either as digitally recorded audio or as synthesized compositions and each type of audio is amenable to procedural manipulation when being played as part of the audiovisual map. At the very least, the playback of audio of either type can be started and stopped as needed, and playback parameters such as amplification, stereo balance, reverb, and playback rate or tempo can be controlled and adjusted.

The use of visual and/or auditory elements may be more or less invariant in a specific map with the user having, for example, no more control than to cause the map to be displayed and the sounds to play (possibly through independent decisions about each). Alternatively, the display of at least some visual elements and/or the playback of at least some auditory elements of the map may be initiated or modified as a result of actions by the user. Audio may be incorporated into the map through the use of controls that determine when or whether the audio is played, or that determine the playback characteristics of the audio such as amplification or stereo balance in response to cursor position, options selected, or other actions by a user working with the map interface. By combining multiple audio recordings through the use of dynamic playback controls, each audio recording could be presented as complementary or oppositional to the others or to elements of the visual design, the difference needing to be deduced from information conveyed within the audio and other cues made available through the design of the map.

The ease with which a user understands the relations between their actions and resulting transformations of the audiovisual map will be important in determining the user's perception concerning the usefulness of the map. In particular, what is to be heard and sound transformations employed during map use may need to be explained through the use of visual elements and/or user training and may need to respect or only carefully circumvent certain expected patterns of sound usage from other, more prevalent, forms of audiovisual media. But sounds may also influence the interpretation of the visual design, possibly providing didactic explanation of the visual design, and thus the overall visual and auditory behaviours of the map must be considered together.[5]

Many types of audio recording carry implications of time and place embedded within them. Recorded narrations are coded by language and dialect, accent, gender, subject matter, and other characteristics that may place the narrator more or less specifically or may indeed place the narrator external to the story (Nichols, 1985). Music can be understood through perspectives on genre, ethnicity, gender, or other characteristics to invoke a sense of time and place or to cite adherence to a specific canon such as Western classical and thereby reference what has by some been interpreted as universalism.[6] Sound effects may be place and/or time

[5] Brauen (2011b, Chapter 4) discusses the influence of sound on the interpretation of visual media.

[6] For a critique of such assumed placelessness, see Leyshon et al. (1998:3–9).

specific, if sufficiently recognizable and assuming that the audience has the knowledge to decode the reference; may be merely suggestive of a type of place (e.g. a windswept plateau); or may be generic and relatively placeless. With all types of audio recording, the characteristics of a sound as perceived by a listener are shaped in part by the acoustic qualities of the space within which that sound was created (Belazs, 1945) but technical manipulation of the sound, including placement and selection of microphones and the use of post-recording technology, may be able to enhance or alter that sense of space. In addition, a recorded audio need not be thought of as statically positioned in space. Transportation technologies and the mobility of audio technologies each on their own enable motion during sound recording and create potential for thinking of a single recording as variable with respect to both time and space (e.g. *Soundtracks* in Table 10.1). Narrative reference can identify multiple places and times along with people, objects and other concepts within a single audio recording.

Turning from discussion of audio as digital media and the implications thereof to questions concerning what sounds and acoustic information could be recorded and used in the design of a map, it is important to note that, as with decisions concerning selection, scale, and framing for a visual map, the selection, sequencing, mixing, and processing of auditory materials seldom would not involve editorial decisions. Even for mapping projects that seek to assemble and organize audio recordings taken in particular locations, decisions concerning where to place microphones and when to record have been made, and the collected recordings are very often specific to both place and time (e.g. the recording of 'Should I Stay or Should I Go?' tagged against Lisbon's Bairro Alto district in *Cinco Cidades*).

When selecting or designing sounds as materials from which cartographic expression of data are to be crafted, rather than merely as the data to be organized within a map, questions to be considered include metaphorical associations between data and the sounds used to express the data, the aesthetics of sounds selected and how they will be used (e.g. frequency of repetition, duration of use, acoustic variations that may be applied through run-time control of the sounds), and possible means for signifying data variations through control of acoustic parameters during playback (Krygier, 1994). This is analogous to abstracting visual variables and their variations for the same purposes but, because of the less common use of sound, a map maker must assume that users are similarly able to dissociate data and the acoustic expression of those data, possibly with the assistance of training and explanation (at least, prior to working with users to understand their reactions to a map design). Regardless of the modality of expression, a map is an abstracting device the operation of which is only completed by acts of imagination. Possibilities for the design of maps as complexes of acoustic and visual materials, and their interpretation and possible uses, remain an area that has received relatively little attention.[7]

When used as part of an audiovisual design, music, voice, environmental sounds, and sound effects could contribute information and affective impact to support narrative intent. Many issues concerning how interactive sound designs could or should work with the

[7] As Cosgrove (2008:168) argued '[n]o spaces can be controlled, inhabited or represented completely [although] the map permits the illusion of such possibilities.' In addition to historical examinations of the roles of maps within society, the linkages between audiovisual media, culture, and meaning (Lukinbeal and Zonn, 2004) and similar linkages when maps are incorporated into such audiovisual media (Conley, 2007) have been studied.

interactive visual design of an online map remain to be studied. The flexibility of a multitrack audio recording and playback system, similar to those developed for narrative film (Altman et al., 2000), will be very useful in allowing a variety of acoustic material to be mixed and varied as a user works with a map. But requirements for the sound design to react to indeterminate actions and timing as a user works with elements of an audiovisual map are very different from those of a film's linear narrative and may be more similar to the sound design requirements of video games (Collins, 2008b). In addition, game sound designs that provide emotional cues, encourage user pacing, or are intended to motivate users to continue to play suggest other motivations for cartographic sound design (ibid.).

10.4 EXAMPLE: AIRBORNE BTEX POLLUTANT EMITTING FACILITIES IN MONTREAL

To present a specific example of an interactive audiovisual map design, a web map prototype was created to:

1. Show facilities in Montreal reporting to Environment Canada's National Pollutant Release Inventory (NPRI) program that may have contributed significantly to aggregate airborne concentrations of benzene, toluene, ethylbenzene, or xylenes (BTEX) during 2008 at specific locations[8] as computed by a simple run-time dispersion model.
2. Use sounds to indicate modelled airborne concentrations for each of the BTEX component compounds relative to specific locations as a user works with the map.

Montreal BTEX Emission Sources 2008, a web application available at http://atlas.gcrc.carleton.ca/montreal_btex, is shown in Figure 10.1, and audiovisual screen captures showing the application in use are included in this book's companion website (http://booksite.elsevier.com/9780444627131/). Brauen (2011a) discussed potential impacts from airborne dispersion of BTEX along with a description of the air pollutant dispersion model used by this map and its strengths and shortcomings for such an application. The discussion here will be restricted to the audiovisual design objectives and approach taken for this prototype.

Airborne dispersion of pollutants from many possible sources (e.g. industry, transportation, agricultural) at levels that are not easily detected by casual observation can have impacts on human and animal health and ecosystem stability and resilience. Therefore, this map was designed to present information concerning the potential presence of airborne pollutants from facilities that report releases to Environment Canada's NPRI program in a manner that is informative while suggesting the uncertainty of the data upon which the application relies and the inherent difficulties of understanding air pollutant dispersion. Air contaminant information is expressed by the following facets of the audiovisual design: synoptic display of facility locations, acoustic expression of modelled airborne pollutant concentrations supported by the visual design, and selectable tabular displays for each reporting facility.

[8] For each of the BTEX component compounds, a minimum threshold set at 1/1000 of estimated background concentration of that compound determined using data from Environment Canada's National Air Pollution Surveillance program stations (Brauen, 2011a) is used to filter out insignificant concentration values.

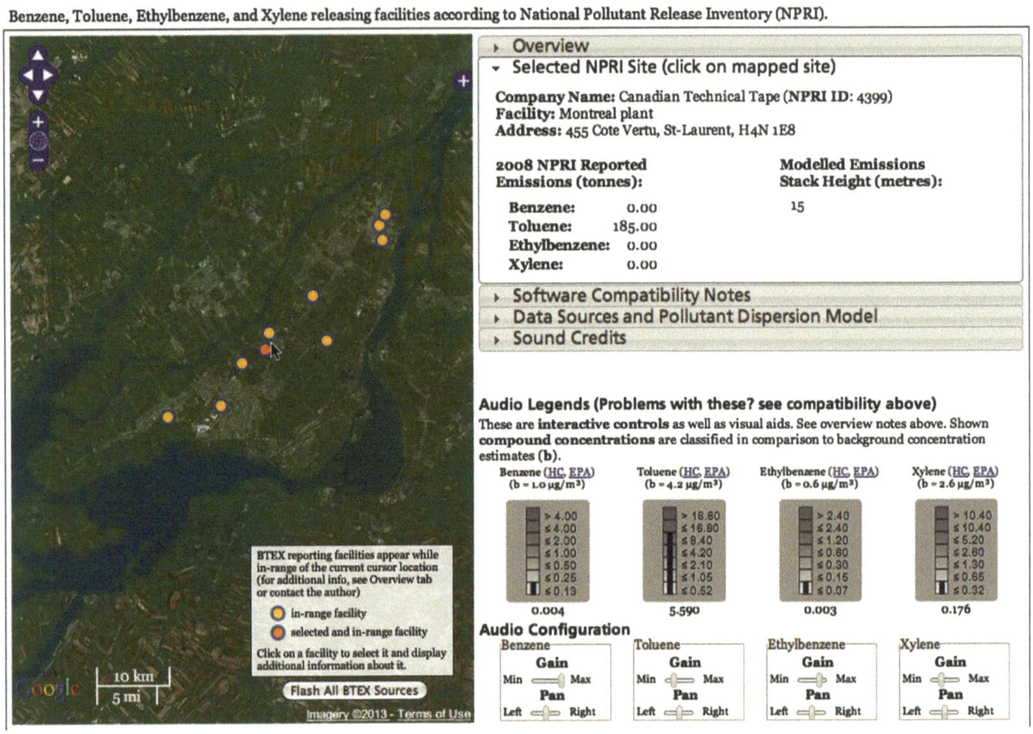

FIGURE 10.1 Montreal BTEX Emission Sources 2008.

Geographic locations as reported to NPRI are synoptically shown by the visual map either because a subset of the reporting facilities would have contributed to the airborne concentration of at least one of the BTEX component compounds during 2008 at the current cursor location set by a user or, temporarily, because a user working with the map clicked on the 'Flash All BTEX Sources' button (bottom right of map in Figure 10.1). In Figure 10.1, a subset of facilities are visible on the map to indicate that these would contribute to aggregate airborne pollutant concentrations at the cursor location according to the application's dispersion model.

Estimated aggregate concentrations of each of the BTEX components at the cursor location are expressed acoustically using a blend of audio loops for each of which amplification is adjusted according to a comparison of the modelled aggregate concentration of the compound from all NPRI reporting facilities against a modelled background airborne-concentration estimate for the same compound (Brauen, 2011a). In Figure 10.1, all of the visible facilities contribute to the aggregate concentrations of at least one of the BTEX component compounds and, therefore, to the set of amplification adjustments for the audio loops associated with the compounds. In addition, Figure 10.1 (lower right) shows a set of dynamic audiovisual legends (Brauen and Taylor, 2008), each of which visually depicts a classification of the concentration level for one of the BTEX components, again in comparison to the estimated airborne background level of that compound. Based on the current cursor position, the dynamic audiovisual legends show that only the modelled aggregate toluene concentration is significant compared to estimated background levels. For all of the other BTEX components, modelled concentrations are all within the lowest possible category tracked by the application (less than one-eighth of estimated background concentrations using a logarithmic classification scale).

A user may selectively display a tabular description for any of the NPRI reporting facilities including details concerning the ownership and address of the facility, the emissions reported by the facility to the NPRI for 2008, and the estimated average stack height used for modelling the emissions from the facility. In Figure 10.1, details for the currently selected facility, shown on the map using a different fill colour compared to all others, are displayed (upper right).

As a user works with the map, cursor movements are continuously tracked as a sequence of locations across the surface of the map, each mapping to a set of updated concentration values for the BTEX component compounds. The concentration values are used to update the audiovisual map iteratively while the cursor remains over the map. Both the acoustic expression of airborne concentrations at the cursor location and the visibility of facilities that reported emissions that contribute to the current modelled aggregate concentrations are thus continuously updated.

Each BTEX component compound is associated with an audio loop containing to varying degrees musical and recognizable anthropogenic-technological sounds. These were selected using criteria that included discriminability and aesthetics but also because the sources of these sounds are rooted in deliberate human activities as are the airborne pollutants from industrial facilities represented by each sound. By contrast, when all of the modelled airborne concentrations, based on cursor location, are classified within the lowest category, the synthesized loops are all muted and the application plays only an environmental recording of an urban wooded area complete with bird calls. Thus, this sound design contains fairly obvious editorial decisions concerning the data mapped (and was critiqued by some attendees at the workshop in which it was initially presented as overplaying the dichotomy between pristine nature and human-caused pollution).

Abrupt changes in the acoustic expression of the estimated airborne concentrations of the BTEX components could cause unpleasant sound artefacts (e.g. clicks and pops). Therefore, amplification levels are always adjusted gradually over time after new target concentration values have been computed, as is common in audio processing for many other forms of media (e.g. film, music). In practice, fairly short fade periods of approximately 50 ms are adequate to avoid unpleasant acoustic events with any except the most abrupt level changes. However, the sound design for the application deliberately lengthens these fade times to a full second to both create relaxed transitions and to suggest the transience, unpredictability, and invisibility of airborne pollutant concentrations and to suggest the uncertainty of the data and modelled concentrations. If a user continues to move the cursor on the map while amplification levels continue to settle from previous moves, it is possible that the amplification levels will never quite express the concentration levels for the latest cursor location selected. In practice, users rarely stay that active for long.

To adjust the visibility of NPRI reporting facilities to show only those that contributed to aggregate levels at the current cursor location, the visual markers for the facilities on the map are designed to allow their opacity to be modified dynamically. To keep the map's visual state consistent with computed and pending acoustic updates, based on user actions, the visibility of each facility marker transitions to a new state by fading from fully opaque to transparent, or vice versa, over a period of 1 s in synchronization with the fading of the acoustic amplification settings. This visual design accommodates the use of sound in the map by relaxing the rate of visual updates to match those used in the sound design. Furthermore, the design of the software that accomplishes the gradual opacity transitions used in this map was both necessitated by and based on the design of the software that manages the gradual acoustic parameter fades. As such, this represents a concrete example of the argument by Théberge (2005) that sound design for cybercartographic maps and atlases should attempt to create complementary acoustic and visual designs rather than taking a comparatively simple but probably less effective approach by attempting to add sound components to a nearly complete visual map.

10.5 CONCLUSION

This research is based on the principle that audiovisual design, although nascent in interactive cartography, can be an effective means of communicating spatial information to a variety of audiences. Hearing and vision frame perceptions differently but through experience people learn to complement one with the other. Each sense also engages attention, judgement, memory, and emotions in different ways (Rodaway, 1994:102–103) based partly on individual learning, experience, and (dis)ability but also on the cultures within which individuals exist (Gibson, 2011; Howes, 1991).

Other media have accepted the need to produce sound designs as part of complex projects. Indeed, many of the common tropes of animated and interactive cartography were foreshadowed by earlier cinematic uses of maps and globes (Caquard, 2009) and these were often accompanied by a film soundtrack. Films and games developed without sound are now very rare. It is possible that the understanding of map making and use as paper-based still dominates so that sound is simply not considered in the development of many projects. It is possible that common tools for producing interactive maps don't make it obvious that including sound is a possibility. It is also possible that designers and potential users of web maps are aware of poor examples of interactive sound designs in computer and web-based application interfaces and these discourage continued patience with the development of new examples, approaches, and technologies. Interactive audiovisual cartography is new and immature but some perseverance with the topic is warranted if one considers the history and lessons of the development of film sound beginning in the nineteenth century and continuing today (LoBrutto, 1994; Stewart, 1980; Weis and Belton, 1985).

Montreal BTEX Emission Sources 2008 is one example of what I refer to as interactive audiovisual design for cartography in addition to the examples outlined in Table 10.1. Cartographic theory and praxis develop in parallel through application on specific projects and digital cartography appropriate for use over distributed networks is a fast-developing domain. Thus digital cartography itself and interactive audiovisual cartography in particular will almost certainly continue to develop through a process of adapting theory and digital media practice from other disciplines to the needs of online map making. Such an approach must be complemented with understandings of cartographic concepts, processes, products, and tools; of tools and processes for working with available visual and acoustic media technologies; of theoretical perspectives out of which data may be created; and of technical skills required to manage, structure, and express those data.

Possibilities for the sound design of a particular audiovisual web map application depend, in part, on the capabilities of the underlying sound subsystem. The cartographic sound subsystem used in this example provides a flexible solution to managing audio clips and streams distributed in real time across the World Wide Web (Caquard et al., 2008) and to applying single-pass digital audio effects to these sound sources under control of parameters adjusted in response to user actions (Brauen, 2011b:181–206). The creative process involved in imagining, searching for and/or recording acoustic materials and in planning, designing and implementing the audiovisual behaviours of the map in relation to user interaction requires effort and skills that may not be familiar to cartographers. Rather than reinventing theory, tools, and methods, sound design for interactive cartography can often adapt methods and theory from other domains including sound design for film (Altman, 1992; LoBrutto, 1994; Ondaatje, 2002; Weis and Belton, 1985) and games (Collins, 2008a,b).

Any user interaction model beyond a simple one in which sounds are directly triggered by interaction with visual elements (e.g. placing the cursor on a visual feature to initiate sound playback) will currently require nontrivial programming to 'connect' the underlying data and its expression. Although the effort to create an audiovisual map may involve additional process steps and considerations, many of these will be similar or exactly the same as those that need to be performed when an interactive data visualization interface is being designed. Data management, structuring, and access challenges must be solved. Process steps to group, accumulate, and normalize values to produce data that can be comparatively expressed will be similar or identical to the steps required in other visualizations. Event handling to detect cursor motion and other events such as keyboard strokes and button clicks to propel the application based on user interaction will be similar. Only the expression of the data through loading of sounds and control of the sound subsystem will be different, and sometimes fundamentally different, than the control of interactive graphics because of the temporality of sound. Finally, because the applications being discussed here are audiovisual, the processing required to produce the graphics will be similar to that done for other visualizations except that, as noted in the previous section, the visual and acoustic interface design should be complementary. If appropriate expertise and tools can be brought to the project, such as for example the sound subsystem discussed here, the additional effort does not have to be prohibitive.

Acknowledgements

This research has been supported, in part, by the Social Sciences and Humanities Research Council (SSHRC).

References

Altman, R. (Ed.), 1992. Sound Theory/Sound Practice, Routledge, New York.

Altman, R., Jones, M., Tatroe, S., 2000. Inventing the cinema soundtrack: hollywood's multiplane sound system. In: Buhler, J., Flinn, C., Neumeyer, D. (Eds.), Music and Cinema, University Press of New England, Hanover, NH, pp. 289–297.

Aprea, D., Antonacci, F., Sarti, A., Tubaro, S., 2009. Acoustic reconstruction of the geometry of an environment through acquisition of a controlled emission. 17th European Signal Processing Conference. EUSIPCO 2009, pp. 710–714.

Arana, M., Martin, R.S., Martin, M.L.S., Aramendía, E., 2010. Strategic noise map of a major road carried out with two environmental prediction software packages. Environmental Monitoring and Assessment 163, 503–513.

Barbagli, B., Magrini, I., Manes, G., Manes, A., Langer, G., Bacchi, M., 2010. A distributed sensor network for real-time acoustic traffic monitoring and early queue detection. Fourth International Conference on Sensor Technologies and Applications, pp. 173–178.

BBC, 2009. BBC Save Our Sounds. Available at: www.bbc.co.uk/worldservice/specialreports/saveoursounds/index.shtml.

Becker, B., 2005. Puget Soundscape. Available at: http://lossycodec.com/soundscape/pugetsound.html.

Belazs, B., 1945. Theory of the film: sound. In: Weis, E., Belton, J. (Eds.), Film Sound: Theory and Practice, Columbia University Press, New York, pp. 116–125.

Binaural/Nodar, 2011. Sound Map of the Paiva River. Available at: http://www.mapize.com/soundmap_paiva/.

Booth, H., Brancasi, J., 2008. Open Sound New Orleans. Available at: http://www.opensoundneworleans.com/core/.

Brauen, G., 2011a. Interactive audiovisual mapping: BTEX emissions from NPRI reporting facilities in Montreal. In: Cartwright, W., Caquard, S., Vaughan, L. (Eds.), Mapping Environmental Issues in the City: Arts and Cartography Cross Perspectives, Springer, Berlin; Heidelberg, pp. 74–108 (Chapter 6).

Brauen, G., 2011b. Toward Interactive Audiovisual Cartography: Motivations, Design strategies, and Methods. Ph.D. Thesis. Ottawa, Carleton University, Ontario.

Brauen, G., Taylor, D.R.F., 2008. Linked audio representation in cybercartography: guidance from animated and interactive cartography for using sound. Revista Brasileira de Cartografia 60 (3), 223–242.

British Library, 2010–2011. Sound Maps. Available at: http://sounds.bl.uk/Sound-Maps.

Butler, T., Whiting, S., 2005. Memoryscape Audio Walks. Available at: www.memoryscape.org.uk.

California Library of Natural Sounds and Oakland Museum of California, 2003. Listening to Nature: A Sound Walk across California. Available at: http://museumca.org/naturalsounds/home.html.

Callard, A., Krughoff, F., Farrow, J., 2006. audioBus. Available at: http://www.andreacallard.com/audiobus/.

Cantizzani, J., Caballero, M., Lastra, M.A., Millán, S., Sanz, P., Bennet, J., Cobussen, M., 2006. Andalucía Soundscape. Available at: www.andaluciasoundscape.net/mapa.

Caquard, S., 2009. Foreshadowing contemporary digital cartography: a historical review of cinematic maps in films. The Cartographic Journal 46 (1), 46–55.

Caquard, S., Brauen, G., Wright, B., Jasen, P., 2008. Designing sound in cybercartography: from structured cinematic narratives to unpredictable sound/image interactions. International Journal of Geographical Information Science 22 (11), 1219–1245.

Cartwright, W., Peterson, M.P., 2007. Multimedia cartography. In: Cartwright, W., Peterson, M.P., Gartner, G. (Eds.), Multimedia Cartography, second ed. Springer-Verlag, Berlin, pp. 1–10 (Chapter 1).

Cartwright, W., Peterson, M.P., Gartner, G. (Eds.), 2007. Multimedia Cartography, second ed. Springer-Verlag, Berlin.

Centeno, H., Sinclair, D., 2008. Mississauga Sound Map. Available at: www.yorku.ca/caseaces/soundmap.

Chapman, D., 2008. Isle of Dogs Music. Available at: http://www.iodmusic.org/youarehear/.

Chesterman, W.D., 1968. Geophysical exploration of the ocean floor using acoustic methods. Contemporary Physics 9 (5), 423–446.

Chicago Public Radio, 2007. Chicago Sound Map. Available at: www.chicagopublicradio.org/Soundmarks_Browse.aspx.

Collins, K. (Ed.), 2008a. From Pac-Man to Pop Music: Interactive Audio in Games and New Media, Ashgate, Hampshire, UK; Burlington, VT.

Collins, K., 2008b. Game Sound: An Introduction to the History, Theory, and Practice of Video Game Music and Sound Design. The MIT Press, Cambridge, MA; London.

Conley, T., 2007. Cartographic Cinema. University of Minnesota Press, Minneapolis.

Cordobés, M., Picanyol, R., 2008. BadiaFonia. Available at: www.badiafonia.net.

Cosgrove, D.E., 2008. Geography and Vision: Seeing, Imagining, and Representing the World. I. B. Taurus, London; New York.

Coulam, D., 2008. Soundlandscapes. Available at: http://www.soundlandscapes.eu/paris%2Dsound%2Dmap.

Cox, T., 2010. Sound Tourism: A Travel Guide to Sonic Wonders. Available at: www.sonicwonders.org.

Cusack, P., 2012. Favourite Sounds. Available at: http://favouritesounds.org.

Dannhorn, A. n.d. *Acoustic Map*. Available at: http://angela-dannhorn.ch/project/acoustic-map.

Dant, A., Davis, T., Gama, V., Gunn, D., 2006. Folk Songs for the Five Points. Available at: www.tenement.org/folksongs/client.

Dant, A., Davis, T., Gunn, D., 2007a. Cinco Cidades. Available at: www.cincocidades.com/en/soundmap/.
Dant, A., Davis, T., Gunn, D., 2007b. Manchester: Peripheral. Available at: http://www.manchesterperipheral.com/phase1/soundmap.
Dein, A., Panetta, F., 2010. Sound Map: The Caledonian Road. Available at: www.guardian.co.uk/society/interactive/2010/apr/26/caledonian-road-sound-map.
Eason, P., Lewis, J., 2007. Soundscape of China. Available at: www.pbs.org/kqed/chinainside/soundmap/index.html.
Ecklund, D. n.d. *Columbia River Sound Map*. Available at: http://www.soundmaps.net/columbiariversoundmap/.
Erkizia, X., Iruretagoiena, O., Hurtado, E., Balderas, X., 2009. Soinu Mapa. Available at: www.soinumapa.net/?lang=en.
European Acoustic Heritage, 2011. European Soundscape Map. Available at: http://map.europeanacousticheritage.eu/.
Fisher, P.F., 1994. Hearing the reliability in classified remotely sensed images. Cartography and Geographic Information Systems 21 (1), 31–36.
Francis, K., 1996. Wula Na Lnuwe'katiyek – an electronic mixed media atlas. MA Thesis. Carleton University, Ottawa.
Freesound Project, 2005. Map of Geotagged Sounds. Available at: http://www.freesound.org/browse/geotags/.
Freesound Project, 2010. Sons de Barcelona. Available at: http://barcelona.freesound.org.
Gartner, G., September 2004. Location-based mobile pedestrian navigation services: the role of multimedia cartography. International Joint Workshop on Ubiquitous, Pervasive and Internet Mapping (UPIMap 2004). Tokyo, Japan, Available at: http://www.ubimap.net/upimap2004.
Gibson, R., 2011. What the eye can hear. Journal of Australian Studies 35 (2), 145–151.
Gwich'in Social and Cultural Institute, 2003. Talking Map Research. Available at: http://www.gwichin.ca/Research/placeNameMap.html.
Harrower, M., 2003. Tips for designing effective animated maps. Cartographic Perspectives 44, 63–65.
Hinde, K., Holroyd, E., 2011. Echo Location: a Sound Map for Bedford UK. Available at: http://echo-location.org/.
Holzer, D., Kolster, S., Boon, M., 2004. Sound Transit. Available at: http://soundtransit.nl.
Hope, W., Sindler, J., Schlaman, K., Lloyd, J., Niemyer, R., Cole, B., Kahn, Z., Mares, J., Sackett, E., 2011. Knox College Sound Map. Available at: http://www.knoxsoundmap.com/index.php.
Hovick, O., Hirt, T., Johnston, M., McGettigan, S., Maccaroni, M., Adams, R., Parry, J., Goldstein, J., Kopel, M., Ellis, B., Muoio, D., Callahan, C., 2012. Ursinus Sound Map. Available at: http://www.ursinusgrizzly.com/ursinus-sound-map/.
Howes, D., 1991. Introduction: 'to summon all the senses'. In: Howes, D. (Ed.), The Varieties of Sensory Experience: a Sourcebook in the Anthropology of the Senses, University of Toronto Press, Toronto, pp. 3–21.
Hu, S., 2003. Web-based multimedia GIS: exploring interactive maps and associated multimedia information on the Internet. In: Peterson, M.P. (Ed.), Maps and the Internet, Elsevier Science, Oxford, pp. 335–344 (Chapter 20).
Kinayoglu, G., 2010. Soundtrack. Available at: http://www.terrasound.org/soundtrack/.
Kiser, S., 2012. Forest Hills Sound Map. Available at: http://spencerkiser.com/storage/processing/forest_hills_sound_map/applet/index.html.
Kite's in the Tree Media, 2012. The Austin Zoo Sound Map. Available at: http://kitesinthetreemedia.com/post/33690684954/the-austin-zoo-sound-map.
Komatsu, T., Mikami, A., Sultana, S., Ishida, K., Hiraishi, T., Tatsukawa, K.-i., 2003. Hydro-acoustic methods as a practical tool for cartography of seagrass beds. Otsuchi Marine Science 28, 72–79.
Kraak, M.-J., Brown, A. (Eds.), 2000. Web Cartography: Developments and Prospects, Taylor and Francis, New York.
Krygier, J.B., 1994. Sound and geographic visualization. In: MacEachren, A., Taylor, D.R.F. (Eds.), Visualization in Modern Cartography, Pergamon, New York, pp. 149–166 (Chapter 8).
Laboratory of Applied Bioacoustics, Technical University of Catalonia, 2010. Listening to the Deep Ocean Environment. Available at: http://listentothedeep.net/acoustics/index.html.
Levine, P., 2006. Signature. Available at: http://paulalevine.net/projects/signature/signature.html.
Levine, P., Lilleman, G., Johnston, R., Redpath, G., Dandan, S., 2004. Baghdad↔San Francisco. Available at: http://shadowsfromanotherplace.net.
Leyshon, A., Matless, D., Revill, G., 1998. Introduction: music, space, and the production of place. In: Leyshon, A., Matless, D., Revill, G. (Eds.), The Place of Music, Guildford, New York; London, pp. 1–30.
Lidbo, H., Franzén, E., Ullberg, H., Stein, M., Liljedahl, J., 2009. Stockholm Sound Map. Available at: http://www.oljudsthlm.se/soundmap/.
LoBrutto, V., 1994. Sound-on-Film: Interviews with Creators of Film Sound. Praeger Publishers, Westport, CT.

locusonus.org, 2006. Locus Sonus. Available at: http://locusonus.org/soundmap.
Louth-Robins, T., 2011. Fleurieu Sound Map. Available at: http://www.tristanlouthrobins.com/fleurieu_soundmap/.
Special issue on cinematic geographies. 2004. In: Lukinbeal, C., Zonn, L. (Eds.), GeoJournal 59, 4.
Madrid Soundscape.org, 2008. Madrid Soundscape Map. Available at: www.madridsoundscape.org.
Mak, A., Chan, S., Yan, Y.C., Chan, R., Ming, W.T., Lin, H., 2009. A Map of Our Own: Kwun Tong Culture and Histories. Available at: http://www.kwuntongculture.hk/en/home.php.
Manovich, L., 2001. The Language of New Media. MIT Press, Cambridge, MA.
MarSensing Lda, 2012. Shipping Noise Underwater Acoustic Map. Available at: http://shippingnoise.com/.
Metzner, J., McAdam, E., 2010. Sound and Story Project of the Hudson Valley. Available at: www.soundandstory.org.
Minard, R., 2010. SoundMap. Available at: http://map.sounding-d.de/map.
Molina, B., Longina, C., González, H., Rodríguez, J.-G., Gómez, J., Sánchez, C.S., 2006. Escoitar. Available at: www.escoitar.org/?lang=en.
Monda, R.L., 2009. Sounds like Staten Island. Available at: www.soundslikestatenisland.com/map/node.
Monmonier, M., 1992. Authoring graphic scripts: experiences and principles. Cartography and Geographic Information Systems 19 (4), 247–260, 272.
Mouafo, D., Müller, A., July 2002. Web-based multimedia cartography applied to the historical evolution of Iqaluit, Nunavut. Proceedings of the Joint International Symposium on Geospatial Theory, Processing and Applications, International Society for Photogrammetry and Remote Sensing, Ottawa, Commision IV, Spatial Data Handling. Available at: http://www.isprs.org/proceedings/XXXIV/part4/pdfpapers/501.pdf.
Müller, J.C., Scharlach, H., 2001. Noise abatement planning: using animated maps and sound to visualise traffic flows and noise pollution. Proceedings of the 20th International Cartographic Conference, vol. I. International Cartographic Association, Beijing, China, pp. 375–385.
murmurtoronto.ca, 2003. murmur. Available at: http://murmurtoronto.ca.
Museum of Modern Art, 2013. Common Senses Sound Map. Available at: http://www.moma.org/moma_studio/soundmap.
Mydlarz, C., Cox, T., Drumm, I., 2009. Sound Around You. Available at: www.soundaroundyou.com.
Netaudio Festival Berlin, 2009. Berlin Wall of Sound. Available at: http://www.netaudioberlin.de/berlin-wall-of-sound/.
Nichols, B., 1985. The voice of documentary. In: Nichols, B. (Ed.), Movies and Methods: An Anthology, vol. II. University of California Press, Berkeley, pp. 258–273.
Noll, U., 2006. Radio Aporee. Available at: http://aporee.org/maps.
Ondaatje, M., 2002. The Conversations: Walter Murch and the Art of Film Editing. Alfred A. Knopf, New York.
Peterson, M.P., 2007. The Internet and multimedia cartography. In: Cartwright, W., Peterson, M.P., Gartner, G. (Eds.), Multimedia Cartography, second ed. Springer-Verlag, Berlin, pp. 35–50 (Chapter 3).
Peterson, M.P. (Ed.), 2008. International Perspectives on Maps and the Internet, Springer, Berlin; Heidelberg; New York.
Poll, A., Feng, S.S., 2009. Sound Seeker. Available at: fm.hunter.cuny.edu/nysae/nysoundmap/soundseeker.html.
Porteous, J.D., Mastin, J.F., 1985. Soundscape. Journal of Architectural and Planning Research 2, 169–186.
Prior, J., 2009. 12 Gates to the City: Acoustic Map. Available at: http://12gatestothecity.com/extras/acoustic-map.
Prior, J., 2013. Introducing the Hebdomadal Sound Project. Available at: http://12gatestothecity.com/2013/01/20/hebdomadal-sound-project/.
Radicchi, A., 2009. Firenze Sound Map. Available at: http://www.firenzesoundmap.org/.
Rawes, I.M., 2010. London Sound Survey. Available at: www.soundsurvey.org.uk/index.php/survey/soundmaps.
Rebelo, P., Chaves, R., Meireles, M., McEvoy, A., Stein, M., 2012. Belfast Sound Map. Available at: http://www.belfastsoundmap.org/.
Ritts, M., Smith, G.J., Imerman, H., Thompson, N., Hunt, M., Ehrlich, A., McMurray, G., Wiernik, N., 2010. Toronto Sound Ecology. Available at: http://torontosoundecology.org.
Roberts, J.M., Brown, C.J., Long, D., Bates, C.R., 2005. Acoustic mapping using a multibeam echosounder reveals cold-water coral reefs and surrounding habitats. Coral Reefs 24, 654–669.
Rodaway, P., 1994. Sensuous Geographies: Body, Sense and Place. Routledge, London.
Rona, P.A., Jackson, D.R., Wen, T., Jones, C., Mitsuzawa, K., Bemis, K.G., 1997. Acoustic mapping of diffuse flow at a seafloor hydrothermal site: Monolith Vent, Juan de Fuca Ridge. Geophysical Research Letters 24 (19), 2351–2354.

Russo, F., Loria, T., Nespoli, G., Stein, M., Lachine, C., Assoiants, A., Cochrane, K., Doshi, S., Finch-Coursey, P., Khazov, O., Lai, R., Lesh, S., Mitchell, J., Migirdicyan, S., Pathmanathan, S., Suarez-Mason, K., 2012. Toronto Sound Map. Available at: http://torontosoundmap.com.

Schafer, R.M., 1977. The Tuning of the World. Alfred A. Knopf, New York.

Seay, J., Cusack, P., Mallozzi, L., Gale, T., Lu, S., Neher, A., 2009. Favorite Chicago Sounds. Available at: http://www.favoritechicagosounds.com/soundmaps.php.

Servigne, S., Laurini, R., Kang, M.-A., Li, K.J., 1999. First specifications of an information system for urban soundscape. IEEE International Conference on Multimedia Computing and Systems (ICMCS'99), vol. II. Florence, Italy, June 7–11, 1999, pp. 262–266.

Shepard, M., Murphy, F., Diesel, B., Flynt, A., Thomas, A., Modi, V., Krishnamurthy, A., 2011. Tactical Sound Gardens. Available at: http://www.tacticalsoundgarden.net/.

Simon Charles Audio, 2012. Dubai Sound Map. Available at: http://scaudio.co/dubai-sound-map/.

Sound@Media, 2010. SeoulSoundMap. Available at: http://som.saii.or.kr/campaign.

Soundwalks.org, 2010. Soundwalks. Available at: www.soundwalks.org/map.

Stanza, 2010. Soundcities. Available at: http://soundcities.com.

Steel, L., Maisonneuve, N., Stevens, M., Ochab, B., Niessen, M.E., Hanappe, P., Bartholomees, S., D'Hondt, E., Kambona, K., Boix, E.G., De Meuter, W., Zaman, J., 2009. NoiseTube. Available at: http://noisetube.net/cities.

Stein, M., 2008. Montreal Sound Map. Available at: http://cessa.music.concordia.ca/soundmap/en.

Stewart, J.G., 1980. The evolution of cinematic sound: a personal report. In: Cameron, E.W. (Ed.), In Sound and the Cinema: The Coming of Sound to American Film, Redgrave, Pleasantville, N.Y., pp. 38–67.

Stollery, P., 2005. Gordon Soundscape. Available at: http://www.gordonsoundscape.net/.

Taylor, D.R.F., 2003. The concept of cybercartography. In: Peterson, M.P. (Ed.), Maps and the Internet, Elsevier Science, Oxford, pp. 405–420 (Chapter 26).

Taylor, D.R.F., Pyne, S., 2010. The history and development of the theory and practice of cybercartography. International Journal of Digital Earth 3 (1), 2–15.

Théberge, P., 2005. Sound maps: music and sound in cybercartography. In: Taylor, D.R.F. (Ed.), Cybercartography: Theory and Practice, Elsevier Science, Amsterdam, pp. 389–410 (Chapter 17).

The Smalls Limited, 2010. The Smalls Street Sounds. Available at: http://content.thesmalls.com/streetsounds/.

Thirion, S., 2007. Cascade on Wheels. Trans Parent Web Design. Available at: http://www.trsp.net/cow/.

Walker, B.N., Lindsay, J., 2006. Navigation performance with a virtual auditory display: effects of beacon sound, capture radius, and practice. Human Factors 48, 265–278.

Weis, E., Belton, J. (Eds.), 1985. Film Sound: Theory and Practice, Columbia University Press, New York.

Wild Sanctuary Inc, 2009. Wild Sanctuary. Available at: http://earth.wildsanctuary.com.

Williams, H., Sumner, R., 2009. Listen to Africa. Available at: www.listentoafrica.com/map.

Yoganathan, N., Jachner, M., Stein, M., 2010. Inukjuak Sound Map. Available at: http://www.inukjuaksoundmap.com/.

CHAPTER

11

A Spatial Typology of Cinematographic Narratives

Sébastien Caquard[1], *Daniel Naud*[2]

[1]Department of Geography, Planning and Environment, Concordia University, Montreal, Quebec, Canada, [2]Département de Géographie, Université de Montréal, Montréal, Quebec, Canada

OUTLINE

11.1 INTRODUCTION

In his essay on the geography of 25 feature films directed by Éric Rohmer in post-war Paris, Richard Misek (2012, 56) argues that in these films 'Rohmer provides a Parisian's view of Paris rather than the more familiar touristic view that we often see in films set in Paris.' This Parisian view is made of cafés, parks, streets, and public transportation, more than of monuments and famous landmarks. Through his analysis, Misek (2012, 65) argues that Rohmer's films can be envisioned as a map of Paris, connecting hundreds of places throughout the city. This argument opens up the possibility of a body of cinematographic work to map places, and make connections between these places. While Misek was addressing the question 'how can films map?' from a systematic analysis of the spatiality of Rohmer's cinematographic production, we propose to address a similar question from a cartographic perspective.

The goal of the research presented in this chapter is to literally map cinematographic narratives in order to contribute to a better understanding of *how films can map*, as well as

Developments in the Theory and Practice of Cybercartography, Second Edition, ISSN 1363-0814
http://dx.doi.org/10.1016/B978-0-444-62713-1.00011-8

to provide a base for mapping narratives in general terms. To explore this issue, we have developed a methodology and a cybercartographic application designed to represent cinematographic narratives. This application has been used to map the geographic structure of 46 contemporary Canadian films and to develop an initial spatial typology of cinematographic narratives.

The first section positions the cybercartographic atlas of Canadian cinema in the context of a growing interest in exploring the multiple relationships that exist between maps and fiction. The spatial structure of Canadian cinematographic narratives are then introduced and discussed in the following section. Finally, in the last section, we propose a spatial typology of the dominant narrative structures that exist in Canadian cinematographic narratives and in films in general.

11.2 THE CYBERCARTOGRAPHIC ATLAS OF CANADIAN CINEMA: INTRODUCTION

The relationships between maps and fiction have stimulated the interest of a growing number of researchers in domains such as literature (Mitchell, 2011; Jenstad, 2011; Piatti and Hurni, 2011; Maleval et al., 2012), cinema (Conley, 2007; Caquard and Taylor, 2009; Aitken and Dixon, 2011; Lukinbeal, 2010; Roberts, 2012; Castro, 2011), painting (Joliveau, 2010; Young, 2011; Pezzutto, 2011), poetry (Horowitz, 2011), music (Cohen, 2012; Long and Collins, 2012), and narratives in general (Caquard, 2013). Across these disciplines, maps and mapping have been used in many different ways that can be grouped into three main categories. There is a large group of scholars in the humanities that has endorsed mapping as a conceptual framework to study narratives and more specifically their spatial dimensions. This is for instance the case of Richard Misek (2012) with his analysis of Rohmer's Paris, as described previously. A second group of scholars is more interested in deconstructing and analysing the presence of maps in narratives and fiction. This is for instance the case of Tom Conley (2007), who addresses the symbolic and political dimensions of maps in films through an in-depth analysis of maps and spatial mechanisms that appear in films throughout the twentieth century. Finally, a third group of individuals is more interested in making maps of narratives in order to reveal some hidden elements of their narrative structures (Bulson, 2007), as well as to study the geography that emerges from these creative works. This approach has been widely inspired by the seminal work of Franco Moretti on literary cartography (1999), and is quite well illustrated by the Literary Atlas of European Literature project (http://www.literaturatlas.eu/en). The *Cybercartographic Atlas of Canadian Cinema* falls within this last category.

The *Cybercartographic Atlas of Canadian Cinema* is a collaborative project between Université de Montréal, Concordia University (Montréal) and Carleton University (Ottawa), involving members from cartography, geography, film studies, and computer sciences. This project, started in 2007, originally had two major goals: (1) to support the development of new ways of designing maps inspired by cinematographic concepts and techniques; and (2) to apply these approaches to mapping Canadian cinematographic territories (i.e. places of film production, film consumption, and cinematographic narratives) in order to contribute to a broader understanding of Canadian cinema and its contribution to the production of an imaginary geography. To reach these two goals, a series of exploratory maps and geovisualizations have been developed over the years (see www.atlascine.org for some examples). This chapter focuses on the cybercartographic application designed to map cinematographic narrative structures.

Mapping cinematographic narratives can be complex both at the ontological level and at the representational level. At the ontological level, it is extremely difficult to clearly define and characterize homogenous spatial units in films in a consistent way, given the level of spatial complexity that cinematographic narratives can reach. For instance, when the action unfolds in one place while two characters are talking about a different place with some inserted shots from different time periods, how do we characterize this/these diachronic and ubiquitous place(s)? At the cartographic level, how can we use the existing cartographic vocabulary to represent these complex spatiotemporal expressions made of sounds, moving images, and subtle references? This becomes even more complicated when we want to map not only these locations, but also the connections that exist between these locations. Indeed, mapping cinematographic narratives cannot be just about mapping the places of the action – which can easily be done with online mapping services such as Google Maps (see for instance http://www.themoviemap.com). It also needs to take into account the relationships and connections that exist between these places in the narrative. Not only places are distorted and produced through the cinematographic lens, but the relationships between these places are also redefined by the film's narrative structure.

These two issues have been addressed in the *Cybercartographic Atlas of Canadian Cinema* (atlascine.org). The ontological one has been addressed through the development of a reading grid and of a methodology inspired from literary studies to extract places from narratives. With this reading grid, any film can be broken down into geographic bits and pieces that can then be rearranged in a geospatial database with various attributes (e.g. the duration of a scene, the level of accuracy of the location, the way the place is materialized in the film). This methodology has been used to extract and map 2200 places from 46 contemporary Canadian movies. Although the mapping of these results provides an interesting perspective on the way Canadian cinemas frame global, national, and local geographies differently (see Caquard et al. (2012) for more details), it did not really address the internal spatial structures of narratives, which is the second major issue in mapping cinematographic narratives: How are these places connected together within a movie and what are the spatial structures produced by these connections?

11.3 THE CYBERCARTOGRAPHIC APPLICATION TO MAP CINEMATOGRAPHIC NARRATIVES

In order to address this second issue, we have developed a specific cybercartographic application. This application has been designed to simultaneously represent the temporal dimension of the unfolding story as well as three spatial dimensions inherent to any type of narrative: (1) the locations of the action; (2) the connections between these locations; and (3) the connections between these locations and other places simply mentioned in the movie.

This application is described in more detail elsewhere (Caquard and Fiset, 2013), but it is important in the context of this chapter to briefly introduce its main technological characteristics. This application is based on Nunaliit, an open source software developed at Carleton University in the context of the cybercartographic project to render geospatial stories (http://nunaliit.org/) (for more details about Nunaliit, see Brauen et al., 2011 and Hayes in Chapter 9 of this volume). The data that drive the map is fetched from a spreadsheet saved in Google Docs, which contains all the data collected with the reading grid. The geographic features generated are then superimposed on a background map provided via OpenStreetMap or Google Maps. This application has been designed to allow any individual to map all sorts of data related to narratives, beyond simply cinematographic ones.

The cybercartographic application renders maps made of lines and points that unfold through time. The points represent the places where the action unfolds. The size of the points represents the amount of time spent in each place throughout the movie. For example, if 40 min of a film take place in Montreal and 20 min in Toronto, the circle representing Montreal will appear twice as big as the circle representing Toronto. Different types of lines are used to represent different types of connections: for example, plain lines represent the displacement of action between two places, while dashed lines represent places that are mentioned while the action unfolds somewhere else. Colours are used to differentiate the way places are materialized in the movie. Red, for instance, is used for places mentioned explicitly, and orange for places that are not explicitly mentioned, but that are implicitly recognizable (see Figure 11.1).

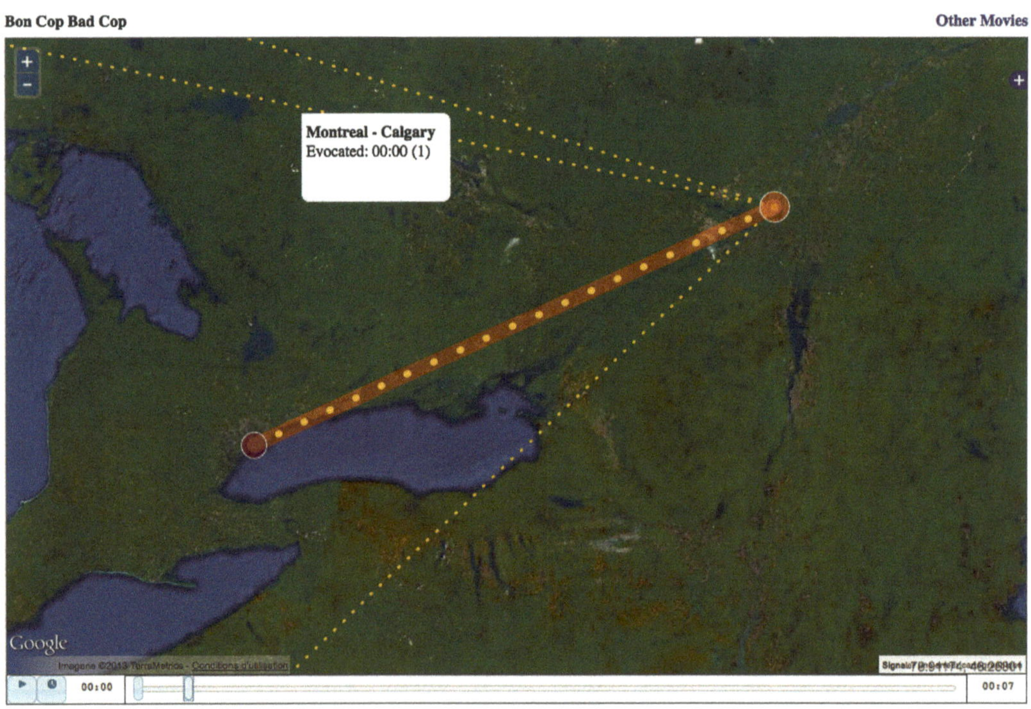

FIGURE 11.1 This example shows the first 7 min of Bon Cop Bad Cop (Dir. Erik Canuel, 2006). So far the action has unfolded evenly between Montreal (orange point in the NE corner) and Toronto (red point in the SW corner). The plain orange line shows that the action has already moved twice between Montreal and Toronto, and the dashed lines connect the action with places that are mentioned in the movie (Note: the working prototype of the cybercartographic application can be accessed here: http://scaquard.classone-tech.com/).

The map is interactive and animated via a play button: the points and lines appear and grow in relation to the unfolding of the narrative. The point representing where the movie starts will appear and grow proportionally to the time spent in this location throughout the movie. If the action moves to another place, then a plain line connects these two places, and the second point grows with time, etc. Meanwhile, dashed lines appear each time a place is mentioned in the narrative (e.g. if the action unfolds in London, Ontario, where a character mentions Paris, Texas, a dashed line will appear to connect these two places). In order to analyse this data, we looked at all the 46 narrative maps produced with this application.

11.4 TOWARDS A SPATIAL TYPOLOGY OF CINEMATOGRAPHIC NARRATIVES

Before looking at the results of our analysis, it is important to mention that the exact location of every scene of a film is not always known. Often we know the city or the community where the action unfolds, but not necessarily the exact neighbourhood or address. This means that the data are often aggregated at the level of the city/community. For instance, if the action navigates between two neighbourhoods of the same city that are not clearly identified, these neighbourhoods will be aggregated at the level of the city. Although this approach does not allow an in-depth analysis of the spatial structure at the local level, it is relevant to developing a general spatial typology of cinematographic narratives, which is our goal here.

There are various ways of looking at the geographic structures of these different movies. One way is to identify them when they first appear throughout the movie. It is indeed interesting to note that the general geographic structure of a movie is often put in place during the first third of a movie. In the first 30 min of about half of the movies studied, the audience has been brought to all the major places where the action will unfold throughout the entire movie. A perfect example is *Ararat* (Dir. Atom Egoyan, 2002), in which the geographic organization of the narration between Toronto, Mount Ararat (Armenia), and New York City is set within the first 15 min of the movie. In movies where new places appear after the first third of the story, it is also interesting to note that most of the time these places appear towards the end of the movie. This is for instance the case in *L'âge des ténèbres* (Dir. Denys Arcand, 2007), in which the main character leaves his depressing suburban life at the end of the movie to regenerate himself in the countryside. In this case, like many others in Québécois cinema, the countryside is used as a place to heal the damages of urban life (see Naud, forthcoming).

This general structure illustrates the fact that dominant spatial settings of films are often loaded at both extremities of the movie: mainly at the beginning – which is expected to set up the action and to provide an 'establishing space' for the story to unfold, as already mentioned by Bordwell et al. (1985, 196) – but also towards the end. What may be more interesting is that new places are rarely introduced in the middle of a film. Throughout the narrative, characters usually keep revisiting the places already introduced in the first section of a movie. Towards the end of the narrative, new places are mobilized to enable characters to reach new dimensions or to escape constraints in which they had been caught throughout the movie. New places are often synonymous with new destinies and new opportunities for characters, as illustrated previously with the example of *L'âge des ténèbres*. In *Guide de la Petite vengeance* (Dir. Jean-François Pouliot, 2006), all the action unfolds in Montreal except for the last 5 min,

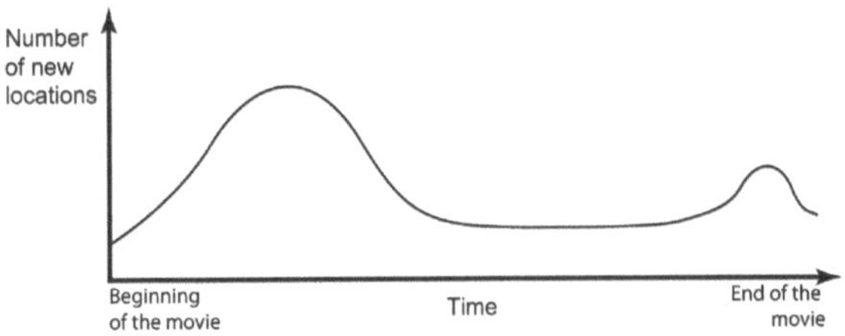

FIGURE 11.2 General structure of the appearance of new places throughout movies. New structures appear mainly in the first third of a movie and, to a lesser extent, towards the end of a movie.

during which the main character moves to Mexico where the movie ends. The first hour and a half of *C.R.A.Z.Y.* (Dir. Jean-Marc Vallée, 2005) unfolds in Montreal, while during the last 30min, Zac, the protagonist, flies to Jerusalem to escape the remorse and the weight of his first homosexual experience before coming back to Montreal. The first hour and a half of *Ma Tante Aline* (Dir. Gabriel Pelletier, 2007) also unfolds in Montreal, while the last 25min take place in Cuba where Genevieve, the main character, can finally (and unexpectedly) resolve family issues. Interestingly enough, all these examples come from Quebecois films (Francophone) while Anglophone Canadian films seem less inclined to mobilize new places towards the end of the story. This could be explained in part by the fact that Quebecois films tend to overly focus on Montreal as the main setting for their actions, which does not leave much room for their characters to escape their destinies. New places open new perspectives and ways of escaping the geographic containment that characterizes the rest of these narratives. Since French Quebecois cinema is more contained spatially than its Anglo Canadian counterpart (Caquard et al., 2012), it is not surprising that it more often mobilizes new places towards the end of its stories, to allow its characters to escape this containment (Figure 11.2).

Another way to look at these spatial structures is to study how each film articulates the three main spatial dimensions described earlier:

- The location of the action, or geographic poles, which ranges from unipolar (i.e. the action unfolds only in one given city/community) to multipolar (i.e. the action keeps on moving between multiple spatial poles or nodes).
- The connection between these locations (i.e. the geometry), which is organized based on the intensity of the connections between the places where the action unfolds (i.e. the number of times the action moves back and forth between the different locations), and materialized by the thickness of the lines connecting these different poles.
- The connection between these locations and other places simply mentioned in a movie, which ranges from none to dozens of places mentioned by the characters throughout the movie.

All the results for the 46 films studied are synthesized in Annex 1. The presentation and discussion of these results follows the level of diversity in terms of the location of the action: From unipolar to multipolar.

Unipolar is the most simplistic spatial form of cinematographic narrative. It can take place in a given neighbourhood – for instance, *Long Life, Happiness and Prosperity* (Dir. Mina Shum, 2002) unfolds only in Vancouver's Chinatown (See Figure 11.3) – or in a city in which specific locations are not clearly identified, as in *Eastern Promises* (Dir. David Cronenberg, 2007), which unfolds in different neighbourhoods in London. Not only is the spatial extent of the

FIGURE 11.3 Example of an unipolar spatial narrative structure with only a couple of places mentioned (dashed lines): Long Life, Happiness, and Prosperity (Dir. Mina Shum, 2002).

action of these movies restricted around one location (or pole), but often very few places are mentioned in these movies. In some of them, such as *La rage de l'ange* (Dir. Dan Bigras, 2006) or *Trailer Park Boys* (Dir. Mike Clattenburg, 2006), no places are even mentioned. Although these movies cannot be defined as 'aspatial', since they are set in a recognizably referential place (Montreal in the case of *La rage de l'ange*; Dartmouth, Nova Scotia for *Trailer Park Boys*), their geographical extent is bounded by the frame of the camera and reduced to its minimum.

Some of these uninodal films are connected to the rest of the world through the evocation of places. In *Eastern Promises* (Dir. David Cronenberg, 2007), the action unfolds in London, but a few other places such as Edinburgh, Amsterdam, Paris, and Russia are mentioned. In these cases, the evocations do not follow any specific direction: they are multidirectional, which is not always the case. For instance, while the action of *Roméo et Juliette* (Dir. Yves Desgagnés, 2006) unfolds in Montreal, all the spatial references made by the characters are directed towards European places (mainly the UK and France). These places serve to anchor the narrative into its original European/British spatial context. Mentioning the places serves to broaden the geographic structure of a movie as well as its historical dimension. Overall, unipolar films offer a restricted geographic extent, not only because of the restricted location of the action, but also because of the limited number of places mentioned.

In bipolar movies, the action unfolds in two main locations. These locations can be nearby – for instance, *Borderline* (Dir. Lyne Charlebois, 2008) navigates between two neighbourhoods in Montreal – or far away as illustrated by *Shake Hands with the Devil* (Dir. Roger Spottiswoode, 2007), in which the action moves back and forth between Kigali in Rwanda and New York City (at the UN headquarters). Although the amount of time spent in each place can be even – for instance, *Ma Fille Mon Ange* (Dir. Alexis Durand-Brault, 2006) unfolds evenly between Montreal

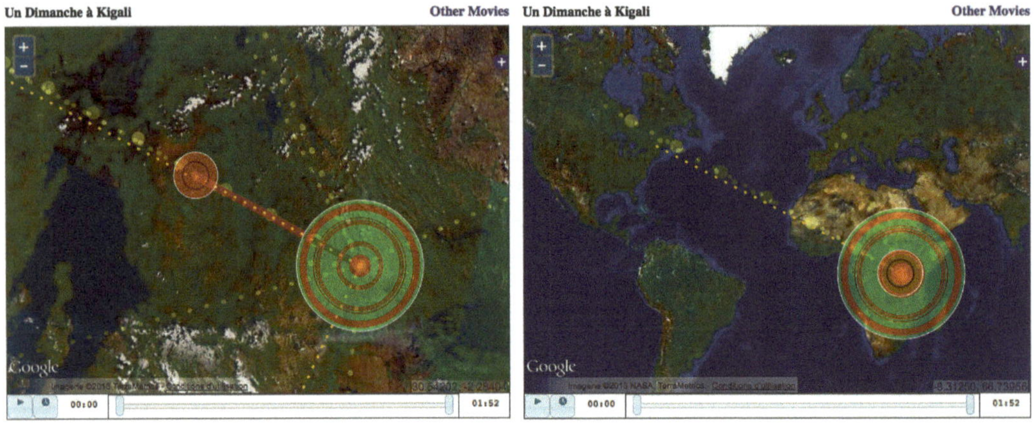

FIGURE 11.4 In *Un Dimanche à Kigali* (Dir. Robert Favreau, 2006), the action unfolds at a local scale between Kigali and Karago, Rwanda (left) but Canadian places are mentioned 15 times in total (right), expanding the geographic frame of the narrative.

and Quebec City – very often the time spent in each place is unbalanced; the action unfolds mainly in one place. In some of these movies very few places are mentioned (e.g. *C.R.A.Z.Y.*), but in general the bipolar movies evocate more places than the unipolar ones. These evocations are also mainly unidirectional and can help to expand the spatial framework of the movie beyond the places where the action unfolds. In *Un Dimanche à Kigali* (Dir. Robert Favreau, 2006), the action unfolds between Kigali and Karago, Rwanda, but Canadian places are mentioned 15 times in total (see Figure 11.4). Beyond their primary functions (e.g. making reference to the origins of the main character), places that are mentioned without being shown contribute to the production of 'nondiegetic places': places that are implicitly present throughout the narrative (via explicit mentioning) but without any visual presence on the screen. This absence/presence contributes to giving a different dimension to these places. *Un Dimanche à Kigali* makes a subtle reference to the absence of Western forces such as Canada during the Rwandan genocide.

Finally, another interesting element to take into consideration when analysing the structure of these bipolar movies is the intensity of the connection between the two places. Overall, the intensity of these connections is less important than it is in general with multipolar films, as will be discussed later. Although the action of some movies such as *Shake Hands with the Devil* moves back and forth between two places, tracing a strong connection between both poles, most bipolar movies will only move a couple of times between the two places. Bipolar films can then be characterized by an unbalanced structure between the two main locations, by a limited number of connections between these poles and by the emergence of 'invisible' implicit poles that are made tangible through their repetitive mentions throughout the movie.

Films structured around three poles can reproduce the structures characterizing the previous categories. Some are mainly unipolar – *Un été sans point ni coup sûr* (Dir. Francis Leclerc, 2008) unfolds mainly in Montreal – while others are mainly bipolar – *It's All Gone Pete Tong* (Dir. Michael Dowse, 2004) unfolds between London and Ibiza before moving to the Fiji Islands at the end. But one main way to categorize films structured around three poles is to differentiate nodal structures from triangular ones.

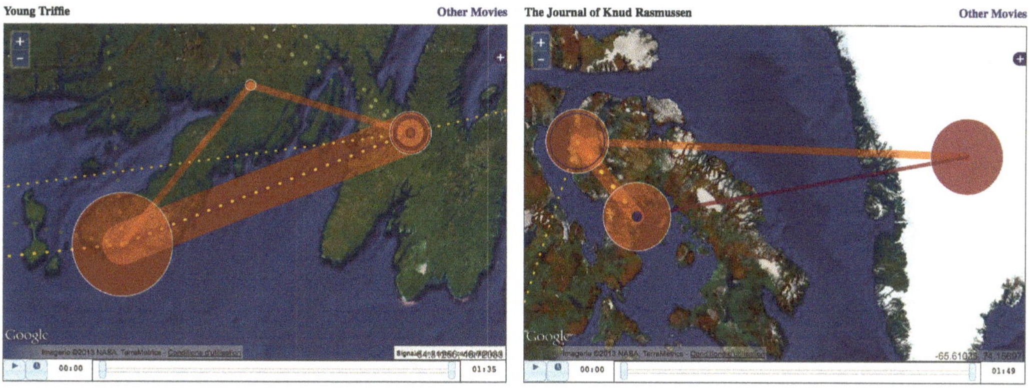

FIGURE 11.5 Two examples of triangular shapes. Young Triffie (Dir. Mary Walsh, 2006) (left) and The Journal of Knud Rasmussen (Dir. Zacharias Kunuk, Norman Cohn, 2006) (right).

In a nodal structure, the narrative is organized around a central place, which is connected to the two other places, while these two later places are not connected to each other. For instance, in *It's All Gone Pete Tong*, London is connected to Ibiza and to Fiji Island, but Ibiza and the Fiji are not connected together. In a triangular spatial structure, the action moves from one place to another, then to the third. This triangle shape appears clearly in films such as *Young Triffie* (Dir. Mary Walsh, 2006) and *The Journal of Knud Rasmussen* (Dir. Zacharias Kunuk, Norman Cohn, 2006) (see Figure 11.5). This triangulated form corresponds to the very basic shape of a circular structure. The narrative moves from place to place without having to always come back to a central node. The spatial flow of the narrative is then circular instead of being centralized around a given node/location.

Multipolar films combine unipolar, bipolar, circular, and nodal structures to generate spatial organizations that can range from mainly unipolar to extremely complex spatial structures. At the unipolar end of the spectrum, we find films like *Maurice Richard* (Dir. Charles Binamé, 2005), which unfolds mainly in Montreal with scenes in Toronto, New York, and Boston (see Figure 11.6). This is the archetypical nodal spatial structure in which the action mainly unfolds in one central node, from which it radiates to a few others. These secondary places are never connected to each other, only to the primary node.

A second recurrent structure is organized around a strong narrative connection between two nodes. For instance, while *Bon Cop Bad Cop* (Dir. Erik Canuel, 2006) unfolds in four main locations, its narrative spatial structure is organized around the recurrent movement between Toronto and Montreal (Anglophone/Francophone). A similar structure appears in *L'âge des ténèbres* (Dir. Denys Arcand, 2007) between Laval and Montreal, as well as in *Radiant City* (Dir. Gary Burns and Jim Brown, 2006) between Calgary and Evergreen. In both cases, the action unfolds at a local scale, mainly between the suburb and the city, and sometimes navigating to surrounding places. This central linear structure is then combined with secondary triangulated narrative structures, illustrating the combination of the basic shapes described earlier (See Figure 11.7). Interestingly enough, both films provide a critical perspective on the suburban way of life: one from a fictional perspective and the other from a documentary point of view.

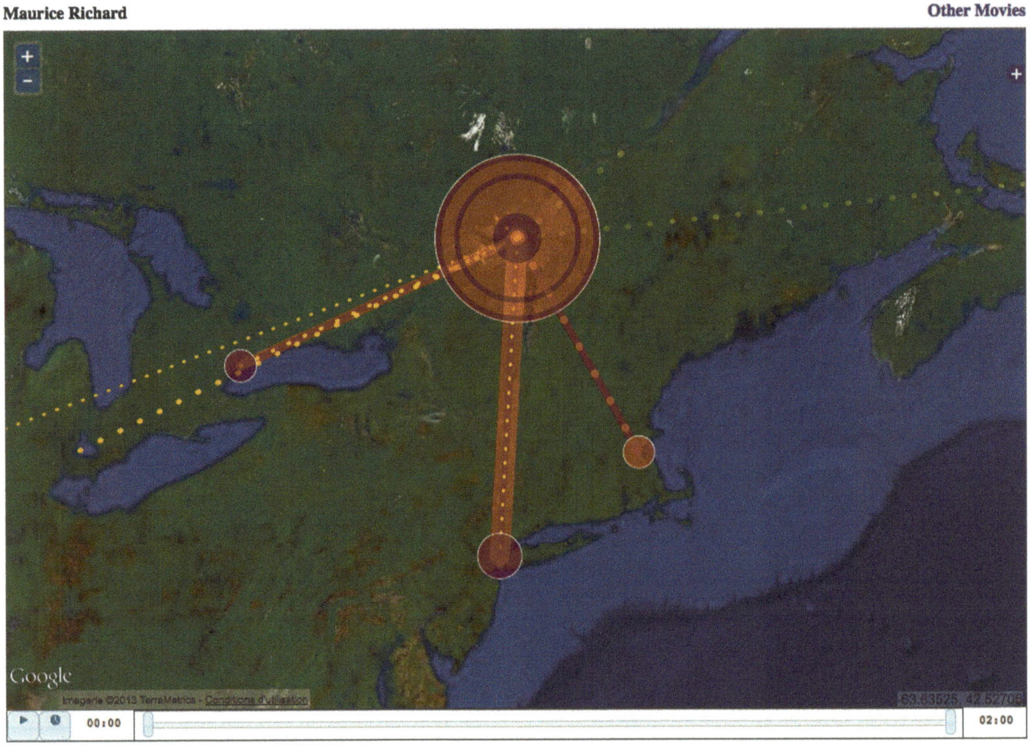

FIGURE 11.6 Example of a multipolar nodal narrative structure that unfolds in several locations but radiates from a central node: Maurice Richard (Dir. Charles Binamé, 2005).

FIGURE 11.7 Two similar complex spatial narrative structures: *L'âge des ténèbres* (Dir. Denys Arcand, 2007) on the left and *Radiant City* (Dir. Gary Burns and Jim Brown, 2006) on the right. In both cases, the action unfolds at a local scale, mainly between the suburb and the city, and sometimes navigating to surrounding places. This central linear structure is then combined with secondary triangulated narrative structures, illustrating the combination of the basic shapes described earlier.

Overall, the spatial structures of documentary films appear to be more complex than that of fiction films (see Annex 1). Most documentaries analysed are in the multipolar category with many locations involved. Although this relationship between documentaries and spatial complexity appears clearly in our results, it cannot be extended to other film genres. Documentary is the only genre that is clearly associated with a specific level of complexity in terms of spatial narrative. This might be explained by the fact that documentary films, much more than fictional ones, need to be grounded in real places. While fictions often favour undefined places to reach a broader audience, documentaries need to be factual and therefore to unfold in well defined places. The spatial complexity of documentary films might then be due to a combination of the investigative nature of this genre, combined with the necessity to explicitly mention the places where the action unfolds. Since these places are explicitly mentioned they can be mapped precisely, while places appearing in fiction films are much more often aggregated. In other words, the general spatial complexity of documentary narratives may be partially due to the fact that documentaries are often more spatially specific than fiction films. This illustrates the stronger relationship that documentaries have developed with referential places in comparison to fictions.

Beyond this relationship, it is important to emphasize that certain topics such as suburbia tend to be addressed throughout a complex spatial structure, as illustrated with *L'âge des ténèbres* and *Radiant City*. Spatial narrative structures are made of a complex set of criteria that may include the film genre and the topic of the film as discussed in this chapter, as well as other issues such as logistical and economical constraints related to film production (Lukinbeal, 2012) and the artistic vision of the filmmaker. Although the cybercartographic application presented in this chapter is far from being perfect, it can definitely contribute to a better understanding of *how films can map* and address the issue of how to map film narratives, and narratives in more general terms.

11.5 CONCLUSION

The goal of the research presented in this chapter was to initiate the development of a spatial typology of cinematographic narratives, using a cybercartographic application dedicated to mapping spatial narratives. Throughout the process of mapping and analysing the narrative structures of 46 Canadian movies, we were able to identify some recurrent narrative forms:

- Unipolar: The stories unfold only in one location/community without much mention of any places at all;
- Bipolar with strong connections between them: the story unfolds back and forth between two locations. In this case the movement between the locations is as important as the places themselves to study the spatial structure;
- Multipolar nodal narrative structure that unfolds in several locations but radiates from a central node; and
- Multipolar circular shape in which the story navigates between three or more locations without necessarily coming back to a central place (note the circular shape is less common with the increasing number of poles).

All these different structures can be combined together to provide complex shapes that become even more complex when combined with the places mentioned throughout the movie. While unipolar movies tend to mention less places, other films with more poles tend to refer to more places, which contributes to ingrain these 'invisible places' into the narrative and to bring them to the map auditorily. It is also important to reaffirm here that even if the granularity of the data collected is not as fine as we would have liked (i.e. some of the local displacements have been aggregated because their exact location within a city could not be identified precisely), the spatial arrangements described here does not seem to be scale-dependent. Some of the most complex spatial structures are set at a local scale while some more basic bipolar structures unfold between continents. In other words, the lack of very precise information associated with certain locations is not necessarily an issue in developing a typology that captures most of the basic cinematographic spatial shapes.

This first attempt in sketching a typology of spatial narratives also emphasizes some of the issues that will need to be addressed in order to deepen the analysis of these shapes. There are three major domains that will need to be further explored in order to improve our understanding of the geography and the meaning of cinematographic narrative spatial structures. First, at the ontological level, we need to refine our definition of narrative cinematographic places. Narrative places have their own units that do not fit with existing geographic and cartographic ontologies. We are currently in the process of redefining what a cinematographic spatial unit could be. Second, we need to improve the cartographic representation of narratives. There is a lot of important information that is not represented with the current cybercartographic application, such as the level of accuracy associated to the location of a narrative spatial unit. We are currently in the process of revising the application to take into account this type of key information. These two domains in which improvement is needed are the key to tackling the third issue, to deepen the analysis of the meanings of these different shapes. We believe that this type of analysis, based on an improved mapping of spatial narratives, could help us and others to better understand the relationships between places and narratives, between spatial narrative structures and narrative contents, and between narratives and our personal and collective spatial imaginations. This research is a modest attempt to start thinking about these questions.

Acknowledgements

The work presented in this paper was supported by the Social Science Research Council of Canada (SSHRC). Thank you to Jean-Pierre Fiset for coding the Cybercartographic application and to Taien Ng-Chan, Tracey P. Lauriault and D.R. Fraser Taylor for their editorial comments.

References

Aitken, S.C., Dixon, D.P., 2011. Avarice and tenderness in cinematic landscapes of the American West. In: Dear, M., Ketchum, J., Luria, S., Richardson, D. (Eds.), GeoHumanities, Routledge, London & New-York, pp. 287–295.

Bordwell, D., Staiger, J., Thompson, K., 1985. The Classical Hollywood Cinema: Film Style and Mode of Production to 1960. Columbia University Press, New York.

Brauen, G., Pyne, S., Hayes, A., Fiset, J.-P., Taylor, D.R.F., 2011. Encouraging transdisciplinary participation using an open source cybercartographic toolkit: the atlas of the Lake Huron Treaty relationship process. Geomatica 65 (1), 27–45. 10.5623/cig2011-006.

Bulson, E., 2007. Novels, Maps, Modernity: The Spatial Imagination, 1850–2000, first ed. Routledge.
Caquard, S., Taylor, D.R.F., 2009. Editorial: what is cinematic cartography? The Cartographic Journal 46 (1), 5–8.
Caquard, S., Naud, D., Gonzalès, R., 2012. Esquisses Géographiques Des Récits Cinématographiques Canadiens Contemporains. Le Géographe Canadien 56 (4), 508–530.
Caquard, S., 2013. Cartography 1: mapping narrative cartography. Progress in Human Geography 37 (1), 135–144.
Caquard S., Fiset J.P., 2013. How Can We Map Stories? A Cybercartographic Application for Narrative Cartography, The Journal of Maps (Published online: 07 Oct 2013).
Castro, T., 2011. La Pensée Cartographique Des Images. Aléas, Lyon.
Cohen, S., 2012. Urban musicscapes: mapping music-making in Liverpool. In: Mapping Cultures: Place, Practice, Performance, Palgrave Macmillan, Basingstoke, pp. 123–143.
Conley, T., 2007. Cartographic Cinema. University of Minnesota Press.
Horowitz, H., 2011. Wordmaps. In: Dear, M., Ketchum, J., Luria, S., Richardson, D. (Eds.), GeoHumanities, Routledge, London & New-York, pp. 107–111.
Jenstad, J., 2011. Using early modern maps in literary studies. Views and caveats from London. In: Dear, M., Ketchum, J., Luria, S., Richardson, D. (Eds.), GeoHumanities, Routledge, London & New-York, pp. 112–119.
Joliveau, T., 2010. The amazing secret map of Mona Lisa/L'étonnant Secret Cartographique De La Joconde. (e)space & Fiction. Available at: http://spacefiction.wordpress.com/2010/03/08/the-amazing-secret-map-of-mona-lisa-letonnant-secret-cartographique-de-la-joconde/.
Long, P., Collins, J., 2012. Mapping the soundscapes of popular music heritage. In: Mapping Cultures: Place, Practice, Performance, Palgrave Macmillan, Basingstoke, pp. 144–159.
Lukinbeal, C., 2012. "On location" filming in San Diego County from 1985–2005: how a cinematic landscape is formed through incorporative tasks and represented through mapped inscriptions. Annals of the Association of American Geographers 102 (1), 171–190.
Lukinbeal, C., 17 June 2010. Mobilizing the cartographic paradox: tracing the aspect of cartography and prospect of cinema. Digital Thematic Education 11 (2), 1–32.
Maleval, V., Picker, M., Gabaude, F., 2012. Géographie Poétique Et Cartographie Littéraire. In: Maleval, V., Picker, M., Gabaude, F. (Eds.), Presses Universitaires de Limoges, Limoges, Available at: http://www.decitre.fr/livres/geographie-poetique-et-cartographie-litteraire-9782842875787.html.
Misek, R., 2012. Mapping Rohmer: cinematic cartography in post-war Paris. In: Les Roberts Mapping Cultures: Place, Practice, Performance, Palgrave-Macmillan, Basingstoke, pp. 53–67.
Mitchell, P., 2011. 'The stratified record upon which we set our feet' the spatial turn and the multilayering of history, geography, and geology. In: Dear, M., Ketchum, J., Luria, S., Richardson, D. (Eds.), GeoHumanities, Routledge, London & New-York, pp. 71–83.
Moretti, F., 1999. Atlas of the European Novel, 1800–1900. Verso.
Naud, D., forthcoming. Les discours ruraux, urbains et périurbains du cinéma québécois, PhD Thesis, Département de géographie, Université de Montréal.
Pezzutto, D., 2011. Leonardo's Val di Chiana map in the Mona Lisa. Cartographica: The International Journal for Geographic Information and Geovisualization 46 (3), 149–159. http://doi: 10.3138/carto.46.3.149.
Piatti, B., Hurni, L., 2011. Cartographies of fictional worlds. The Cartographic Journal 48 (4), 218–223. http://doi: 10.1179/174327711X13190991350051.
Roberts, L. (Ed.), 2012. Mapping Cultures: Place, Practice, Performance, Palgrave Macmillan, Basingstoke.
Young, S.S., 2011. The earth exposed. How geographers use art and science in their exploration of the earth from space. In: Dear, M., Ketchum, J., Luria, S., Richardson, D. (Eds.), GeoHumanities, Routledge, London & New-York, pp. 183–186.

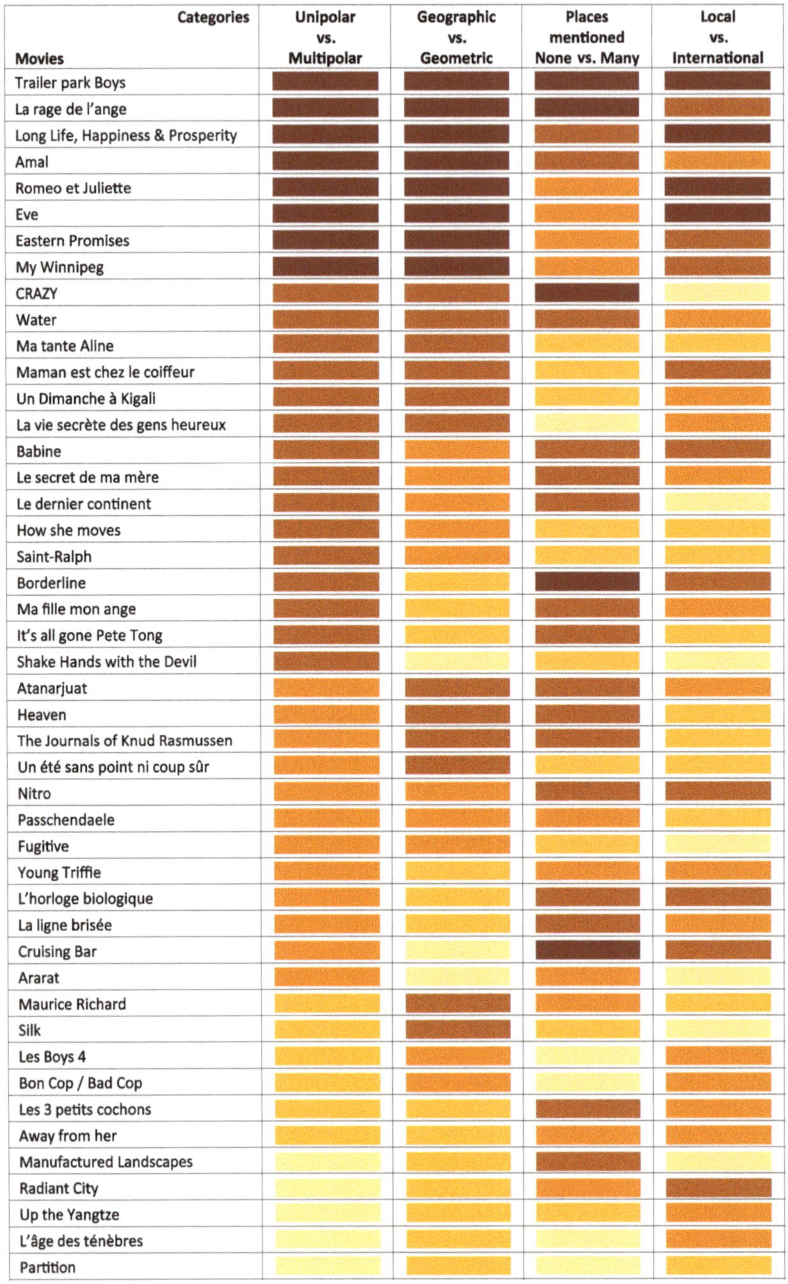

FIGURE A11.1 The different films studied in this project have been organized in this grid based on four criteria: (1) Unipolar vs. Multipolar: The locations of the action range from unipolar (one location) (dark) to multipolar (many locations) (light); (2) Geographic vs. Geometric: The unfolding of the action ranges from no connections between locations (Geographic) (dark) to many displacements between these locations (Geometric) (light); (3) Places mentioned: None vs. Many. The number of places simply mentioned throughout the movie ranges from none (dark) to many (light); and (4) Local vs. International. The action unfolds at different scales: From very close locations (e.g. within a neighborhood) (dark) to across continents (light).

CHAPTER

12

Pilot Cybercartographic Atlas of the Risk of Homelessness

Tracey P. Lauriault

National Institute for Regional and Spatial Analysis (NIRSA), National University of Ireland at Maynooth, Maynooth, Co. Kildare, Republic of Ireland; Member of the Geomatics and Cartographic Research Centre (GCRC), Carleton University, Ottawa, Ontario, Canada

OUTLINE

12.1 INTRODUCTION

The *Pilot Cybercartographic Atlas of Risk of Homelessness* (See Figure 12.1) (http://gcrc.carleton.ca/homelessness) maps risk of homelessness indicators in an interactive fashion using timelines, unique interactive visualization techniques, text, charts, and instructional videos. *Atlas* modules include the visualization of indicators across time at three scales: Canada, 23 municipalities, and three featured cities/metropolitan areas. The City of Toronto module showcases longitudinal data about the City's aging social housing stock; the City of Calgary module

Developments in the Theory and Practice of Cybercartography, Second Edition, ISSN 1363-0814
http://dx.doi.org/10.1016/B978-0-444-62713-1.00012-X

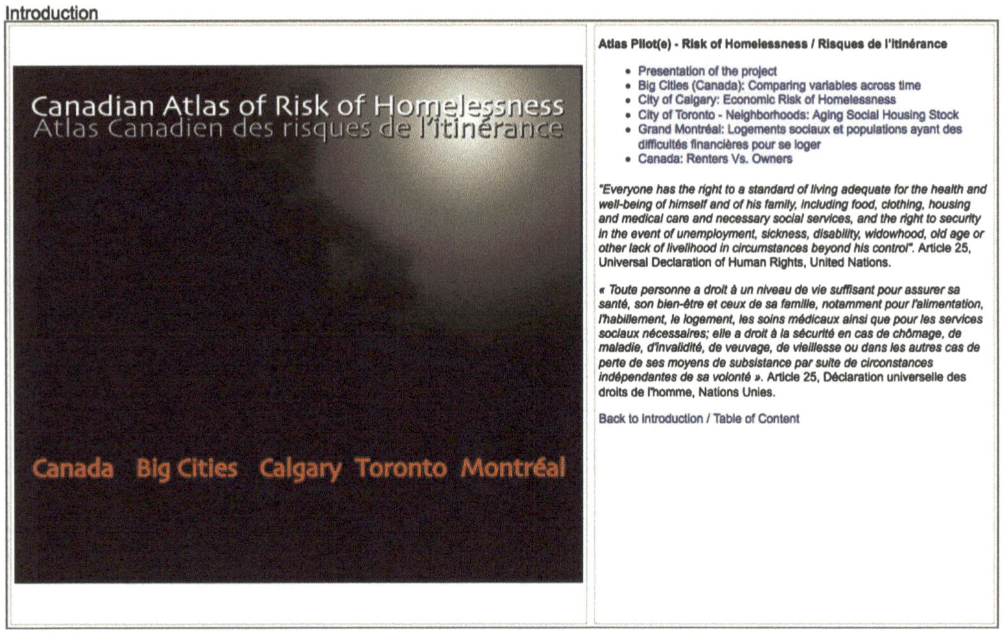

FIGURE 12.1 Home page of the Pilot Atlas of the Risk of Homelessness.

features poverty and the disproportionate spending on rent while the Communauté métropolitaine de Montréal (CMM) tells the story of social housing availability, affordability, and types of social housing programmes. The Big Cities module applies a graphomap visualization technique (Caquard and Fiset, 2008) to engagingly depict risk indicators at three time-intervals for 19 of Canada's largest cities. Alternatively, the *What is wrong with this map?* module discusses issues related to Statistics Canada data, what happens when city boundaries change, and the how easy it is for geospatial data novices to erroneously map census data in the absence of local knowledge.

The *Atlas* was intended as a pilot to demonstrate to a variety of stakeholders and decision-makers that cartographers, geographers, and social policy subject matter specialists can work together to create engaging, understandable, and useful visual artefacts to explain the structural issues of the risk homelessness. This was an unconventional approach for the social sector, as normally resources are limited, restricting advocates and decision-makers to the use of commonplace communication methods such as text based reports, newsletters, and presentations.

The intended users of the *Atlas* are the public and decision-makers. The motivation was to show that homelessness is not only about an individual's life trajectory but a collective Canadian story created by changes in the political economy. Additionally, the aim was to show that there is no uniform picture of the risk homelessness and it is for this reason that indicators were mapped at local scales. By mapping small geographies, it was possible to show considerable variation between and within some big Canadian cities. Moreover, the *Atlas* was designed to illustrate the potential of the Cybercartographic Atlas Framework to the social sector and municipalities to promote this way of conceptualizing, visualizing, accessing, and disseminating social policy indicators. More importantly, the objective was to advance a focus on prevention by featuring the impact of past policy interventions. An unintended outcome of the *Atlas* has been its appeal for educational purposes.

The *Atlas* was launched in 2009 using an early version of the Nunaliit Cybercartographic Atlas Framework (see Chapter 9). It was designed as a 'living atlas' where modules could be added should new partners join the project. Many datasets were used and sources include city or metropolitan region base maps, census, Federation of Canadian Municipalities (FCM) Quality of Life Reporting System (QoLRS) indicators and social housing provider data collected from a number of providers. This information and methodological notes are available to users in the metadata located in the right pane of the *Atlas* for each module. The *Atlas* includes timelines, interactive maps, and data visualizations. The Cybercartographic Atlas Framework has evolved since the 2009 launch. However, it consistently continues to facilitate visual story telling as opposed to simply data delivery.

12.2 THE RISK OF HOMELESSNESS IN CANADA

It is generally understood that homelessness is related to structural factors associated with becoming shelterless. Homelessness is not an issue that is simply associated with personal issues such as addiction or mental health. For example, a household that pays 50% of its gross income or more on housing is considered at risk of homelessness. This same population is also considered to be in a core need of housing. Households spending more than 30% of their gross income on housing are also at risk of becoming homeless but are not considered to be in core need. There are many others at risk factors such as: being in arrears with rent, living in precarious housing, experiencing serious health issues, family conflict, and/or domestic abuse. Economic factors include households relying on fixed incomes such as social assistance, disability support programmes, employment insurance or seniors' pensions. In the latter cases, the portion allotted for housing is far below the value of market rents. Along with the maintenance and the lack of construction of new social housing, affordable rental accommodations and the supply of Rent Geared to Income (RGI) units have not kept pace with growing need. The low vacancy rates in many cities compound the issues as this affects the availability and price of rental accommodations.

Job security is another factor. Many people only work part-time, seasonally, or juggle several part-time jobs to make ends meet while others work full-time earning minimum wages, which have not increased with inflation and rents. Low education levels are also associated with low-income jobs and/or lower income categories. Family status such as single-parent families, particularly female led, are also considered at greater risk of homelessness.

In many cities, visible minorities, the aboriginal population, or immigrants are also at risk of homelessness. In Halifax, for example, '18% of persons of Aboriginal identity, 19% of visible minorities, 22% of nonofficial language speakers and 30% of recent immigrants pay more than 50% of their income on rent. This is much higher than the 9% average for the population as a whole' (Halifax Regional Municipality, 2004: 24). Family status, visible minorities, aboriginal people, and newcomers face a number of structural barriers to entry in the job market or access to services such as a lack of affordable childcare spaces, discrimination, lack of Canadian experience, or an inability to speak an official language. The imminent homeless may also include those who may be in the process of losing their homes because they are being evicted or their rental accommodations are being converted into condominiums – a growing phenomenon in boom towns such as Calgary. This affects a large number of renter seniors as well as people being discharged from a public institution (e.g. hospitals, prisons, mental institutions, etc.) who have nowhere else to go. Some of the people on social housing registry waiting lists and those who use food banks also fall into this risk category. Finally, there is a category that falls into the zone of at-risk of homelessness, imminent homelessness and homeless and these are the invisible homeless. The invisible homeless are individuals or families who stay with family and friends for lengthy periods of time because they cannot access or maintain their own housing or are experiencing difficult circumstances involving housing instability. This is a difficult population to track.

The risk of homelessness is complex, multilayered, and multifaceted. The *Pilot Atlas of the Risk of Homelessness* does not represent all of these issues as that would require a massive data collection and standardization endeavour. The *Atlas* project does however illustrate some of the factors just discussed.

12.3 INDICATORS OF THE RISK OF HOMELESSNESS

Indicators of absolute homelessness are generally available (e.g. Number of homeless shelter users). However, comparable indicators of the risk of homelessness had to be developed. These indicators were constructed by a working group of the FCM QoLRS in 2002. A literature review was conducted to determine the types of indicators currently being used in Canada, combined with an examination of what data QoLRS municipalities were already tracking (FCM, 2002a,b, 2004a,b). The working group developed a short list of potential indicators, discussed options among all of the 24 QoLRS members, and assessed their validity as well as data availability, comparability, data access and cost (see Table 12.1). These indicators are imperfect but nonetheless do provide insight on issues related to housing insecurity such as income, affordability, family circumstance, and the supply of housing. In addition, it was fully recognized by the working group that the emphasis on structural issues negates the personal circumstances and behaviour of individuals. These indicators have not been statistically assessed, nor is any one indicator determinative. However, together, they provide a barometer of causes and conditions associated with the issue. In 2013, most of these data are routinely collected and reported by the QoLRS in the Municipal Data Collection Tool, and publicly available in the FCM online visualization tool municipaldata.ca. Many of these indicators are featured in the *Atlas*.

12.4 REASONS FOR MAPPING THE RISK OF HOMELESSNESS

When the *Atlas* was created, homelessness and at risk of homelessness data and information were mostly represented in static time contingent reports or PDF maps and documents that do not generally include maps and a few thematic online maps (Lauriault and Taylor, 2009). By 2013, not much has changed, although there are more data portals disseminating

TABLE 12.1 QoLRS Indicators of the Risk of Homelessness

Shortlist of Quality of Life Reporting System Risk of Homelessness Indicators	
Readily Available Indicators	**Difficult to Obtain**
Number of rent geared to income (RGI) units (available from most cities)	Number of people on social housing registry lists (not collected consistently or comparable between cities)
Population in core housing spending more than 50% of income on rent (Custom data order Canada Housing and Mortgage Corporation (CMHC) and StatCan)	Number soup kitchen and food bank users (data not collected consistently if at all)
Population spending more than 30% of income on rent (census)	Number of economic evictions (data not collected consistently)
Number of people living on fixed government incomes (census)	Provision of mental health care in a community (difficult to assess)
Incidence of low income (census)	
Vacancy rates (CMHC)	
Unemployment rate (census)	
Number of rental housing starts (census)	
Number of lone parent families (census)	

related data (e.g. Homelessness Hub), some open data advocates are collecting data and mapping them (e.g. Montréal Ouvert Logement Insalubre Coalition), and the FCM QoLRS is now available in an interactive online format, although without maps. Reports and data are generally very location specific and are most often found after digging through municipal and community-based organizations' databases or websites. As discussed, at risk of homelessness variables change over time, and patterns vary according to region, province, city, and within cities. Furthermore, while it is known that social housing and the risk of homelessness has a geographic dimension as seen in the distribution of affordable and social housing and shelters, this geographical variability is rarely seen in maps (Communauté métropolitaine de Montréal, 2005).

The ability to access this type of knowledge in maps provides the public with a better understanding of the issue both in their communities and in Canada as a whole. This also provides policy and decision makers the information they need to geographically target their interventions.

In addition, the transmission of structured knowledge on the topic of homelessness is difficult to communicate, as the public does not fully appreciate the nature, complexity and variability of this problem. New technologies, including multimedia cartographic techniques, such as those shown in the *Atlas*, are therefore seen as ways to improve the organization and transmission of this knowledge and to engage people to examine the issue in new ways.

Historically, maps were created to inform and also reform (e.g. John Snow Cholera maps); they are shaped by the social world and reshape it. In the case of this *Pilot Atlas of the Risk of Homelessness*, the intention was to create an atlas to inform public policy, shift perceptions on the issue of homelessness, and demonstrate a new way to collaboratively work towards achieving its prevention.

12.5 TRUSTED PARTNERSHIPS

The Geomatics and Cartographic Research Centre (GCRC) partnered with the FCM QoLRS and in particular with a fledgling working group on data visualization and with the QoLRS housing experts (FCM, 2002a,b, 2004a,b, 2008, 2009). Lauriault worked closely with these teams as a consultant from 2002 onward.

Outreach to QoLRS members was done to seek *Atlas* module creators. Two QoLRS cities and one metropolitan area came forward. GCRC researchers with city and metropolitan area officials brainstormed ideas for modules and discussed potential datasets that could help with telling those stories. Officials discussed these ideas with their colleagues (e.g. analysts, data creators, geographic information system experts), local housing authorities, and community groups. Once a general module story was developed, the acquisition of datasets and a geographic unit of significance to their specific public was chosen.

Both the City of Toronto and the City of Calgary contributed their neighbourhood boundary files and the CMM contributed its 82 municipality geography files. These types of files are only available from cities. These were shared with the GCRC on the understanding that they would only be used for the *Atlas* and would be disseminated further. These types of sharing agreements were common prior to the advent of open data initiatives and open licenses. Access to these data was always negotiated individually with specific officials with whom one has a trusted and personal relationship. It was also agreed that full data citation, the names of those who collaborated on the project, would be provided, and that any publications would acknowledge contributors. This information is captured in the metadata and data sources for each module on the right-hand pane.

Each module was iteratively designed with officials and GCRC researchers. With each iteration, changes were made to the maps, text, etc. based on suggestions, comments, or the emergence of new ideas. This is important as GCRC researchers are not experts on the subject, do not know the specificities of the local context nor how these stories might be received by stakeholders and the public locally. Local knowledge from experts to validate the maps and representations are key to the cybercartographic method. As the process unfolded content congealed and satisfied the design, aesthetic, messaging, and representation ideals of all those involved. A deep appreciation of local knowledge and expertise was an early part of the cybercartographic process and this is discussed in more detail in Chapters 18 and 19.

12.6 ATLAS MODULES

Users arrive at the *Atlas* Introduction module as seen in Figure 12.1. The text on the map links to thematic modules. The modules are also available from the text window on the right of the map and the text window provides a variety of administrative informations.

A series of tabs at the bottom can be selected to change how the data are viewed. The scroll bar on the right pane enables the reading of the accompanying text, which includes hyperlinks to information outside the *Atlas*. Maps are also interactive, with pop-up windows providing additional information about map features. A region can be selected and will remain so as the user navigates across topics in the tabs, which allows for the comparison between geographic variables or the same variable across time. The user can also zoom in and out to see the data at different scales.

12.6.1 Communauté métropolitaine de Montréal Module

The CMM Atlas module tells the story of social housing and housing affordability for lower-income renting populations in 82 municipalities. As seen in Figure 12.2, the demand for and the supply side of social, affordable and different types of housing supplied by different social providers and programmes are part of this module. This module shows the distribution of these data in both real numbers and proportionally for all CMM municipalities. The CMM contributed these data to the project and crafted the text to accompany the interactive map. The indicators mapped are RGI, the supply of social housing by type of programme, affordable rental accommodation, and households, which spend more than 50% of their income on rent. The data were collected from the census, the Société d'habitation du Québec and the Canada Housing and Mortgage Corporation (CMHC). The CMM joined the project as officials wanted to see what type of unique maps could be made with their data. The CMM services mostly French speakers and it is for this reason that this module is in French.

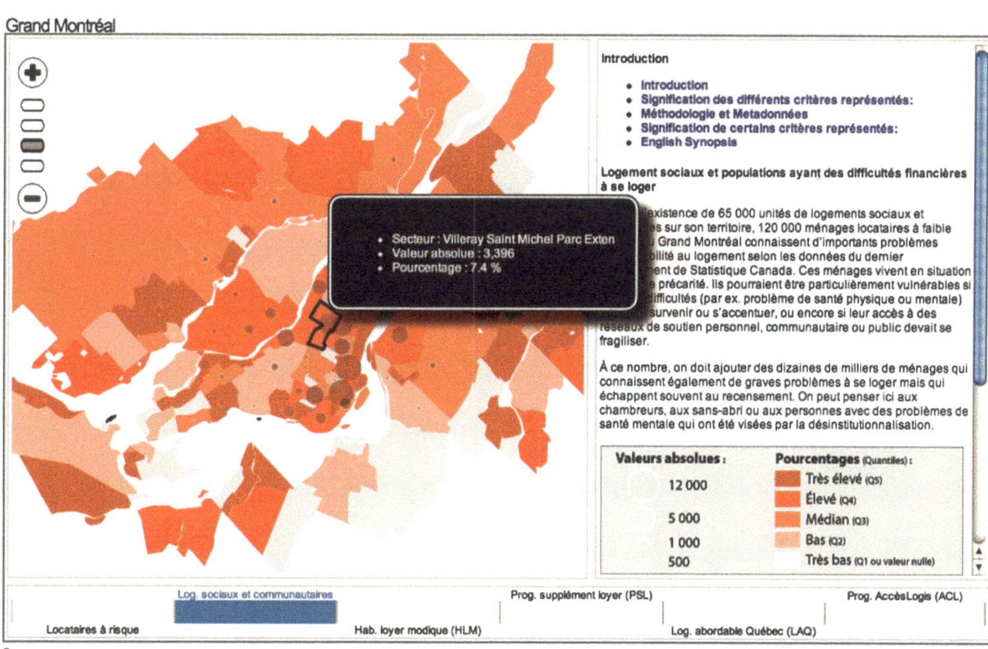

FIGURE 12.2 CMM social housing and those who spend more than 50% of their income on rent Module.

12.6.2 The City of Toronto Module

The City of Toronto module tells the story of the city's aging social housing stock. The City is faced with the daunting and expensive task of managing a large inventory of social housing that is aging, in need of repair and requires refurbishing to meet current energy efficiency and heating standards. The City also has a large number of people on its social-housing waiting list and cannot meet increasing demand for this type of shelter. Not only does the City need new social housing to be constructed, it also needs to pay for the maintenance and upkeep of its current stock. Canadian big city mayors were lobbying the Federal government at the time for the financial support to construct new social housing across Canada (FCM, 1999). A request for additional support to pay for maintenance was also part of that lobby. The City contributed its neighbourhood base files and the Toronto Community Housing Corporation (TCHC) shared their housing data. The TCHC dataset includes the location, name, number of RGI units and the year of construction for all city-managed social housing. The data were aggregated by city neighbourhood and by decade as seen in Figure 12.3. The dark green circles represent the decade selected and the lighter green represents the construction of social housing prior to that. The first iteration as seen in Figure 12.4 included the location of social housing by building and the decade of construction. This was considered to be useful but less informative than the former (Figure 12.3) and also very slow to load on the Internet at the time. Both are made available to users. The story quickly becomes obvious as the construction boom for social housing appears in the 1960s and 1970s, with much stock built earlier, and drops in the 1990s, and tapers to only one new construction in 2003. Viewers therefore get an ageing stock story, background information on the age of stock, and associated maintenance

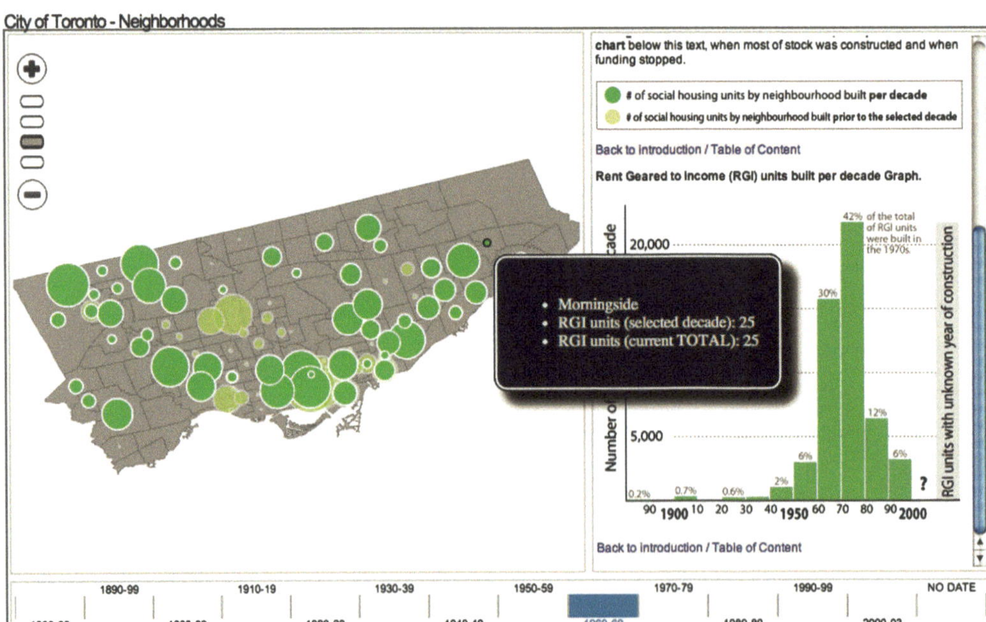

FIGURE 12.3 City of Toronto Ageing Social Housing Stock by Neighbourhood Atlas Module.

issues. The maps also reveal some of the political changes in the priorities for social housing in the City of Toronto, which were directly related to national social housing policy funding priorities that dwindled at the turn of the millennia.

12.6.3 The City of Calgary Module

The City of Calgary module tells the story of the economic risk of becoming homeless. The variables shown are Low-Income Cut Offs, which is a measure of poverty in Canada (StatCan, 2009) and also those who spend 30% or more of their income on rent (close up view Figure 12.5). Figure 12.6 keeps the same local selected but provides a full view of the City and shows the 30%+ variable. The City wanted to see if these economic risk data could be rendered in such a way as to demonstrate income variability within City of Calgary neighbourhoods. Often, it is perceived that neighbourhoods have contiguous population groupings, while the reality, as seen in these maps, is that neighbourhoods depending on their historical evolution and where they are located vary in terms of their composition demographically and economically. The GCRC Researchers interpolated social data using Dissemination Area (DA) geographies (StatCan, 2012), the smallest geographic unit publicly available from StatCan, into isolines and superimposed the Calgary neighbourhood boundaries. Unfortunately, 1991 and 1996 Census data were collected at Enumeration Area geographies (StatCan, 2013), which are not compatible with 2001 and 2006 DA geographies. Otherwise more time points would have been added. Isolines are a unique way to visualize socioeconomic data and show dynamic change at a small scale. This map

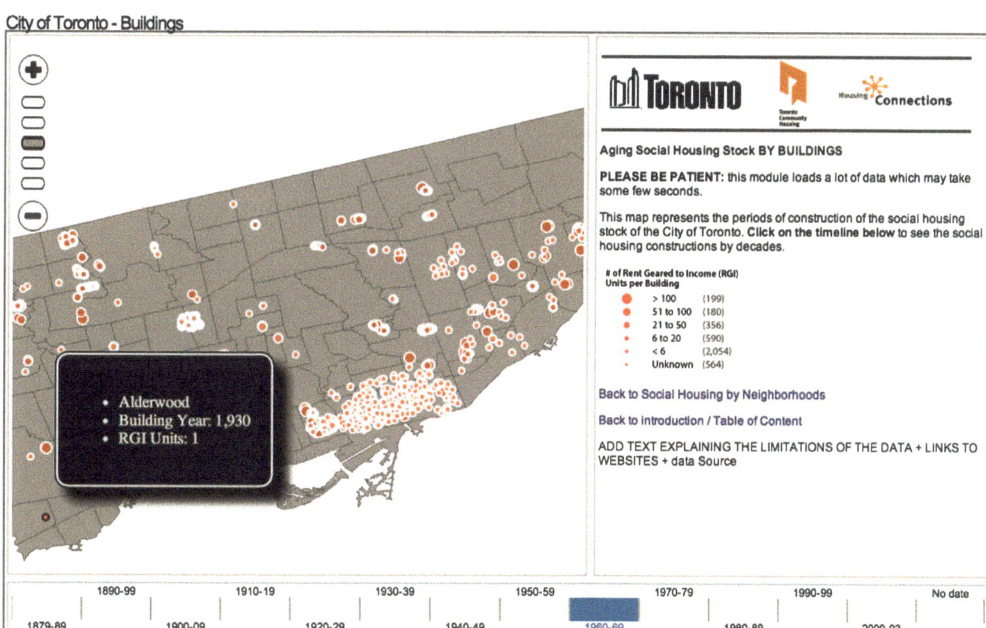

FIGURE 12.4 City of Toronto Ageing Social Housing Stock by Building Atlas Module.

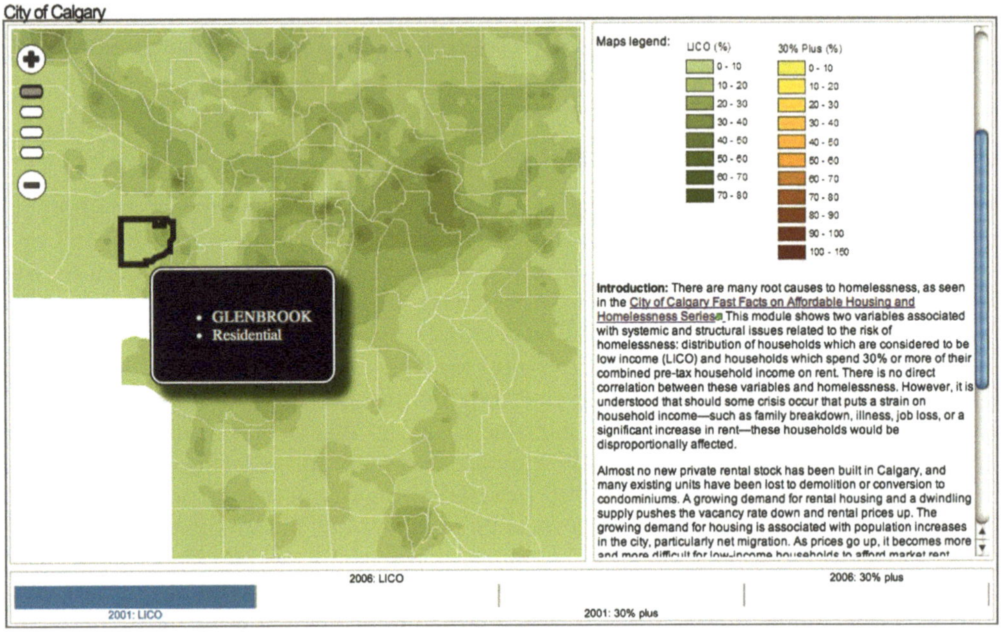

FIGURE 12.5 City of Calgary Economic Risk Atlas Module (detailed view).

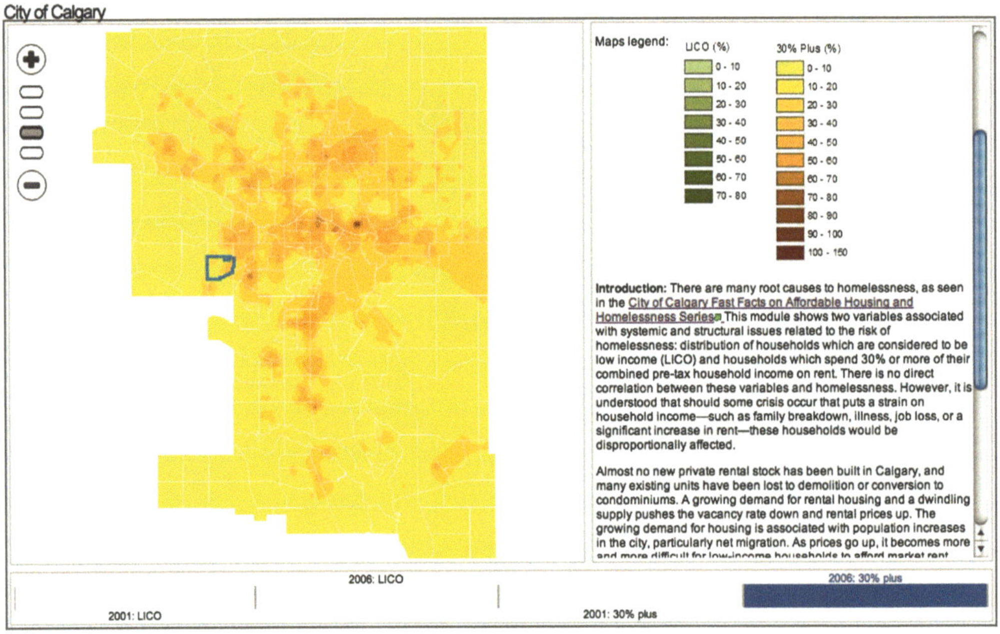

FIGURE 12.6 City of Calgary Economic Risk Atlas Module.

was most interesting since local Calgarian experts had never seen their neighbourhoods in this way, and early interactions with the maps yielded much interesting discussion. The maps were also used to determine the location of new community outreach projects.

12.6.4 Big Cities Module

The Big Cities module uses a GraphoMap visualization technique as seen in Figure 12.7(a)–(c), which depicts 11 FCM risk of homelessness indicators at three time-intervals. The GraphoMap visualization was designed by Dr Sebastien Caquard (Caquard and Fiset, 2008), and it abstracts Canada's geography into a 180° semi circle with QoLRS cities located relationally and distance wise from east to west. Circles proportionally represent the value of a particular variable in real numbers. For instance, the circles on the Vacancy Rate indicator represent the Total Number of Rented Dwellings for a particular municipality and year. The position of the point on the radial line works as follows: the closer a point is to the centre of the semicircle, the higher the risk of homelessness for that particular variable. For example, the location of the proportionally sized dots on the Lone Parent Families Spending variable represents the Percent of Lone-Parent Family Households spending 50% or more of Household Income Spent on Rent for a given year as a proportion of the total number of Lone-Parent Family Renters that same year. The higher the proportion of these Lone Parent Households being over extended, the higher the risk of homelessness for that year. A tutorial video accompanies the GraphoMap to explain how to read the visualization. Interactivity and a well-designed visualization can make accessible great complexity relatively easily when compared to data tables on multiple pages in a PDF report.

(a)

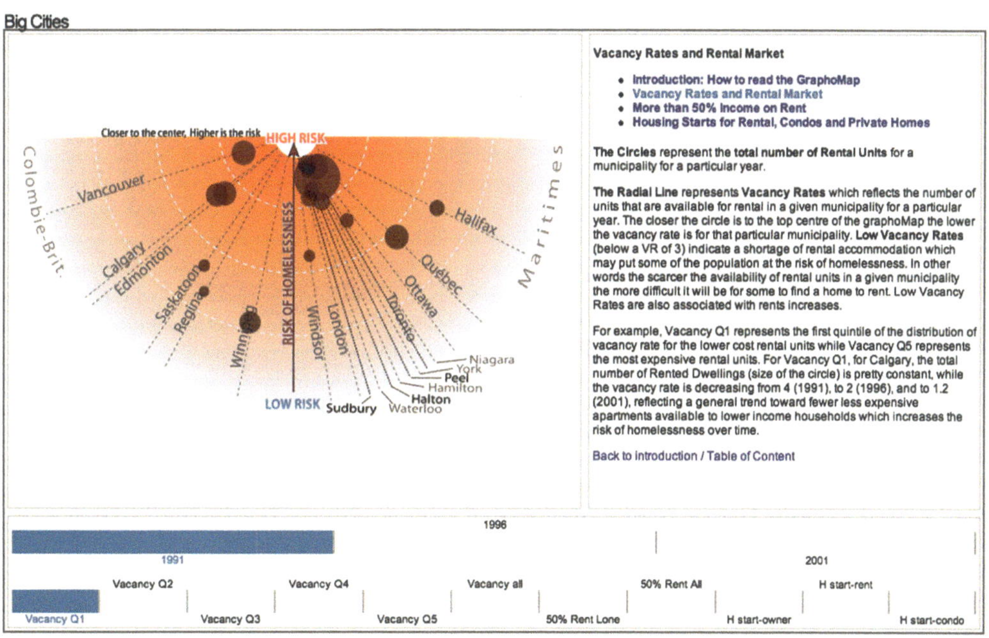

FIGURE 12.7 (a)–(c) Big Cities module view of the 1991, 1996, and 2001 Vacancy Rates for the most affordable rental accommodations.

(b)

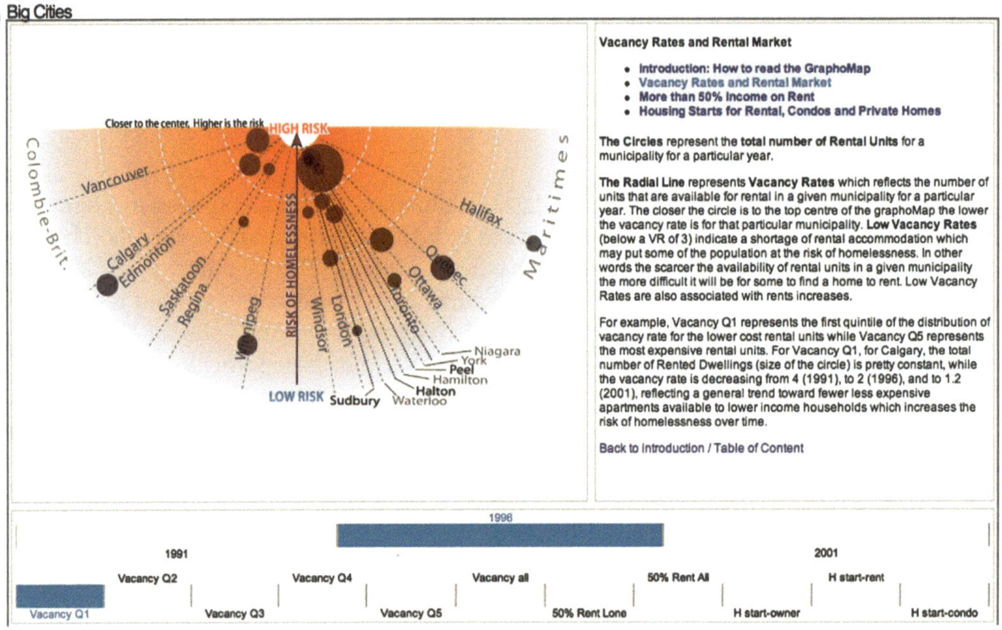

(c)

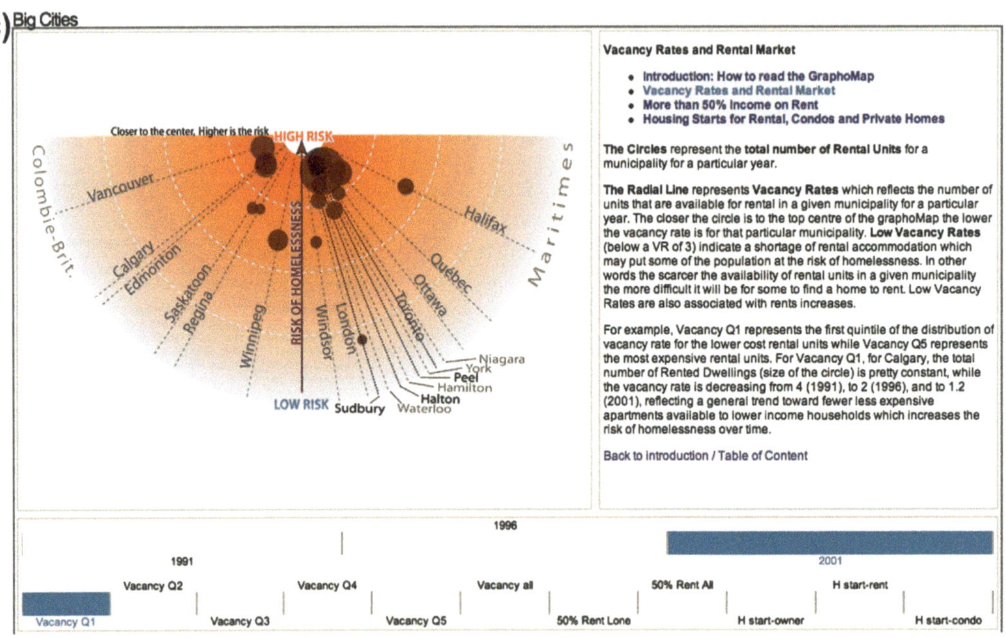

FIGURE 12.7 Cont'd

12.6.5 The Canada Module

The Canada Module as seen in Figure 12.8 was created to assess data variability of Census Subdivisions (CSDs) in Canada across time. CSDs are the Statistics Canada geographic units that represent the administrative boundaries of cities and municipalities. The points on the map represent the 24 QoLRS cities. Initially, researchers wanted to develop a rate of change map series to demonstrate change in terms of units available for sale and for rent. Maps, Data, and Government Information Centre Data Liberation Initiative census data were used to create these maps. The Rate of Change methodology developed is useful, albeit difficult to understand. Irrespective, since CSD geographies have radically changed across time due to the wave of amalgamations across the country and with the advent of Nunavut, researchers concluded the results of this experiment to be inaccurate. The cost of acquiring data adjusted to post amalgamation geographies was too prohibitive again precluding the telling of the story with a longer timeline. At the time, the joke was that it might be necessary to mortgage a house to afford to study homelessness in Canada. The map was kept in the *Atlas* as a cautionary tale for other researchers and to discuss issues of data access in Canada. Since the launch of the *Atlas*, StatCan changed its policy and now disseminates census data at no cost, although custom orders requiring geography adjustments or cross-tabulations are still cost prohibitive. Also, the census was cancelled in 2011, making it now impossible to do any up-to-date and meaningful time series.

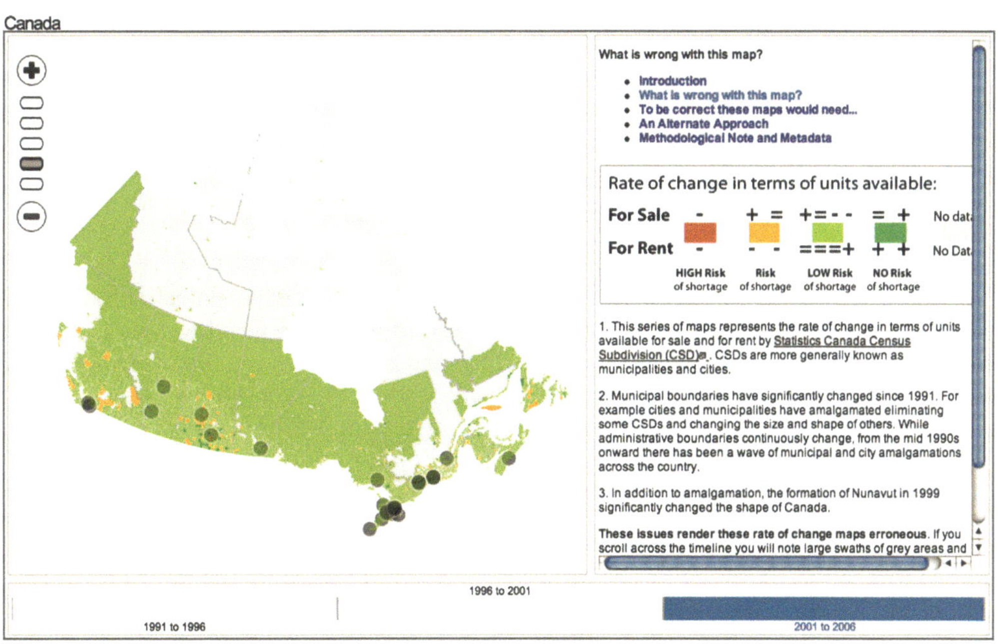

FIGURE 12.8 Canada Module.

12.7 CONCLUSION AND RECOMMENDATIONS

Atlas users have uniformly been surprised at how easy it is to see trends and interrelations between the variables depicted in the *Atlas*. Unfortunately, many of the datasets included in the *Atlas* are notoriously difficult to access, very expensive to purchase or under restrictive licensing, making it cost prohibitive and difficult to expand or reproduce. This type of work is therefore too expensive for nonprofitable and charitable organizations to carry out. Beyond data and infrastructure, skilled personnel are also expensive. Because this project was housed in a university research centre, it was possible to mobilize some highly skilled expertise, to create the GraphoMap (Figure 12.7(a)–(c)), the City of Toronto module (Figures 12.3 and 12.4), the CMM modules (Figure 12.2), and the Calgary module (Figures 12.5 and 12.6). The *Atlas* is a multidisciplinary endeavour, which is a key element of the practice and theory of Cybercartography, but this takes time, which many organizations do not have. Yet, it is this way of thinking and way of working that creates an environment conducive to the reception of critical feedback, iterative design and flexibility.

The social sector is often accused of being slow in terms of innovation and technology adoption through no fault of its own. Resources are scarce, mandates limit what is to be done and the geography of action, and skill sets are attuned to innovative social practices but not necessarily technology. In addition, there are no non governmental organizations in Canada that have a national scale mandate to do this kind of national and localized research. Some provinces and cities do some mapping work, but jurisdictional mandates delimit the scale of analysis. Questions arise as to who in Canada can do multiscalar, multisectoral, and multithematic analysis on the issue of housing and homelessness as well as poverty and social issues in general. This *Atlas* marks the beginning of a conversation about finding ways to facilitate the dissemination of information on these important issues in an engaging, accessible, understandable, and dynamic way. It is hoped that sociotechnological funding can be made available to build atlases and data infrastructures to enable this type of analysis.

The *Pilot Atlas of The Risk of Homelessness* was created to help change perceptions about the issues of homelessness and exemplifies collaboration in action (Lauriault and Taylor, 2005). In addition, it inspired other researchers, public-sector officials, and educators in the social sector to consider visualizing their research findings and data in a new way. The FCM QoLRS Team used this *Atlas* to inform public policy locally and the visualization working group began exploring ways to incorporate some of these ideas in the QoLRS, which is now an online interactive report. Data access, geography, and cost issues remain a serious impediment to this type of research outside of an academic setting. This *Atlas* however showed what is possible. The advent of open data has helped alleviate some of the costs of regular data, but as of yet these programmes have not released sought after administrative data on the topic of homelessness such as those collected nationally by the Homeless Families Information System (HIFIS) (Lauriault, 2013).

Unexpectedly, the School of Social Work at Carleton University has, since the *Atlas* was launched, asked that it be discussed in their 3rd or 4th year methods classes while a student from that programme, who participated in one of the sessions, decided to pursue her post graduate work on the examination of mapping and social issues as a result. While students from that programme are primarily trained for front line work, they found it interesting to examine social issues in this way. The *Atlas* was also extensively discussed in each of the partner cities, and the City of Calgary module was actually used to inform a public spending policy decision.

12.7.1 Recommendations

The following are recommendations that were made to Human Resources and Skills Development Canada (HRSDC) in the final report (Lauriault and Taylor, 2009):

- Support research projects that have to undertake multidisciplinary, multiscalar, multisectoral, and cross-institutional collaborations of this kind. The issue of homelessness crosses many boundaries and it is critical that those who deliver programmes and collect data can pool resources and work together.
- Encourage government agencies to consider using cybercartography and cybercartographic atlases as methods and practices for disseminating the research results they have funded (e.g. Homelessness Partnering Secretariat).
- Provide support for projects to allocate a small portion of their research towards the creation of atlas modules or to map their findings.
- Support negotiating access to national public statistics and other public datasets at no cost. Open data programmes have helped, but these types of social indicator data have not made their way into these portals.
- HRSDC should work with government statistical agencies to collect data relevant to the issues and to make these standard products available to the sector at no cost.
- Encourage greater allocation of resources towards the production of mapping service and technologies for the social sector.

Overall, the *Atlas* met its objectives, remains very popular and continues to get discussed and showcased. New modules have not been added since its launch, and this is primarily attributed to the difficulty in acquiring funding in the social sector for this type of application. Online applications are being created as part of App Contests and Hackathons in civil society open data movements across Canada. However, the creation of methodologically rigorous atlases, that are scalable and supported by a well-resourced infrastructure, have not yet emerged in those circles. Sociotechnological approaches to the resolution of issues in the political economy are not yet commonplace, but this is evolving, as seen in the creation of many of the traditional knowledge atlases that have since been developed at the GCRC.

Acknowledgements

This project was funded by the Homelessness Partnering Secretariat, Human Resources and Skills Development Canada (HRSDC). Thanks also go to those who contributed to the creation of the *Atlas*: Dr Sebastien Caquard, Christine Homuth, Amos Hayes, Jean-Pierre Fiset, Dr Glenn Brauen, Michel Frojmovic, Barbara George, Harvey Low, Sylvia Novac, Natalie Chicoine, Philippe Rivet, Michel Rochefort, Dr Sharon M. Stroick, Derrek Cook, Wendy Watkins, Joel Rivard and David Sharp.

References

Caquard, S., Fiset, J.P., 2008. Un cyber cartogramme source ouverte et interopérable pour l'analyse de données spatiotemporelles complexes et diverses. Revue des Nouvelles Technologies de l'Information (RNTI). Sélection des meilleurs articles de SAGEO 2007.

Communauté métropolitaine de Montréal, 2005. L'Accès à un logement de qualité pour chaque ménage de la Communauté de Montréal. Orientation de la Communauté métropolitaine de Montréal en matière de logement social et abordable. Rapport du ler décembre. Available at: http://www.cmm.qc.ca/fileadmin/user_upload/documents/CMM_orientations_logement_social_abordable.pdf.

Communauté métropolitaine de Montréal, 2007. Analyse de la répartition spatiale de l'offre et de la demande de logements sociaux et abordables sur le territoire de la Communauté métropolitaine de Montréal. Available at: http://www.cmm.qc.ca/fileadmin/user_upload/documents/RapportINRS_Analyse_repartition_spatiale_logement_sociaux.pdf.

FCM, 1999. Three Articles on the Announcement of Big City Mayors on Homelessness Crisis in Canada. Available at: http://action.web.ca/home/housing/resources.shtml?x=66831&AA_EX_Session=5990d138d82a48bc6124f2a3f7d46e35.

FCM, 2002a. The FCM Quality of Life Reporting System. Technical Report on Measuring Homelessness. Annex 4 Survey Feasibility Analysis. Results of a Survey on Data Availability and Indicator Preferences for the Measures of the Changing Face of Homelessness and Relative Homelessness. (Unpublished Report).

FCM, 2002b. Findings and Recommendations. The FCM Quality of Life Reporting System. Technical Report on Measuring Homelessness. Annex 4 Survey Feasibility Analysis. Results of a Survey on Data Availability and Indicator Preferences for the Measures of the Changing Face of Homelessness and Relative Homelessness.

FCM, 2004a. Incomes, Shelter and Necessities Report (http://www.fcm.ca//CMFiles/nov1720041VJU-3272008-2356.pdf) and 2008-Trends & Issues in Affordable Housing & Homelessness Report (http://www.fcm.ca//CMFiles/qol20081VVM-3272008-3162.pdf).

FCM, 2004b. QOL Issues Report #1. Incomes, Shelter and Necessities. Technical Annex. (Unpublished Report).

FCM, 2008. Trends & Issues in Affordable Housing & Homelessness.

FCM, 2009. Immigration & Diversity in Canadian Cities & Communities.

FCM, Quality of Life Reporting System. Available at: http://www.fcm.ca/home/programs/quality-of-life-reporting-system.htm.

Halifax Regional Municipality, January 2004. Homelessness in HRM: A Portrait of Streets and Shelters. Planning and Development Services.

HIFIS, 2013. Homeless Individuals and Families Information System Initiative Page. Available at: http://hifis.hrsdc.gc.ca/index-eng.shtml.

Lauriault, T.P., 2013. A Fed Open Data Ask!. Datalibre.caAvailable at: http://datalibre.ca/2013/07/02/a-fed-open-data-ask/.

Lauriault, T.P., Taylor, D.R.F., 2005. Cybercartography and the new economy: collaborative research in action. In: Taylor, D.R.F. (Ed.), Cybercartography: Theory and Practice, Elsevier, Amsterdam, pp. 181–210.

Lauriault, T.P., Taylor, D.R.F., 2009. Homelessness Knowledge Development Program, Homelessness Partnering Secretariat. Final Report on the Pilot Cybercartographic Atlas of the Risk of Homelessness. Available at: https://gcrc.carleton.ca/confluence/download/attachments/2850828/FINAL+REPORT_LauriaultTaylor_RiskofHomelessnessAtlasPilot-2.pdf?version=1.

StatCan, 2009. Definition of LICO. Available at: http://www.statcan.gc.ca/pub/75f0002m/2009002/s2-eng.htm.

StatCan, 2012. Definition of Dissemination Area. Available at: http://www.statcan.gc.ca/pub/75f0002m/2009002/s2-eng.htm.

StatCan, 2013. Definition of Enumeration Area. Available at: http://www12.statcan.ca/English/census01/products/reference/dict/geo024.htm.

CHAPTER

13

Mapping *Views from the North*: Cybercartographic Technology and Inuit Photographic Encounters

Carol Payne[1], *Amos Hayes*[2], *Sheena Ellison*[3]

[1]Art History, Carleton University, Ottawa, Ontario, Canada, [2]Geomatics and Cartographic Research Centre (GCRC), Carleton University, Ottawa, Ontario, Canada, [3]Institute for Comparative Studies in Literature, Art and Culture, Carleton University, Ottawa, Ontario, Canada

OUTLINE

13.1 INTRODUCTION

Views from the North (http://viewsfromthenorth.ca/index.html) links communities from across Nunavut and generations of Inuit through photographic images, words, and technology. A collaboration between the Inuit training program Nunavut Sivuniksavut (NS), the Library and Archives Canada (LAC) and Carleton University's Geomatics and Cartographic Research Centre (GCRC), it is lead by Carleton University Art History faculty member Carol Payne as Principal Investigator and funded under a 2009–2013 SSHRCC Northern Communities grant. Sheena Ellison has worked as a Research Assistant for the project since 2011 and Amos Hayes has overseen the development of the cybercartographic atlas. We have also been aided by, among others, Spencer Stuart, Hellin Alariaq, Suzanne Crowdis, and Clare McKenzie. Like many of the initiatives discussed in this volume, *Views from the North* employs cybercartographic technology to serve an indigenous group and to reach across the great expanses of the North.

Developments in the Theory and Practice of Cybercartography, Second Edition, ISSN 1363-0814
http://dx.doi.org/10.1016/B978-0-444-62713-1.00013-1

At the core of *Views from the North* are a series of intergenerational Inuit encounters over archival photographs. NS students work as researchers for the project conducting oral history interviews with elders from their home communities about archival photographs depicting those same communities from the collection of LAC. Across the territory of Nunavut, the project has been sparking conversations and connections since 2005. In Kugluktuk, Pangnirtung, Rankin Inlet, Gjoa Haven, Pond Inlet, Coral Harbour, and many other communities, elders and youth are gathering together over photographs to forge intergenerational bonds and pass on traditional knowledge, and those conversations are being extended through a web-based cybercartographic atlas, where elders' recollections, students' perspectives, archival photographs, and new images made by youth are being disseminated across Nunavut and beyond.

This chapter will introduce *Views from the North* as a whole, outlining its methodology, the cybercartographic website (http://www.viewsfromnorth.ca), the perspectives of participating elders and youths, and discuss the ability of cybercartographic technology to reach and involve Inuit communities. Although presenting this collaborative visual repatriation project as a whole, this chapter will specifically discuss the use of cybercartographic technology as a vehicle for reaching and involving Inuit communities. We will argue that this technology complements Inuit culture values in providing a collaborative, interactive, and interdisciplinary forum for multigenerational groups of Inuit participants to engage with Inuit history through traditional oral sharing of information prompted by archival images.

13.2 STARTING THE CONVERSATION: NS

Views from the North engages students from the Ottawa-based training program NS and has been developed in collaboration with NS faculty Murray Angus and Morley Hanson. Housed in a former bank on Rideau Street, NS is a lively school and a centre of activity among Ottawa's large Inuit community. Open areas and classrooms are filled with the sounds of drumming, hip-hop, throat singing, electronics, and concentrated discussions about culture and politics.

NS was founded in 1985. Initially, it was mandated to prepare Inuit youth to participate in the land claims negotiations that would result in the establishment of the Territory of Nunavut in 1999. Ottawa, as the seat of Canada's federal government and the largest Inuit population in the south, was a logical venue for training students to negotiate for Inuit culture. Students in the first few years of the programme were in their later twenties and many became involved in the development of Nunavut (Angus and Hanson, 2011). Today, however, NS is a more youthful environment. Most NS students are in their late teens or early twenties and come to Ottawa from across Nunavut's 26 small communities. The majority of students stay at NS for eight months, although a few return for a second year.

If NS was founded with the goal of political advocacy, today it is guided in part by a mission to redress the grave situation facing many Inuit youth now. Among Inuit youth, rates of substance abuse and suicide far exceed national and international averages. A recent study conducted by the Government of Nunavut, for example, showed that young Inuit men in their teens and twenties are roughly 10 times more likely to commit suicide than people in the rest of the country (Government of Nunavut, The Working Group for a Suicide Prevention

Strategy for Nunavut, Qaujijausimajuni Tunngaviqarniq). As Murray Angus and Morley Hanson, founding NS faculty members have described:

> ... the depth of insecurity that many youth feel about their identity, both as Inuit and as young individuals in the larger Canadian society. Having to negotiate two cultural worlds at home, they have often found themselves forsaking Inuktitut language and traditional skill development as they follow the school system for its prospect of employment and "better life" at the end. They often doubt their own ability to live up to the image of what it means to be a real Inuk, and find themselves even questioning the value of trying to do so. At the same time, with limited life experience beyond their small communities, they have little confidence in being able to measure up to the image of the successful individual, as portrayed by media and institutions in their community. Caught between these two images, and developing proficiency in neither, they are at risk of becoming marginalized, of becoming spectators in the world played out before them (Angus and Hanson, 2011, 42–44).

To aid students caught between two value systems, NS offers them a supportive educational environment that blends traditional Inuit knowledge with practical life skills and academics. Classes in Inuit history, land claims settlements, and the Inuktitut language are taught alongside those in computer skills and English as well as guidance in navigating adulthood. Reflecting Inuit cultural values, for example, the school emphasizes collaboration with students working together as a community and learning from elders. Indeed, among the music and the young adult voices, those of elders can be heard regularly at the school. Visits from leaders like Mary Simon and instruction in drumming and sewing by elders from the North are highlights of the school year. It is the school's emphasis on traditional knowledge that makes NS a distinctively Inuit institution and its pedagogical approach particularly innovative. And in turn these innovations have made NS and its students Inuit success stories.

In stressing Inuit traditions and knowledge passed down by elders, NS focuses on *Inuit Qaujimajatuqangit* (IQ), or, as it is familiarly known across the north, 'IQ' (Stevenson, 2006, 177; Graburn, 2006, 146)· The nature and importance of local and traditional knowledge are discussed in several chapters of this book. An indigenization of intelligence quotient, 'IQ' is defined broadly as Inuit 'traditional knowledge'. The scholars Frédéric Laugrand and Jarich Oosten stress that 'IQ is not something abstract and separated from the context in which it is produced. It is always related to the present and updated' (Laugrand and Oosten, 2011, 136). The Government of Nunavut has codified IQ into six types of knowledge (Government of Nunavut, Department of Human Resources):

- *Pijitsirarniq*: Concept of serving;
- *Aajiiqatigiingniq*: Consensus–Decision Making;
- *Pilimmaksarniq*: Concept of Skills and Knowledge Acquisition;
- *Qanuqtuurungnarniq*: Concept of Being Resourceful to Solve Problems;
- *Piliriqatigiingniq*: Concept of Collaborative Relationship or Working Together for a Common Purpose; and
- *Avatimik Kamattiarniq*: Concept of Environmental Stewardship.

These concepts and values are at the core of pedagogy at NS and are often conveyed through elders' examples. As former student Kevin Iksiktaaryuk of Baker Lake described his experience at NS during the 2008–2009 academic year:

> I came to appreciate ... elders. I spoke to elders and actually listened to them what they were saying, how it actually related. Before I came to NS I didn't know anything about IQ principles and learning about them and then actually putting them in to play in my head (Iksiktaaryuk, 2009).

As part of their efforts to engage youth with elders, NS faculty members Angus and Hanson began collaborating with LAC in 2001 on an undertaking known as 'Project Naming'. For this initiative, NS students originally researched some of the thousands of archival photographs depicting previously unnamed Inuit in the LAC collection. Their research comprised asking elders to identify individuals in the photographs (LAC, Project Naming). Subsequently, identifications have been largely done on the web through Project Naming's extensive website, developed by archivist Beth Greenhorn. To date, Project Naming has been successful in identifying hundreds of Inuit in its collection and, through the web, reaching out to Inuit communities across the north (Payne, 2006; Alia, 2008).

13.3 VIEWS FROM THE NORTH

Views from the North, which was initiated in 2005, is intended to complement the pedagogy at NS and to extend Project Naming's emphasis on identifications by including photo-based oral history interviews. In this way, the project reflects the primacy of oral tradition within Inuit culture while, in more practical terms, enabling a more extensive, personal and less structured flow of information to be handed down.

The methodology that guides this project is termed 'visual repatriation', the recovery and reclamation of visual or material culture made by or depicting Aboriginal peoples undertaken by the same Aboriginal groups or in collaboration with outside researchers. Ann Fienup-Riordan first coined the concept in her collaborative research conducted with Yup'ik elders (Fienup-Riordan, 1998, 1999). Today, it reflects a widespread turn towards collaborative methodologies in recent anthropological research. Though extensive visual repatriation work has been conducted on clothing design, handcrafts, utensils, and other materials, photography holds a special place within this method. Presumptions about photographic indexicality – the photograph's seeming ability to offer a direct, unmediated imprint of the past – have lent camera-made images an authority among indigenous groups and western researchers alike. Since the late nineteenth-century, for example, photography has been an important tool in anthropological fieldwork. From Edward S. Curtis to Franz Boas and Marius Barbeau, the camera has been a central means of gathering cultural data (Pinney, 2011). By the mid-twentieth century, many anthropologists turned to 'photo elicitation', the use of photographs during fieldwork interviews with informants as a means to prompt recall of the past (Collier, 1967). Visual repatriation, in effect, indigenizes photo elicitation because the indigenous group depicted uses the recollections sparked by the photograph. In important visual repatriation work conducted by Elizabeth Edwards, Roslyn Poignant, Allison K. Brown, and Laura Peers among others (Edwards, 2003; Poignant, 1996; Peers et al., 2006), the very groups depicted in anthropological photographs reclaim and renarrate these images. In all of this work, the western or southern academic researchers take on the roles of facilitators and collaborators and the archive is seen as a flexible tool that can be shaped by the indigenous viewer. As Elizabeth Edwards and Christopher Morton have observed, 'there is an increasing concern for instance to constitute photographs as contested sites of encounter and cultural exchange even within asymmetrical power relations' (Edwards and Morton, 2009, 4).

At the same time, this project draws on the important research currently being conducted into the revitalization of the Inuktitut language and oral culture. Oral tradition is vital to Inuit culture. As the anthropologists Béatrice Collignon and Michèle Therrien have explained, 'Inuit claim that *the reach of our tongue, the spread of our language determines our homeland...*' (Collignon and Therrien, 2009).

The Iqaluit Oral History Project at Nunavut Arctic College, initiated in 1994, and extensive work by Jacques-Louis Dorais, Therrien, Collignon, John Bennett, and Susan Rowley, among others, has developed vital resources about Inuit linguistics and practical curricular material (Nunavut Arctic College, 2007; Laugrand and Oosten, 2011; Dorais, 2010; Bennett and Rowley, 2004). *Views from the North* contributes the approach of photo-based oral history to the revitalization of Inuit oral tradition (Freund and Thomson, 2011).

As a visual repatriation project employing photo-based oral history, *Views from the North* facilitates Inuit communities to reclaim and reinterpret archival photographs of their home communities. The photographs used are in the collection of LAC and, for the most part, depict Inuit in what is now Nunavut during the 1950s and 1960s. This was a period of great disruption in the North with increased efforts at assimilation by the federal Canadian government and religious groups (Tester and Kulchyski, 1994). The establishment of residential schools and the movement away from a traditional nomadic life date from these years. As a result, the images – although largely made by the federal Canadian government and intended to show the implementation of its policies in a positive light – depict a harsh episode in recent Inuit history.

For the project, NS students are hired as researchers to interview elders each in his or her own home community about photographs depicting that same community. Participating students arrange to interview an Elder or groups of Elders; almost all approach a parent, grandparent, another relative or a neighbour, who in turn are also paid for their time as researchers. The students report back that the Elders are consistently delighted to be asked by their own children, grandchildren, and community members to talk and acknowledge the implicit respect this request confers. To further acknowledge their role and contributions, Elders are also paid as researchers.

These small groups of Elders and youths sit down together usually in the elder's home over a cup of tea and the photo album. The photographs break the ice between elders and youth, who often feel intimidated speaking with elders. In almost all interviews, the elders recognize family members and old friends. They are encouraged to keep photographic prints; indeed, extras are printed so that there are ample copies. If the photographs ease the conversation, they do not constrain it. The images provide starting points for conversations that reflect the rich and dramatic narrative flow of Inuit oral tradition. As noted previously, the interviews to date have emphasized community history and 'traditional' Inuit life including hunting, dog teaming, and handcrafts (Payne, 2011).

The photo-based interviews are always personalized. Elders draw in their young interviewers, pausing to relate how the youths are related or connected to the people in the images. In effect, the conversations are both about the Elders' perceptions of the past but also include the youths in a cultural continuum. Among recent interviews conducted in December 2011 to January 2012, for example, Arqnarulunnguaq Arqna Audlaluk and Pauline Akeeagok spoke with Arqna's father and aunt, who talked about their experience of being relocated to Grise Fiord decades earlier (Audlaluk, et al., 2012). Janet Evic of Pangnirtung interviewed her brother-in-law's father, Jarloo Akulujuk, who talked about whaling (Evic, 2012). Jennifer Kilabuk of Iqaluit interviewed her grandfather only days before he passed away. She spoke to Sheena Ellison about the value that interview holds for her family (Kilabuk, 2012). And in Kugluktuk, in Nunavut's far west, student Katrina Hatogina sat down with elder Bernadette Elgok, her best friend's grandmother. In the course of chatting and leafing through the photo album, Bernadette recognized herself modeling fox fur in a photograph from the 1960s. As Katrina told Sheena Ellison:

> When [Bernadette] saw herself, she was like "Boy, I was such a beautiful young lady, and I noticed that she felt good about it, after seeing herself. I told her afterwards, I'm like, "You're always pretty!" She was really smiling as I walked out (Hatogina, 2012).

The interviews are often filled with the pleasure and focused attention of discovery or rediscovery; at other times they are solemn and punctuated by silences, particularly when the traumatic events of the 1950s and 1960s come into focus. Indeed, as noted elsewhere, the students' experiences call to mind the concept of 'post-memory' as coined by Marianne Hirsch. Like the children and grandchildren of Holocaust survivors, these students have lived their lives under the shadow of the events marking the previous generation of Inuit (the establishment of residential schools and the demise of nomadic life, key among them). These were not the students' own direct experiences and yet those earlier traumas also impact their lives profoundly (Hirsch, 1997; Payne, 2011).

Sitting together over photo albums, Elders and students reframe these archival images. What had been intended as promotional images of government-led assimilation programmes or ethnographic photographs have instead become the impetus for a renewal of Inuit cultural knowledge, community history, and intergenerational bonds. In this way, through these simple encounters (a series of conversations over old photographs), cultural regeneration is fostered.

The practical methodologies reflect some of the key principles of IQ. The project provides a framework for students to work collaboratively with elders and peers. (Often students conduct their interviews together in small groups of two or three.) But others from their home communities get involved, too. Parents, neighbours, or other family members help to arrange the interviews and translate from the Inuktitut if necessary. Students often note the joy that the elders display at their desire to learn about traditional culture – an aspect of the value *Pilimmaksarniq*.

In addition to the interviews, the students are asked to take their own photographs of their communities for the atlas. Many students offer contemporary views of the same vantage points found in the LAC archival photographs, documenting the change in infrastructure and landscape over time. Others demonstrate a local pride, expressing their desire to show others how beautiful their communities and the surrounding natural resources are. Some choose to focus on the people in their communities: the elders they are interviewing or their family and friends, while still others want to depict their favourite or most memorable places so as to share their own personal experience of their communities. The students' photographs create a bridge between the LAC photographs and the principles of IQ because they provide the students with another platform to learn, share, and engage with traditional culture and traditional knowledge in a collaborative way.

Often times the same photographs are sent up North more than once with subsequent students or groups of students, giving other elders the opportunity to recount different stories in relation to each specific image. In this way, the interviews demonstrate how oral history functions in contradistinction to conventional empirical histories: individuals' responses to the past may vary and yet those responses all remain valid.

13.4 CYBERCARTOGRAPHIC ATLAS

But those conversations held across a cup of tea and a photo album in communities across Nunavut are also reverberating across the North and beyond through the *Views from the North* cybercartographic atlas developed by Carleton University's GCRC. Once the students' interviews have been completed and they have made their own photographs, these files are uploaded onto the cybercartographic atlas where the public can access them as well as the archival photographs, which prompted the initial interviews. The cybercartographic atlas will also be linked to the LAC's *Project Naming* site, thereby, reaching a wider audience and joining with information at the LAC. On the *Views from the North* atlas, visitors can browse archival photographs, read new identifications of people in the images, listen to audio recordings of the students' interviews with elders as well as students' own perceptions, and see new photographs made by students, as well as engaging in the process by contributing new information.

Like other projects discussed in this volume, the *Views from the North* atlas uses *Nunaliit*, a Cybercartographic Atlas Framework technology that has been developed under the auspices of the Social Sciences and Humanities Research Council of Canada (SSHRCC) funded Cybercartography and the Initiative on the New Economy (INE) project (Taylor and Caquard, 2006). One of the key innovations of this open-source atlas framework technology is its accessibility: not requiring extensive specialized equipment, anyone with an internet connection can access the site. It also enables information to be input in various forms (digital files with photographs or videos, text, etc.) and languages. Such flexibility and accessibility are crucial to Views from the North, which seeks to reach out to communities across Nunavut. The framework is more fully discussed in Chapter 9 of this volume.

The *Views from the North* cybercartographic atlas uses mapping both as an interface and a metaphor to link Inuit across a vast geographical span and across generations. Upon entering the site, viewers are presented with an interactive map of Nunavut. 'Interactive points', representing and placed at the cartographic location of each of the Territories 26 communities, allow the visitor to virtually enter that community's images and interviews. By organizing the larger research project and the cybercartographic atlas by specific communities, *Views from the North* is able to emphasize local histories and render the information in highly specific and personalized terms. Scrolling over the 'interactive points' reveals the name of the community and access to the sidebar where more detailed community information and the documents pertaining to that community can be found. These documents include: archival photographs; NS students' own photographs and photo essays; the students' recorded interviews with elders; transcripts of the interviews with elders; and follow-up interviews conducted by Carleton University researchers with NS students in order to enter the youths' perceptions on the record. Users can match photographs to the information provided by the elders in the interviews by descriptions, and in some cases, by reference to the LAC photograph accession numbers. The stories recounted during the interviews come to form a part of the photographic

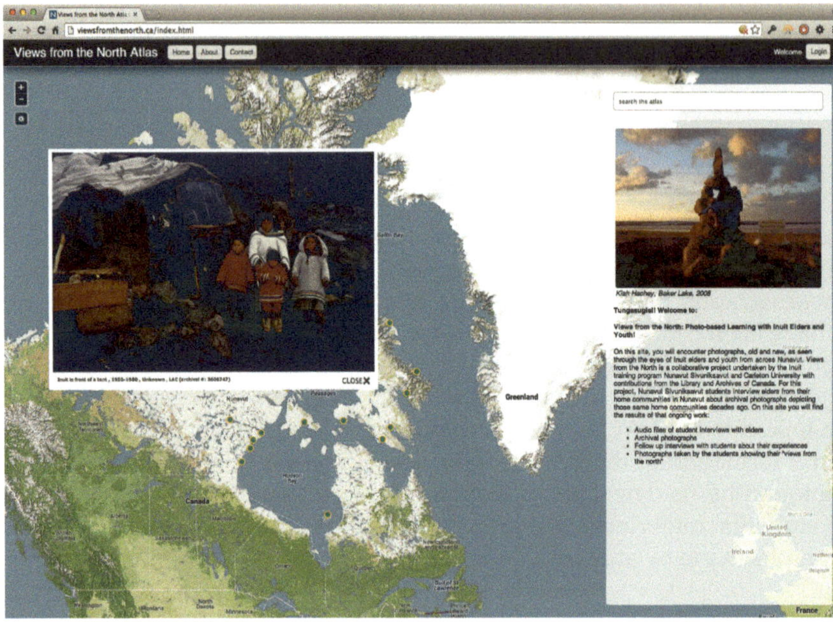

FIGURE 13.1 Views from the North cybercartographic atlas, Carleton University, Geomatics and Cartographic Research Centre http://viewsfromthenorth.ca/index.html

FIGURE 13.2 Gar Lunney (photographer), Governor General's Northern Tour. Group of Eskimos watch square dance performed by Eskimos in Northern Affairs garage for Governor General Vincent Massey, during his visit to Frobisher Bay, N.W.T. Governor General's Northern Tour. Group of Eskimos watch square dance performed by Eskimos in Northern Affairs garage for Governor General Vincent Massey, during his visit to Frobisher Bay (Iqaluit), N.W.T. March 1956. Accession no. 1971–271, MIKAN no.3198738, National Film Board of Canada Still Photography Division fonds, Library and Archives Canada.

records of the communities through their connection to *Views from the North* and LAC's *Project Naming*, allowing elders to self-narrate a visual record that has a history of invasiveness and imposition.

The web has been employed to reach the projects' key participants and audience – Inuit in the Nunavut communities depicted. Through those linked sites, cultural knowledge and youth engagement are disseminated across Nunavut and beyond, allowing this information to be modified in a dynamic model of research dissemination that reflects the flow of Inuit oral tradition. Indeed, as many Inuit acknowledge, interactive web-based technology are particularly effective tools for geographically dispersed but internet-connected communities across the north. According to celebrated Inuk filmmaker Zacharias Kunuk, for example, the web is 'the most important media tool of the twenty-first century to protect Inuit language and culture' (Kunuk, 2008). The format of the cybercartographic atlas, particularly, is a well-suited format to disseminate and engage Inuit in multilayered information, allowing it to be accessed in relation to the geographic location of the community it pertains. Student Jennifer Kilabuk, for example, noted the value of her experience, which permitted her to hear stories from her grandfather's life that had never heard before. Having continued access to the interview, she told us, made the experience even more important. Jennifer was excited to share her grandfather's stories and memories with other members of her family and community (Kilabuk, 2012). Cybercartographic technology allows the values of IQ to be disseminated widely and in a format that reaches out to technology savvy youths.

In photo-based conversations between elders and youth and through the cybercartographic site, archival photographs are being used in this project to link past and present and in doing so strengthen Inuit cultural heritage. Through its intergenerational, collaborative, interactive, and interdisciplinary format, the cybercartographic atlas complements the values of IQ.

References

Alia, V., 2008. Names and Nunavut: Culture and Identity in the Inuit Homeland. Berghahn, New York and Oxford.

Angus, M., Hanson, M., 2011. The new 'Three Rs'. Our Schools/Our Selves Special Issue: The Voice of Nunavut. Ottawa: Canadian Centre for Policy Alternatives. 20 (4) Summer: 31–51.

Audlaluk, A.A., Akeeagok of Grise Fiord, P., 2012. In conversation with Sheena Ellison, Ottawa.

Bennett, J., Rowley, S. (Eds.), 2004. Uqalurait: An Oral History of Nunavut, McGill-Queen's University Press, Montreal and Kingston.

Collier, J., 1967. Visual Anthropology: Photography as a Research Method. Rinehart & Winston, New York.

Collignon, B., Therrien, M., 2009. Orality/Oralité in the 21st Century Inuit discourse and practices. Proceedings of the 15th Inuit Studies Conference 2006, Paris, Inalco. Available at: http://www.inuitoralityconfrence.com.

Dorais, L.-J., 2010. The Language of the Inuit: Syntax, Semantics and Society in the Arctic. McGill-Queen's University Press, Montreal.

Edwards, E. (Ed.), 2003. Talking visual histories: introduction, Peers, L., Brown, A.K. (Eds.), Museums and Source Communities: A Routledge Reader, Routledge, London and New York, pp. 83–99.

Edwards, E., Morton, C., 2009. Introduction. Photography, Anthropology, and History. Ashgate, Farnham, Surrey pp. 1–25.

Evic, J., 2012. In conversation with Sheena Ellison, Ottawa.

Fienup-Riordan, A., 1998. Yup'ik elders in museums: fieldwork turned on its head. Arctic Anthropology 35 (2), 49–58.

Fienup-Riordan, A., 1999. Collaboration on display: a Yup'ik Eskimo exhibit at three national museums. American Anthropologist 101 (2), 339–358.

Freund, A., Thomson, A. (Eds.), 2011. Image and Memory: Oral History and Photography, Palgrave Press, London.

Geomatics and Cartographic Research Centre. 2013. Views from the North. Cybercartographic Atlas. Available at: http://viewsfromthenorth.ca/index.html.

Government of Nunavut, Department of Human Resources. Available at: http://www.gov.nu.ca/hr/site/beliefsystem.htm.
Government of Nunavut, The Working Group for a Suicide Prevention Strategy for Nunavut, Qaujijausimajuni Tunngaviqarniq. Available at: http://www.gov.nu.ca/suicide/SP%20WG%20discussion%20paper%20E.pdf.
Graburn, N., 2006. Culture as narrative. In: Stern, P., Stevenson, L. (Eds.), Critical Inuit Studies, pp. 139–154.
Hatogina, K., 2012. In conversation with Sheena Ellison, Ottawa.
Hirsch, M., 1997. Family Frames: Photography, Narrative and Postmemory. Harvard University Press, Cambridge.
Iksiktaaryuk, K., 2009. In conversation with Carol Payne, April 20, 2009, Ottawa.
Kilabuk, J., 2012. In conversation with Sheena Ellison, Ottawa.
Kunuk, Z., 2008. Keynote address. Inuit Studies Conference. University of Manitoba, Winnipeg, Manitoba.
Laugrand, F., Oosten, J., 2011. 'From the Classroom to the Tundra': The transfer of Inuit qaujimajatuqangit between Elders and Youth 'out of the land'. Our Schools/Our Selves Special Issue: The Voice of Nunavut. Ottawa: Canadian Centre for Policy Alternatives. 20 (4) Summer: 135–150.
Nunavut Arctic College, 2007. Surviving in Different Worlds: Transferring Inuit Traditions from Elders to Youth. Nunavut Arctic College, Iqaluit.
Payne, C., April 2006. Lessons with Leah: re-reading the photographic archive of nation in the National Film Board of Canada's Still Photography Division. Visual Studies 21 (1), 4–22.
Payne, C., 2011. 'You hear it in their voice': photographs and cultural consolidation among Inuit youths and elders. In: Freund, A., Thomson, A. (Eds.), Image and Memory: Oral History and Photography, Palgrave Press, London, pp. 97–114.
Peers, L., Brown, A.K., Members of the Kainai Nation, 2006. 'Pictures Bring Us Messages': Photographs and Histories from the Kainai Nation. University of Toronto Press, Toronto.
Pinney, C., 2011. Photography and Anthropology. Reaktion Books, London.
Poignant, R., 1996. Encounter at Nagalarramba. National Library of Australia, CanberraProject Naming, Library and Archives Canada. Available at: http://www.collectionscanada.gc.ca/inuit/index-e.html.
Stevenson, L., 2006. The ethical injunction to remember. In: Stern, P., Stevenson, L. (Eds.), Critical Inuit Studies: An Anthology of Contemporary Arctic Ethnography, University of Nebraska Press, Lincoln, pp. 168–183.
Special issue on cybercartography. 2006. In: Taylor, D.R.F., Caquard, S. (Eds.), Cartographica 41 (1).
Tester, F., Kulchyski, P., 1994. Tammarniit (Mistakes): Inuit Relocation in the Eastern Arctic, 1939–1963. University of British Columbia Press, Vancouver.

CHAPTER

14

The Creation of the Inuit *siku* (Sea Ice) Atlas

Gita J. Ljubicic[1], *Peter L. Pulsifer*[2], *Amos Hayes*[3], *D. R. Fraser Taylor*[3]

[1]Department of Geography and Environmental Studies, Carleton University, Ottawa, Ontario, Canada, [2]National Snow and Ice Data Centre (NSIDC), University of Colorado, Boulder, USA, [3]Geomatics and Cartographic Research Centre (GCRC), Carleton University, Ottawa, Ontario, Canada

OUTLINE

14.1 INTRODUCTION

For Inuit in northern communities, sea ice is a part of daily life. All but one of the communities in the territory of Nunavut, Canada, are coastal communities, with the ocean being frozen for 6–8 months of the year. Therefore, the freezing, presence, and thawing of sea ice drives seasonal patterns of land and ocean use for people and animals. The sea ice is an important travel and

Developments in the Theory and Practice of Cybercartography, Second Edition, ISSN 1363-0814
http://dx.doi.org/10.1016/B978-0-444-62713-1.00014-3

hunting platform, and thus plays a strong role in local diets, recreation, livelihoods, health, and cultural values (Krupnik et al., 2010). Through years of experience in travelling and hunting on the sea ice, Inuit elders and hunters are local sea ice experts. They have learned about the relationships between winds, weather, tides, currents, and sea ice in order to travel safely and provide sustenance for their families (Aporta, 2002; Nichols et al., 2004; Gearheard et al., 2006; Laidler and Elee, 2008; Laidler and Ikummaq, 2008; Laidler et al., 2008; Krupnik et al., 2010). Much of this detailed Inuit knowledge is not written down, but it has been passed on orally over generations and through sharing of experiences. In the Baffin Island communities of Cape Dorset, Igloolik, and Pangnirtung, Nunavut, local sea ice experts wanted to work with researchers to document their knowledge, in order to: (1) share their knowledge with Inuit youth; (2) enable more Inuit knowledge and northern content to be incorporated into the northern education system; and, (3) share their knowledge more broadly with scientists and the general public. This was the impetus for developing the *Inuit siku (sea ice) Atlas* (http://sikuatlas.ca), which was facilitated by a long-term working relationship with these three Baffin Island communities, as well as by funding and technical expertise supported through the Government of Canada International Polar Year (IPY, 2007–2008) Science Program funding for The Inuit Sea Ice Use and Occupancy Project (ISIUOP).

In this chapter, we will begin by providing some background context on how this project evolved, moving from an effort to document Inuit knowledge and uses of the sea ice towards an atlas-building activity. We will then present the key components of the *siku Atlas*, providing an overview of key content elements, integrated with the rationale for why and how they were developed. This will include highlights of how the Cybercartographic approach facilitated improvements in sharing results with more direct Inuit voices, interaction, and context. Finally, we will identify the technical innovations that emerged from addressing community interests in atlas development, which also contributed to advancement in some aspects of the overall Nunaliit framework shared between various atlases (see Chapters 9 and 17).

14.2 BACKGROUND

14.2.1 Initial Community-Based Research

Although *siku Atlas* development was facilitated by project collaborations through ISIUOP, and funding from the IPY program, this project began several years earlier as part of Ljubicic's (née Laidler) doctoral research with the Nunavut communities of Cape Dorset, Igloolik, and Pangnirtung (Laidler, 2007). As part of her PhD in Geography, she sought to understand 'Ice, Through Inuit Eyes', in other words attempting to characterize the importance, uses, and changes of sea ice around these three communities from Inuit perspectives. Starting with preliminary research visits in the fall and winter of 2003/2004, she spent nearly 9 months in these communities over the course of two years. Through interviews (including participatory mapping), the sea ice trips, and focus groups, Ljubicic sought to document Inuit knowledge of:

- sea ice freeze-thaw processes;
- Inuktitut sea ice terminology;
- the influences of winds and currents on sea ice formation and movement;
- indicators of sea ice travel safety;
- the importance of sea ice for hunting and marine mammals; and,
- local observations of sea ice changes and associated implications for Inuit communities.

The results of this collaborative, community-based sea ice research have been shared through her doctoral thesis (Laidler, 2007), journal articles (Laidler and Elee, 2008; Laidler and Ikummaq, 2008; Laidler et al., 2008, 2009; Ford et al., 2009), and book chapters (Laidler and Elee, 2006; Laidler et al., 2010). Various outreach materials were also made available in the partner communities, such as research trip reports, information pamphlets, summary results reports, posters, radio shows, and public community meetings (see www.straight-upnorth.ca for examples). However, over the course of this research, Inuit Elders continued to emphasize their interest in having more Inuit-centred and Inuktitut educational materials for Inuit youth, but also Inuit wanted to share their knowledge more broadly with scientists and the public. Therefore, options to make the research results more accessible were explored, including for cultural heritage, teaching and learning, and language preservation purposes (see Chapter 20). The IPY funding, announced in 2005, provided an important opportunity to collaborate with other researchers and communities with similar interests.

14.2.2 Atlas Collaboration Supported by ISIUOP

In 2006, several researchers working on similar sea ice projects across the Baffin Region came together to apply for IPY funding to expand and better coordinate their work through the formation of the ISIUOP. Dr Claudio Aporta (Carleton University) led this initiative, and funding was secured for ISIUOP from 2006 to 2011. There were four subprojects contributing to the broad ISIUOP objectives of:

1. documenting and mapping sea ice conditions, uses, and hazards;
2. evaluating sea ice changes (and local impacts of change) over time;
3. investigating new technologies to map and/or represent the dynamic ice environment;
4. developing community-based monitoring/ice safety programmes or material; and,
5. developing education materials for Inuit youth, researchers, and governments.

For a full overview of the ISIUOP, associated subprojects and related objectives, please refer to Aporta (2011). To develop the *siku Atlas*, participants from subprojects one and four joined forces to build on the research Ljubicic had done previously on sea ice conditions, uses, and hazards. Subproject one was led by Dr Gita Ljubicic (Carleton University), and subproject four was led by Dr Fraser Taylor (Geomatics and Cartographic Research Centre, Carleton University). The goal of this partnership was to develop an online, interactive atlas of the sea ice as an educational resource to share the knowledge, stories, maps, language, and lessons from years of working with Inuit sea ice experts in Cape Dorset, Igloolik, Pangnirtung. Some of this atlas development process has already been detailed in Pulsifer et al. (2010, 2011); however, those publications discussed the *Atlas* during its evolution. This chapter provides a description of *Atlas* development rationale and content using the final, publicly available *siku Atlas* (http://sikuatlas.ca).

14.3 BUILDING AND LAUNCHING THE *INUIT SIKU (SEA ICE) ATLAS*

Building on the work done since 2003 (Laidler, 2007), additional verification workshops were undertaken in the fall of 2007 and spring and fall of 2009 in all three communities as part of ISIUOP research activities. In addition, three MA students at Carleton University expanded upon earlier sea ice research in each community between 2008 and 2009. In this process, Ljubicic and the students worked closely with a local researcher in each community, as well as collaborated with local schools, government departments, Inuit organizations, hunters and trappers organizations, Hamlet Councils, and other local organizations. The entire *Atlas* was built on a set of research and development principles, including: (1) community partnership; (2) preservation; (3) appropriate use of qualitative methods; (4) information content over presentation; (5) interoperability; (6) iterative design and development; (7) use of open-source technology; (8) capacity building; and (9) constant dialogue (Pulsifer et al., 2010). In total, 120 interviews (including participatory mapping) were conducted, along with 20 sea ice trips, and 13 focus groups and workshops (2004–2009). From this vast resource, a selection of recorded stories in the form of maps, audio, video, pictures, and text were compiled in a way that encouraged different types of interactive learning. In order to incorporate this material into a cybercartographic atlas of Inuit sea ice knowledge and use, specialized data collection and conversion considerations had to be taken into account to support effective data modelling and publishing, including: (1) participatory mapping using hard copy 1:250,000 topographic maps; (2) digitizing map data to account for individual attribution and contextual stories; (3) Global Positioning System data conversion for mapping alongside participatory map data; and (4) transcription, translation, and data-format conversion of digital media recordings (audio, video, and photographs) (Pulsifer et al., 2010). Through this process, ways to tailor Atlas development towards an Indigenist data-management approach were explored, with specific considerations for: (1) the approach to knowledge documentation; (2) system development; (3) the representation of relationships; (4) the recognition of Indigenous languages as living processes; and, (5) data sharing practices (Pulsifer et al., 2011). Making this material available online also meant it could be shared across Nunavut (and beyond), and it could be updated more easily. Eventually, the goal is that it can also be modified and updated by Nunavut community members and Inuit experts themselves. Therefore, the *siku Atlas* became an important way to share research results, and to connect research and education initiatives (see Chapter 20).

The public launch of the *siku Atlas* was celebrated on March 17, 2011, with simultaneous community events held in Cape Dorset, Igloolik, and Pangnirtung, Nunavut, as well as at Carleton University in Ottawa, Ontario. These celebratory gatherings were organized to recognize the hard work and important contributions of Inuit elders, northern community members, community and university researchers, funders, and other supporting organizations over the four years of *Atlas* development. These celebrations brought together over 40 people in Cape Dorset, over 60 in Igloolik, over 50 in Pangnirtung, and more than 70 in Ottawa. At these events, northern community members and interested members of government, Inuit organizations, and the public, could: learn about the Atlas, try it out, and enjoy country food and other refreshments along with a number of door prizes. Much positive feedback was received, with elders being happy to finally see their contributions more broadly available. Along with the launch events, several local elder-youth workshops were hosted at each community high school. Through these launch events, and earlier informal consultations, we received a valuable feedback that helped to refine the final Atlas product. There was also room for improvement, and with new funding and/or community partnerships it may be possible to continue to improve the *Atlas* based on community, educator, and researcher recommendations. For a full description of consultation feedback, the educational potential of the *Atlas*, and efforts to integrate it as a resource for Nunavut high school curriculum, please refer to Chapter 20.

14.4 THE *INUIT SIKU (SEA ICE)* ATLAS

The research activities mentioned above were used to develop the vision for the *Atlas*, in terms of: what content was included, how it was organized, the kinds of media to interact with, an emphasis on oral Inuktitut, and as many visual representations of the sea ice as possible. The aim was to provide various multimedia and multisensory ways of engaging with, and learning about, the material in order to facilitate different types of learning, including:

- the *Atlas* framework that helps bring together and represent different knowledge systems;
- the written text is mainly English (since scientific knowledge and formal education focuses more on textual communication);
- the audio is mainly Inuktitut (since Inuit knowledge and ways of learning are based on oral culture);
- many visuals (photos, video, maps, graphics) help to maintain context (they provide examples for those who are unfamiliar with sea ice topics); and
- interactive elements are meant to promote independent exploration and learning using the *Atlas*.

This *Atlas* was created as an educational resource for northern schools, youth, and scientists, who wish to better understand Inuit perspectives on the sea ice environment. The primary emphasis is placed on local Inuit sea ice experts sharing their stories and experiences, and trying to share research results in an accessible, interactive, culturally appropriate, and educational manner. Because of limited bandwidth across Nunavut and other technical considerations, the aim was to keep the *Atlas* simple yet effective, in order to ensure the smoothest and most reliable access possible on the existing northern internet networks (see Chapter 9). Therefore, the Inuit *siku* (sea ice) Atlas (http://sikuatlas.ca) can be used to learn about:

1. Inuit knowledge of sea ice features, uses, hazards, and changes;
2. Nunavut community elders, hunters, and researchers involved in the project;
3. maps of Inuit sea ice knowledge and use;
4. Inuktitut sea ice terminology in different seasons;
5. The Floe Edge Service – satellite imagery available for communities across the Canadian North; and
6. background on the larger sea ice project and researchers involved.

This section provides more detail on each of the main components of the *Atlas* to provide a sense of content, functionality, and rationale for development. But prior to this, it is important to understand why the *siku* Atlas is organized as it is.

14.4.1 Atlas Structure

The Atlas has been organized based on contributing communities. While there are some generic sections, including the welcome page, sea ice background, and project background, the majority of content falls within community-specific pages that can be navigated from the main tabs along the top of the site. This is because of the importance of acknowledging the local geographic variations in sea ice conditions and dynamics, people who contributed to the project, and local Inuktitut dialect (Laidler and Elee, 2008; Laidler and Ikummaq, 2008; Laidler et al., 2008). In the early consultations, it was evident that community members were most interested in learning about ice conditions around their home, and from the Elders and hunters they are familiar with (and are often related to). While some are also certainly interested in geographic and cultural comparisons, creating a navigational structure based on community seemed to match best with the community-specific nature of the research, and also served to provide as much relevant local context as possible. In addition, to help with *Atlas* navigation and using the full functionality, help sections and overview videos are provided for each of the main components of the *Atlas*, to aid users learn to effectively explore relevant content. Within each community-specific *Atlas* section is an introduction to the community and the relative local importance of sea ice, a section describing all the people involved in the project, an interactive sea ice map, a discussion of sea ice changes, a section on Inuktitut terminology, and access to locally relevant Floe Edge Service information. Below, only the key components of the *siku Atlas* are described (including the Sea Ice, People, Sea ice Map, and Terminology pages). Readers are encouraged to explore the *Atlas* online in order to experience the full *Atlas* content and functionality (http://sikuatlas.ca).

14.4.2 Key Components of the siku Atlas

14.4.2.1 Sea Ice

The 'Sea Ice' section of the Atlas provides some general background on Inuit knowledge of the sea ice based on commonalities between all communities (Figure 14.1). This is a basic introduction to commonly recognized sea ice features from Inuit perspectives, to help novices gain some important context prior to getting into community-specific sea ice terminology or map interpretation. The interactive sea ice graphic is made to look like a generic icescape in an attempt to place key features in context. Users can use their mouse to hover over the text to see a picture and brief description of each ice feature, or they can learn in more detail by clicking on the text (which then leads them to a new page with more detailed interactive information). Also, as a part of this important introduction to the sea ice, *Atlas* users can access background information through the 'Learn more' topics related to:

- Inuktitut sea ice terminology – to learn about the value of understanding Inuktitut sea ice terminology, including some important Inuktitut sea ice terms that are commonly used across many Inuit communities;
- Inuit knowledge – to gain more context around the unique perspectives Inuit have of the sea ice;
- working together – to gain more perspective on the challenges and opportunities of bringing Inuit and scientists together to learn about the sea ice; and
- the Floe Edge Service – to see an overview of how this Service provides satellite imagery of sea ice in an accessible format to Inuit communities across Nunavut.

Having this general sea ice background page was important, to provide a basic foundation for those who may be unfamiliar with sea ice environments. Even for those who are knowledgeable about the sea ice from a scientific perspective, it helps provide a familiar starting point, along with opportunities to learn more about various ice conditions that are important

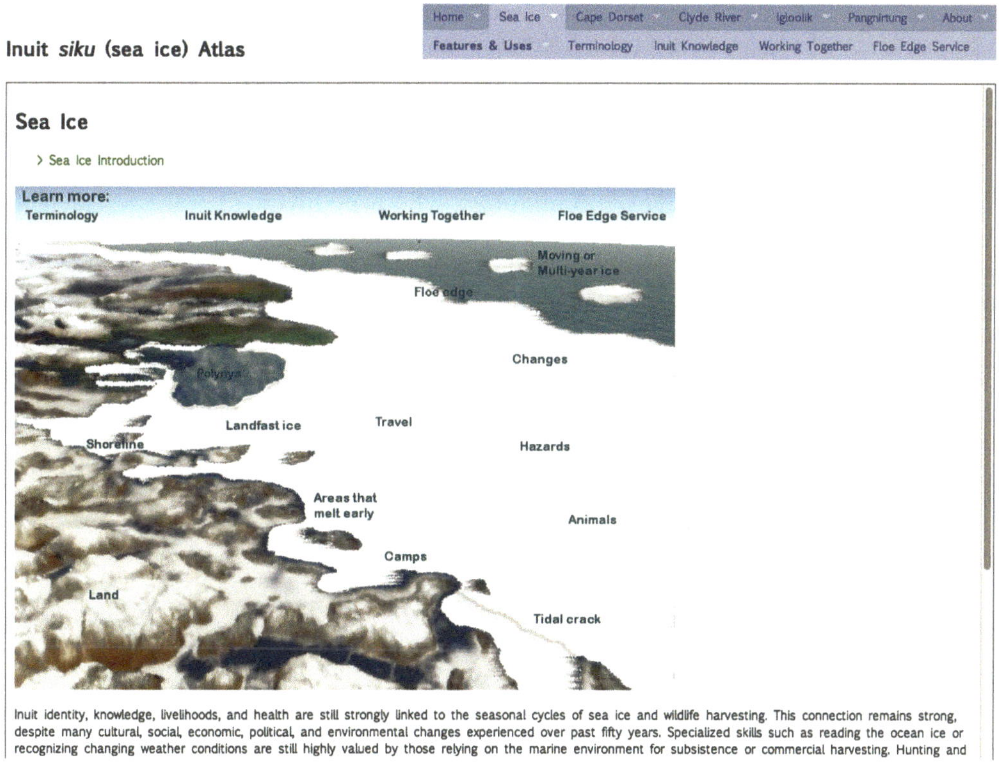

FIGURE 14.1 Screen capture of the Sea ice page in the *Inuit siku (sea ice) Atlas.*

to Inuit. Furthermore, for *Atlas* users who may be unfamiliar with the importance of Inuktitut terminology, Inuit knowledge, and the challenging legacy of scientists and Inuit working together, this background helps provide a common starting point to more effectively interpret information shared in community-specific sections of the *Atlas*. Although there are Floe Edge pages associated with each community-specific section, the background in this general section helps to understand how this programme works, and in community-specific pages users can find out more about accessing and interpreting satellite images of the sea ice around their particular community of interest.

14.4.2.2 People

Within Inuit communities, it matters who you learn from. Inuit knowledge is highly contextual, experiential, and has traditionally been communicated orally to pass on through generations (Krupnik et al., 2010). The Elders we worked with have long-term experiential knowledge, as well as understandings of teachings passed down from their Elders, and are thus well known and respected within their communities. In addition, we worked with active hunters who were engaged frequently with sea ice travel and/or hunting. These individuals are considered as Inuit sea ice experts within their respective communities, and were recommended to us for their particular specialized expertise related to travel or hunting techniques, Inuktitut sea ice terminology, understanding of sea ice formation, decay, and dynamics, or specific geographic areas (usually related to traditional seasonal family travel, camp, and hunting grounds prior to moving into settlements) (Laidler, 2007). The majority of Elders wanted to be identified in sharing their knowledge, so that other community members would understand some of the personal context of the knowledge they were sharing. Therefore, the development of the 'People' section was very important for each community section of the *Atlas*. People from each community, who were involved in the research, were identified in these sections. This was our way of recognizing local experts and researchers who made important contributions to the *Atlas* project. Initially, the page comes up as a list of names, but clicking on any name serves to expand the content to show more about that person (Figure 14.2). This emphasis on highlighting all the individual contributors to the *Atlas*, in their varying capacities, in each community was critical for several reasons, including to:

- identify all the local experts who participated in interviews, focus groups, sea ice trips, mapping sessions, and terminology workshops;
- appropriately acknowledge the local sea ice experts who shared their knowledge in the *Atlas* (i.e. they are referred to by name throughout various sections);

FIGURE 14.2 Screen capture of the Igloolik People page in the *Inuit siku (sea ice) Atlas*.

- have photos and a short biography so that *Atlas* users can learn more about their personal and cultural context (and be able to cross-reference from relevant audio/video clips); and
- provide the metadata to indicate the authenticity and reliability of the information provided.

Audio/video clips are also used throughout the community pages so that Elders and hunters can share their stories in their own voices, or demonstrate specific conditions on the sea ice. In particular, there are important compilations about the importance of the sea ice to each community as part of the opening page for each community section.

14.4.2.3 *Sea Ice Map*

An important aspect of the *siku Atlas* is the geographic representations of Inuit knowledge of the sea ice. This spatial information was digitized from the participatory mapping sessions done as part of most interviews, and some workshops. The maps are within the community-specific sections, and thus their opening extent is limited to the region around each community that was the focus of mapping the local sea ice uses (i.e. the scope of the basemaps used). However, satellite data from Google Earth is used in the *Atlas* as backdrop to the features mapped in each community, so users can zoom and pan to see any specific region of interest. The box in the top right of the map window enables users to select the sea ice features of interest, such as tidal cracks, floe edges, areas of open water year-round (polynyas), areas that melt early in the spring, travel routes (Figure 14.3), ice ridges, and camps. Once a particular sea ice layer is activated, some general background on the relevant sea ice feature is provided in the window to the right of the map, as well as the features becoming visible on the map. Users can then hover over a feature to confirm which feature they are viewing, and access a general feature description (if available). If users later click on a feature of interest, even more information is accessible to the right of the map, where additional descriptions are available along with more context about when the sea ice feature was documented (i.e. during what event, on what date, and which people were involved). However, we also felt that it was important that users were able to search relevant information of interest, either based on a particular person (e.g. a family member), or event (i.e. interview, focus group, sea ice trip), or to find associated media files that might provide added context about the feature or person. Therefore, in the navigation frame to the top right of the map, a search function is included to maximize the ability to retrieve information of interest from the database driving the map functionality. Thus, not only does this mapping section highlight spatial aspects of Inuit knowledge of the sea ice, but it also helps to search the entire scope of the *Atlas* project to seek other relevant contributions.

This mapping functionality is very powerful in recognizing the vast extent of Inuit sea ice knowledge and use, and it maximizes the ability to query information in the database to retrieve relevant contextual information. However, it also raises a number of challenges related to effectively representing peoples' knowledge and engagement with the sea ice environment. For one, there were limitations in mapping extent based on the original 1:250,000 maps used in participatory mapping sessions (Pulsifer et al., 2010). Therefore, where some travel routes or sea ice features may seem to abruptly end, this is not a reflection of the limit to a person's knowledge, but an indication of the boundary of the maps used in interviews. As such, the maps available through the *Atlas* can be a useful resource, but can by no means be

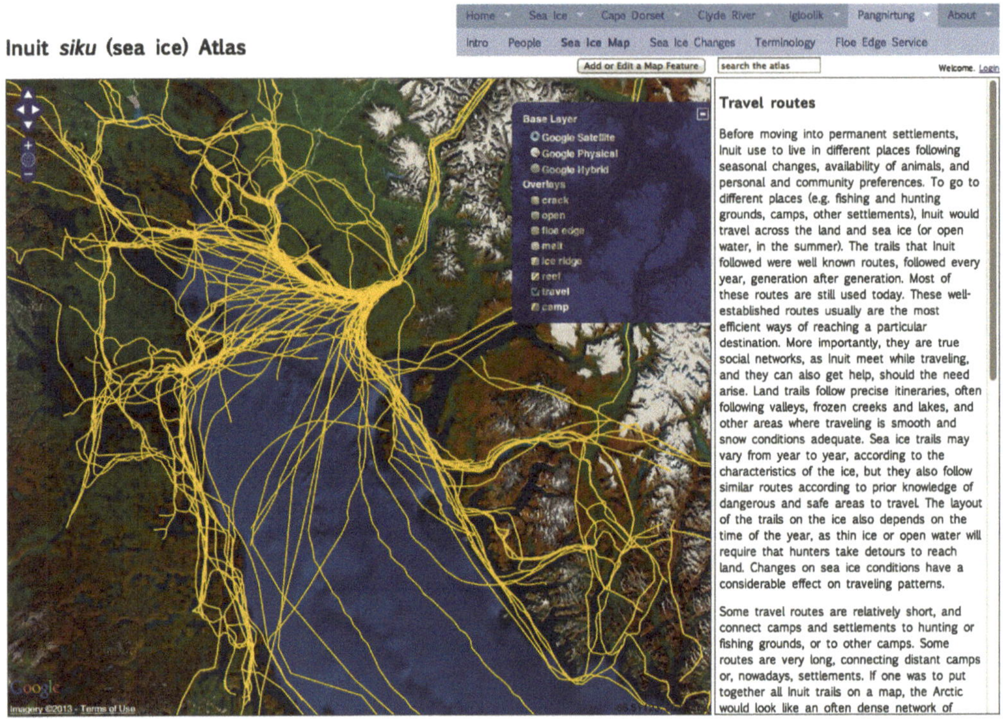

FIGURE 14.3 Screen capture of the Pangnirtung Sea ice Map page in the *Inuit siku (sea ice) Atlas* (showing Travel routes).

considered a comprehensive documentation of all the sea ice features and uses in partner communities (nor are they at sufficiently fine scale to be used for navigational or safety purposes, these are for broad reference only). Secondly, while efforts were made to incorporate temporal elements (e.g. seasonal uses/features, or time references for sea ice changes), inevitably when recording information on maps there is an effect of rendering knowledge or features static (Pulsifer et al., 2010). Further challenging this issue is that the sea ice itself is constantly moving and changing, even when landfast (i.e. attached to the shore). Sea ice is dynamic by nature, and mapping efforts could never presume to effectively represent such seasonal and interannual variation. This was emphasized over and over by hunters and Elders during the mapping process (i.e. even though they were indicating certain lines, points or polygons on a map, they were noting only *general* feature locations, highlighting that the position may change each year depending on weather, temperature, ocean, and current conditions).

14.4.2.4 *Terminology*

Much emphasis was placed on documenting Inuktitut sea ice terminology in each community, as it quickly became clear that to understand Inuit knowledge of the sea ice leads to a need to learn about Inuktitut terminology (Laidler and Elee, 2008; Laidler and Ikummaq, 2008; Laidler et al., 2008). Terminology lists were previously developed, as were posters showing seasonal connections and sea ice feature pictures, but the *Atlas* enabled an exciting opportunity to develop a more interactive way of learning about terminology. Especially important was having the

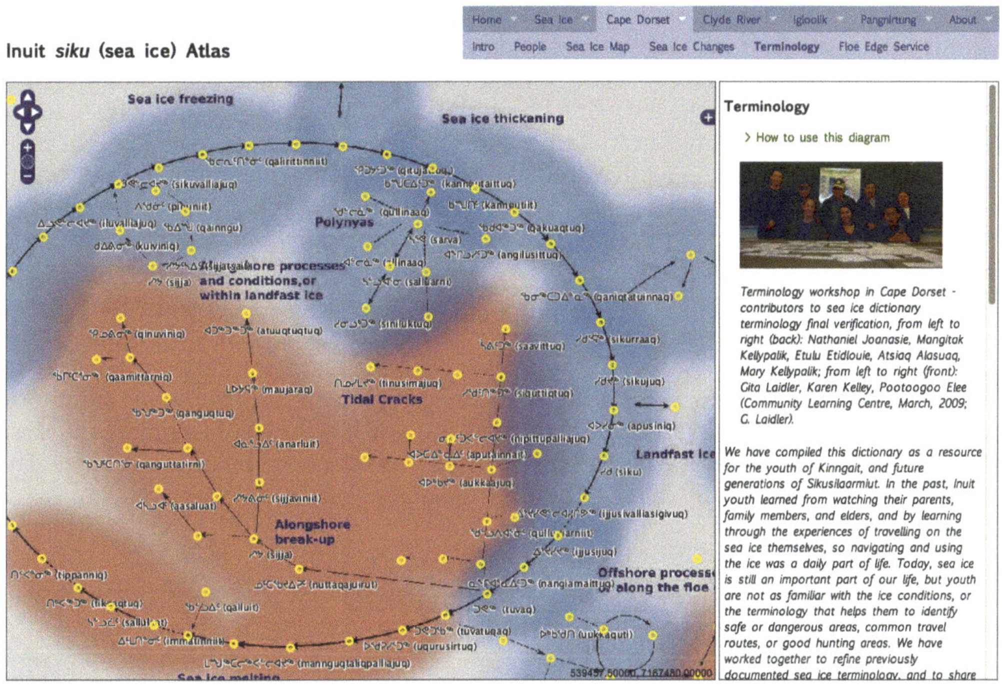

FIGURE 14.4 Screen capture of the Cape Dorset Inuktitut terminology page in the *Inuit siku (sea ice) Atlas*.

accurate pronunciation of Inuktitut terms (so users can hear and learn the terminology, instead of just reading), and to have accompanying photos wherever possible. In addition, it was critical to include an indication of seasonality in formation and melting processes. Typically, terminology documentation results in a dictionary-type compilation of important listings of Inuktitut terms. However, if organized alphabetically, there is no sense of interaction between ice conditions/dynamics (as reflected in particular terminology), nor a sense of seasonal evolution. Therefore, the goal with this 'Terminology' portion of the *Atlas* was to develop a visual representation (using a map metaphor) to show the seasonal sea ice cycle (Pulsifer et al., 2011). Because the Inuktitut terminology was at various stages of verification within each community in 2009, the Cape Dorset terminology page is most developed because it was closest to being finalized at the time of *Atlas* development.

Along with the guidance of five key Elders, work was conducted to compile all terminology references from all interviews in Cape Dorset, and to represent the seasonal processes of freezing and thawing of the sea ice around the community. To do this, a circular diagram was created, where the blue background shading represents freezing stages, and the red background shading represents melting (Figure 14.4). Furthermore, general sea ice freeze/thaw stages are used to label key areas around the diagram. However, the diagram itself is developed to highlight individual Inuktitut sea ice terms in relation to one another, within the seasonal ice cycle. Users can zoom in to see detailed terms, with more labels appearing the closer the zoom (so as not to overcrowd the diagram when zoomed out to the full extent). Users can also pan around the diagram, so that they can follow the links when zoomed in. Furthermore, by hovering the cursor over individual Inuktitut terms users hear the Inuktitut pronunciation of the word, and by clicking on the term it will open a picture of the ice feature/process, full description (Inuktitut audio and text, as well as a written English translation), and associated World Meteorological Organization (WMO, the standard for operational and scientific sea ice monitoring) terminology, where relevant. This link to WMO terminology was important to identify areas of commonality between Inuktitut and scientific terms, as well as to highlight the many cases where there are no equivalent WMO terms, and thus the importance of learning from Inuit about specific localized ice conditions/dynamics. It was an interesting challenge to come up with the Inuktitut definitions of terminology, since, for native Inuktitut speakers, the word itself is the definition (if the various components making up the term are understood). However, for Inuit youth, or others trying to learn Inuktitut, they may not know the full meaning of root words or additional descriptive elements added to the roots, and thus it was important to have an alternate Inuktitut description to explain the sea ice condition/dynamic.

While each of these sections describing key *Atlas* components help to highlight important content and rationale for *Atlas* functionality, there are also a number of technical and design strategies and innovations that emerged through the process of creating the *siku Atlas*. These are discussed below, especially in relation to the data model and user interface.

14.5 TECHNICAL AND DESIGN STRATEGIES AND INNOVATIONS

14.5.1 Background

A primary attribute of the concept of Cybercartography is the use of maps and various forms of media to supportive a narrative, the telling of a story (see Chapter 1). The *siku Atlas* tells the story of the sea ice from the perspective of Inuit experts. The development of the *Atlas* is part of another story told through a number of chapters in this book: the evolution of the Nunaliit atlas development framework. Nunaliit provides a story-telling platform through its features, while each atlas developed drives a change that continues the story of Nunaliit's development. Using the narrative method, data elements can be seen as representations of story elements including, characters, events, settings, scenes, plot lines, etc. (see Chapter 9).

At a high level, the goals of the *Atlas* guided the technological development. The intention was to establish an information system that could represent Inuit knowledge and stories of the sea ice with as much contextual information as possible. The goal was to provide the ability to speak about general themes across communities without losing community-specific attributes of the knowledge related to unique land, sea and icescapes, local terminology, and individual expertise.

The primary innovations in Cybercartographic story-telling through technology and design strategies brought about through the *siku Atlas* include: (1) the establishment of a database model focused on flexible representation of relationships between data elements, and related improvements in data management functions; and, (2) an expanded user interface.

14.5.2 Data Model

The development of the *siku Atlas* moved towards seeing data elements as individual representation objects that can be related in many different ways depending on the story being told (i.e. the context). While mainstream relational database management systems (RDBMS) support the modelling of relations, these relations are typically established at the outset of an information system project with the result being a 'schema' upon which an application is built. These schemas encode knowledge of the type of data being stored, related data constraints (e.g. numerical ranges allowed), and relationships between data elements. If our knowledge remains constant, or a finite set of stories are being told, then this schema-based approach works well. However, in the domains of sea ice, the Arctic environment, and Inuit culture where knowledge is dynamic and many different stories need to be told, an RDBMS schema-based approach can be limiting. Major changes to the schema can be difficult, time-consuming, expensive, and can result in the addition of bugs and system instability.

At the outset of development, the very complex nature of the *siku Atlas* content pushed the boundaries of the RDBMS approach then being used within Nunaliit. The *Atlas* required representing relationships between and among many geographic features, people, sea ice concepts and terminology, events (e.g. interviews, sea ice trips, workshops, mapping sessions), roles (e.g. sea ice expert, interviewer, interpreter), and other elements. Early in the development, a schema would be established only to realize that a new story and related elements developed for the *Atlas* required schema modifications. A more flexible, adaptable approach was needed. The result was the development of the 'sparse relations table' (henceforth 'the Table'), led by co-author Hayes.

In this context, the term 'sparse' refers to the information technology usage meaning that an entity can expand as needed. The term is often used in reference to data volumes, which can expand as additional storage space is required. Here, the term is used to describe a table that represents relations across the *Atlas*. Relationships are modeled in the Table, which can be expanded without modifying the structure of other existing tables (e.g. sea ice features, people). The application software focuses on accessing the Table providing a 'single point of contact' and reducing effort when change is required. Figure 14.5 presents a simple example of the Table structure. A detailed discussion of the data model developments resulting from the *siku Atlas* can be found in Pulsifer et al. (2011).

The sparse relations model approach allowed *Atlas* authors to connect many different story elements in a variety of contexts. For example, as previously discussed within the sea ice mapping section (and drawn upon in various representations throughout *Atlas* sections), people can be represented in relation to their community, the interviews that they participated in, the geographic features attributed to them, or the role(s) that they played in the project such as a sea ice expert, interpreter, or both. At the same time, events can be discussed with respect to when they occurred, who participated, and where they were held. All of these relationships are captured in the Table. In this way, no single concept (i.e. database table) is primary in the model, providing the flexibility needed to deal with the complex subject matter. The stories told dictate the relationships established between data elements, rather than a data schema dictating the stories that can be told.

The Table also provides significant contextual information that assists in data documentation and preservation activities. As a result of queries applied through the Table, it is known who contributed data, where it was collected, when, and how. This information can be used to develop metadata and archival packages. Moreover, the information speaks to the authenticity of the data, which in turn relates to data authority. In effect, a form of 'living metadata' is developed, where knowledge and experience are seen through maps and media and trust relationships are documented, for example the relationships established between

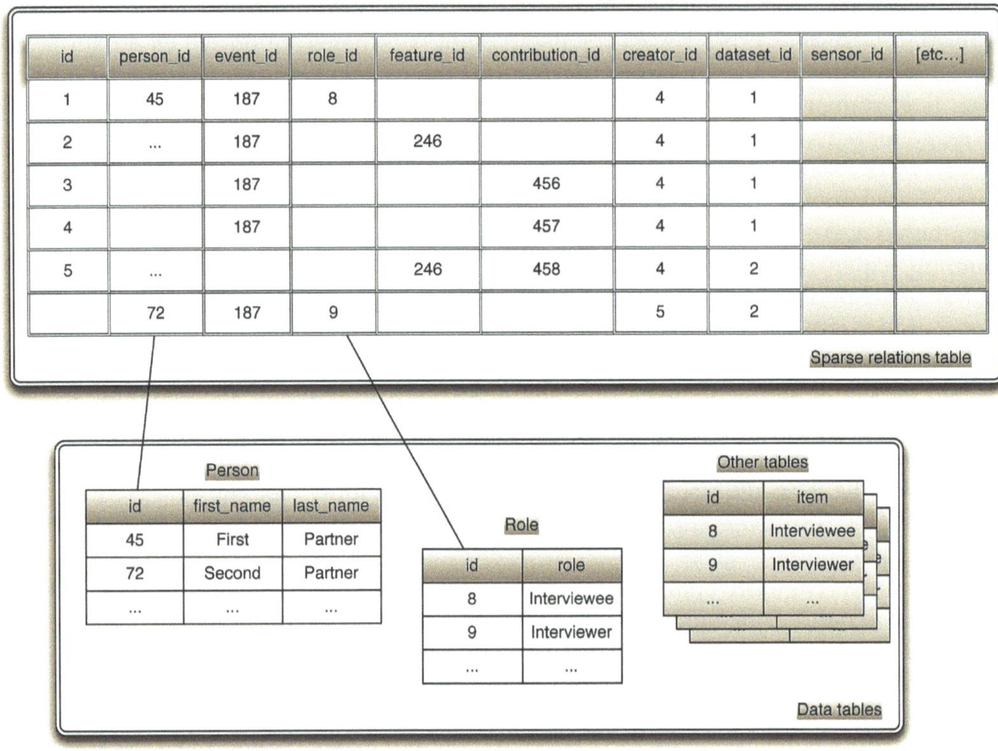

FIGURE 14.5 Many types of relationships can be established between two or more tables using the sparse relations table developed for the *Inuit siku (sea ice) Atlas. Source: Pulsifer et al. (2011: 117).*

researchers from outside the community and local experts. Thus, the *siku Atlas* led to new forms of representation and data management (Pulsifer et al., 2011).

Although the Table approach was effective for the *Atlas*, it was still not ideal. As stories and relationships increased in complexity, the Table and related software required significant development and maintenance effort. Subsequently, the schema-less document-oriented approach described by Hayes and others in Chapter 9 was adopted to help tell new stories.

14.5.2.1 *User Interface*

The *siku Atlas* database and the Table are the foundation of a rich user experience. A number of the atlases described in this book focus on a particular topic or type of feature (e.g. place names) and are well served by a lean user interface. The *siku Atlas* required much more in the way of multifaceted access to data. Thus a broad and deep interface was developed as demonstrated in the previous sections of this chapter. To develop and maintain such an interface in an efficient manner required the use of an underlying database. Many of the relationships presented through the *Atlas* interface are contained in the Table, and the content experienced is stored in the database. For example, the People section previously discussed connects a person's biography to media representing that person, while the various Sea ice

Map sections present people in relation to geographic features, all supported by the previously described data model.

In addition to browsing previously modeled relations or stories, the search and filtering features allow a user to create new stories based on their own interests and experience, perhaps a young person wanting to know more from an older relative. Another case is the researcher interested in observations across communities. The database allows such cross-community access to information, as do the maps. While a map may start out focused on a particular community, space is represented as continuous and can be 'panned' as the user wishes. This fits with Inuit cultural and usage concepts where there are some recognized bounded territories, but generally the landscape is seen as continuous (and extending across the sea ice), with overlapping and negotiated usage patterns. In the *Atlas*, many communities share the same map as they do the land or sea.

The ability to store not only relationships but also textual and media content in the same database provides a foundation for new ways of representing information. The 'spoken map' capabilities developed initially through the *Kitikmeot Place Name Atlas*, described in Chapter 15, provided a starting point for the development of the Sea ice Terminology map previously described. The *siku Atlas* brought the previously used audio-enhanced mapping to representation of conceptual space that helps users to better understand sea ice dynamics, concepts and related terminology.

The development of the *siku Atlas* implemented and inspired a number of technical innovations that have contributed to a growing interest in establishing new and appropriate ways of representing Inuit knowledge specifically, and Indigenous knowledge in general (Christie, 2004; Dyson and Leggett, 2006; van der Velden, 2010; Pulsifer et al., 2012).

14.6 LESSONS LEARNED

Overall, efforts to develop the *Inuit siku (sea ice) Atlas* were a valuable learning experience. Through this, a number of important lessons were learned, and are discussed here as a means to conclude this chapter.

First is the importance of maintaining an emphasis on relationality, from a few different angles. Relationality, as in relationships between people, is at the foundation of research within an Indigenous research paradigm (Louis, 2007; Wilson, 2008; Pulsifer et al., 2010; Chapter 17 in this volume). Therefore, the relationships between researchers, partner communities, and the key local sea ice experts are of paramount importance in order to facilitate the primary research, which provides the content to be shared within the *Atlas* framework. Without strong, mutually beneficial, and respectful working relationships built on trust and accountability, community members would not be comfortable sharing their knowledge in this online public forum. We have prioritized relationships throughout the *Atlas*-creation process, which at times means taking longer to develop functional prototypes or update certain components, but in the end we hope the result is much more representative and reflective of community goals in sharing their knowledge. In addition, great efforts to ensure the preservation of relationality (i.e. between local experts, the knowledge they shared, and the geographic, cultural, language, or interpersonal context within which that knowledge was imparted) were made through the development of a more flexible data model that can account for such relations.

As such, the relational emphasis is not only between people, but also in the entire approach taken to data management, which supports the *siku Atlas* (Pulsifer et al., 2011). This is certainly challenging when attempting to comply with open data standards and representing complex qualitative data relationships, but worthwhile in order to suit the interests and priorities expressed by the communities.

Second, and intertwined with relationality, is the crucial element of engaging in ongoing community consultations and verification. This was a highly iterative process within the project, in order to not only gain and maintain support for *Atlas* development, but also to ensure that the content was accurately representing the original intent and context of knowledge shared as best as ability permitted. The challenge involved was the issue of gaining feedback without a prototype version of the *Atlas*. Early rounds of verification focused on the results of earlier community-based collaborative research, and then as different elements of the *Atlas* were developed there were several rounds of discussions and improvements made as a result. It became quite clear that without an actual example of how the *Atlas* would look and function, it was difficult for community members to provide feedback. However, with each subsequent version more specific comments were made, which in turn made it easier to direct efforts at improvement. Ultimately, a broad community consultation was initiated in the winter of 2011 prior to the official *Atlas* launch, which provided invaluable feedback (Chapter 20).

Third, both aspects of relationality and community consultations highlight the incredible time commitments involved in developing an *Atlas* in a collaborative fashion. These are crucial to create something that is representative and useful, but it is just as important to recognize that this is a long-term process and to plan accordingly. When considering the development of an atlas project, it is recommended that the atlas not simply be seen as an end product. Indeed, it is the process of building research relationships, conducting the research (which is the foundation of the *Atlas*), conceptualizing the *Atlas*, and actually building the database structure and interactive online interface, which is the most intensive and time-consuming (as well as rewarding). However, without an emphasis on process (especially iterative processes of engagement to facilitate improvements and meet community needs), the atlas is less likely to: (1) contain the most relevant information; (2) be available in an accessible format; (3) be useful for the communities involved; and/or (4) be supported or maintained in an ongoing fashion.

Fourth, although already contained in the description above, it is worth specific mention that the efforts to conceptualize atlas structure, function, and content are far greater than the actual *Atlas'* technical development and implementation. Many hours were spent developing conceptual diagrams for the kinds of information people wanted to share, using various interfaces and structures, and these continued to change as the technology improved and as community feedback was received. To effectively tell a story through an atlas, much of the thought goes into framework design, which in turn influences the database structure, all of which has to be tailored to the particular goals for the *Atlas* in an iterative manner. Perhaps this becomes quite evident as readers browse the various atlas examples throughout this book, but we wish to make explicit mention here for those who may be considering developing their own atlases – that the majority of the work is up front, before starting to see anything available online (and even what is shared online is only representing a fraction of the overall content and effort from within a particular project).

Fifth, considerations for technical expertise are imperative when considering developing an atlas. Although the Nunaliit framework has been developed and improved over the years to enable user-contributed content (to support atlases as living resources), we relied heavily on technical staff with the specialized expertise to support *Atlas* development efforts to represent complex relationships, and share dynamic content, in a more narrative format. This was challenging in a number of ways. One is that it is difficult to secure funding to specifically support a technological development, so this had to be funded through various different grants supporting atlas projects – all with their own needs and specification, thus dividing the time of the technical experts on hand. Another is that it means that those working most closely with developing the conceptual model for the *Atlas* cannot directly implement a lot of the ideas without help, or specialized training. Therefore, a team effort is certainly required, and the composition of team members and relative time commitments are thus important considerations when initiating an Atlas project.

Sixth, in order to support the longevity of the *siku Atlas*, it must become a living resource. As mentioned earlier, this is facilitated by the development of user-contributed content options, as well as all the above-mentioned considerations for relationality, consultations, time investments, and technical expertise. But for this *Atlas* to truly be a living resource over the long term (i.e. not limited by project-specific funding), it should be housed and maintained at an institution (ideally an Inuit organization, or northern government department) that can incorporate such learning resources into their relevant programming. In the context of Nunavut, the process of exploring options with the Nunavut Arctic College is ongoing, especially with the long-term vision of having a Nunavut Atlas to which all relevant atlases could contribute (see Chapters 9, 10, 13, 15, 16, 17, 18, 19, and 20).

In the end, it is hoped that we have done justice to the knowledge and wisdom shared by Inuit Elders and hunters in Cape Dorset, Igloolik, and Pangnirtung, Nunavut. The depth and richness of their knowledge can never be fully or adequately represented in an atlas framework; however, it is hoped that the *Atlas* does help to meet their objectives of sharing their knowledge in a more broadly accessible form, especially for Inuit youth and for incorporation into the Nunavut education system (Chapter 20).

References

Aporta, C., 2002. Life on the ice: understanding the codes of a changing environment. Polar Record 38, 341–354.

Aporta, C., Taylor, D.R.F., Laidler, G.J., 2011. Geographies of Inuit sea ice use: introduction. The Canadian Geographer 55 (1), 1–5.

Aporta, C., 2011. Shifting perspectives on shifting ice: documenting and representing Inuit use of the sea ice. The Canadian Geographer 55 (1), 6–19.

Christie, M., 2004. Computer databases and aboriginal knowledge. Learning Communities: International Journal of Learning in Social Contexts 1, 4–12.

Dyson, L.E., Leggett, M., 2006. Towards a metadesign approach for building indigenous multimedia cultural archives. In: Attwater, R., Merson, J. (Eds.), Proceedings of the 12th ANZSYS Conference: Sustaining Our Social and Natural Capital, 3–6 December 2006, Katoomba, NSW, Australia, ISCE Publishing, Mansfield, MA.

Ford, J.D., Gough, W.A., Laidler, G.J., MacDonald, J., Irngaut, C., Qrunnut, K., 2009. Sea ice, climate change, and community vulnerability in northern Foxe Basin, Canada. Climate Research 38, 137–154.

Gearheard, S., Matumeak, W., Angutikjuaq, I., Maslanik, J., Huntington, H.P., Levitt, J., Kagak, D.M., Tigullaraq, G., Barry, R.G., 2006. "It's not that simple": a collaborative comparison of sea ice environments, their uses, observed changes, and adaptations in barrow. Alaska, USA, and Clyde River, Nunavut, Canada. AMBIO 35, 203–211.

Krupnik, I., Aporta, C., Gearheard, S., Laidler, G.J., Kielsen-Holm, L. (Eds.), 2010. SIKU: Knowing Our Ice, Documenting Inuit Sea ice Knowledge and Use, Springer, Dordrecht.

Laidler, G.J., 2007. Ice, through Inuit eyes: Characterizing the importance of sea ice processes, use, and change around three Nunavut communities. Unpublished Ph.D. Thesis, University of Toronto, Department of Geography.

Laidler, G.J., Elee, P., 2006. Sea ice processes and change: exposure and risk in Cape Dorset, Nunavut. In: Oakes, J., Riewe, R. (Eds.), Climate Change: Linking Traditional and Scientific Knowledge, Aboriginal Issues Press, University of Manitoba, Winnipeg, pp. 155–175.

Laidler, G.J., Elee, P., 2008. Human geographies of sea ice: freeze/thaw processes around Cape Dorset, Nunavut, Canada. Polar Record 44 (228), 51–76.

Laidler, G.J., Ikummaq, T., 2008. Human geographies of sea ice: freeze/thaw processes around Igloolik, Nunavut, Canada. Polar Record 44 (229), 127–153.

Laidler, G.J., Dialla, A., Joamie, E., 2008. Human geographies of sea ice: freeze/thaw processes around Pangnirtung, Nunavut, Canada. Polar Record 44 (231), 335–361.

Laidler, G.J., Ford, J.D., Gough, W.A., Ikummaq, T., Gagnon, A.S., Kowal, S., Qrunnut, K., Irngaut, C., 2009. Travelling and hunting in a changing Arctic: assessing Inuit vulnerability to sea ice change in Igloolik, Nunavut. Climatic Change 94, 363–397.

Laidler, G.J., Elee, P., Ikummaq, T., Joamie, E., Aporta, C., 2010. Mapping Inuit sea ice knowledge, use, and change in Nunavut, Canada (Cape Dorset, Igloolik, and Pangnirtung). In: Krupnik, I., Aporta, C., Gearheard, S., Laidler, G.J., Kielsen-Holm, L. (Eds.), SIKU: Knowing Our Ice, Documenting Inuit Sea ice Knowledge and Use, Springer, Netherlands, pp. 47–81.

Louis, R.P., 2007. Can you hear us now? Voices from the margin: using indigenous methodologies in geographic research. Geographical Research 45 (2), 130–139.

Nichols, T., Berkes, F., Jolly, D., Snow, N.B., The Community of Sachs Harbour, 2004. Climate change and sea ice: local observations from the Canadian Western Arctic. Arctic 57, 68–79.

Pulsifer, P.L., Laidler, G.J., Taylor, D.R.F., Hayes, A., 2010. Creating an online cybercartographic atlas of Inuit sea ice knowledge and use. In: Krupnik, I., Aporta, C., Gearheard, S., Laidler, G.J., Kielsen-Holm, L. (Eds.), SIKU: Knowing Our Ice, Documenting Inuit Sea ice Knowledge and Use, Springer, Dordrecht, pp. 229–254.

Pulsifer, P.L., Laidler, G.J., Taylor, D.R.F., Hayes, A., 2011. Towards an Indigenist data management program: reflections on experiences from the Inuit sea ice use and occupancy project. The Canadian Geographer 55 (1), 108–124.

Pulsifer, P.L., Gearheard, S., Huntington, H.P., Parsons, M.A., McNeave, C., McCann, H.S., 2012. The role of data management in engaging communities in Arctic research: overview of the Exchange for Local Observations and Knowledge of the Arctic (ELOKA). Polar Geography 35 (3–4), 271–290. http://dx.doi.org/10.1080/1088937X.2012.708364.

van der Velden, M., 2010. Design for the contact zone. In: Sudweeks, F., Hrachovec, H., Ess, C. (Eds.), Proceedings of the Seventh International Conference on Cultural Attitudes towards Communications and Technology, 15–18 June, Murdoch University, Vancouver/Murdoch, pp. 1–18.

Wilson, S., 2008. Research is Ceremony: Indigenous Research Methods. Fernwood Publishing, Black Point, Nova Scotia.

CHAPTER

15

The Kitikmeot Place Name Atlas

Darren Keith[1], Kim Crockatt[1], Amos Hayes[2]

[1]Kitikmeot Heritage Society, [2]Geomatics and Cartographic Research Centre (GCRC), Carleton University, Ottawa, Ontario, Canada

OUTLINE

15.1 THE KITIKMEOT HERITAGE SOCIETY

The Kitikmeot Heritage Society (KHS) was incorporated in 1995 as a volunteer-run heritage organization with an elected board of directors made up primarily of Elders from the community of Cambridge Bay, Nunavut. Cambridge Bay is located on the Southeast corner of Victoria Island in the Kitikmeot Region of the Nunavut Territory. The community is a regional centre with a mixed population of Inuit beneficiaries to the Nunavut land claim, and other Canadians. From its inception, the KHS was originally concerned with local Cambridge Bay heritage, and while that continues to be a focus of many of its projects, with time the organization has expanded its mandate to serve the region as a whole (See Map of the North in the introduction).

The KHS is based in the May Hakongak Community Library and Cultural Centre – a facility which it helped build in 2001; the centre consists of a library with a reference section on regional history and culture, an archives, and significiant floor space committed to museum quality cultural and archaeological exhibits (Figure 15.1).

Developments in the Theory and Practice of Cybercartography, Second Edition, ISSN 1363-0814
http://dx.doi.org/10.1016/B978-0-444-62713-1.00015-5

FIGURE 15.1 May Hakongak Centre.

The KHS defines its mandate as follows:

The KHS preserves, promotes, and celebrates the history, culture, language, and diversity of the people of the Kitikmeot region. It does this by:

- Collecting and archiving the oral histories of the elders;
- Preserving archaeological sites through oral history and archaeological study;
- Collecting historical information and archives;
- Using the information collected to promote and preserve Inuinnaqtun and Inuktitut;
- Developing educational and learning materials; and
- Promoting and celebrating the cultures and traditions of the people.

Key to the operating philosophy of the KHS is to try whenever possible to take the results of cultural preservation projects, and transform them into vehicles for the promotion and transmission of Inuit culture, history, and language. Through its own projects and through collaboration with other organizations and academic institutions, the KHS has accomplished this goal through: museum exhibitry; print publications; audio disc production; documentary film production; map design and printing; website publication, and; academic publications.

15.2 THE KITIKMEOT ATLAS PROJECT

The KHS began the systematic collection of the Inuktitut place names of the land-use area of the people of Cambridge Bay in the year 2000 (Keith and Angulalik, 2000). Several Cambridge Bay Elders gathered over several days to review topographic maps of the area and identify the locations of all the traditional Inuktitut place names they could remember for the area. Many of the places had oral traditions associated with them such as: the meanings of the names; descriptions of the significant subsistence resources available at these places; accounts of historical events that occurred there in the lifetime of the Elder, or

FIGURE 15.2 Mapping in the Field.

in the lifetimes of their predecessors; and occasionally, mythological events that occurred there. These oral traditions were recorded on video with the intent of preserving them for future generations. This initial recording effort by Darren Keith and Martha Angulalik resulted in a written report, archived videotapes and a first draft of printed 1:250,000 topographic maps that were freely distributed to interested residents of Cambridge Bay. Over the years, the Cambridge Bay place name maps were continually reviewed and added to by Cambridge Bay Elders, and further video recordings were done of oral traditions associated with named places. The Cambridge Bay work was accomplished with the help of several people over the years, including Martha Angulalik, Emily Angulalik, Gwen Angulalik, and Julia Ogina. This body of work was submitted to the Territorial Toponymist of the Government of Nunavut and resulted in 299 place names becoming official names of Nunavut and Canada in September 2012 (Figure 15.2).

The utility of the exercise soon became evident with the popularity of the place name maps locally, and individuals from adjacent communities expressed interest, and so the KHS expanded its commitment to include the recording of traditional place names for other communities in the Kitikmeot region. Building on a Parks Canada place name recording project completed in 1996 by Darren Keith and the late Martha Dwyer with the Gjoa Haven-based Utkuhiksalingmiut population, the Kitikmeot Atlas project started recording in Gjoa Haven in 2001. The local researchers in Gjoa Haven who contributed greatly to the project were Louis Kamookak, whom had already started recording names on maps, and Jerry Arqviq. Pelly Bay followed in 2003 with local researcher Marie Kakkiarniut.

Over the course of the Kitikmeot Atlas Project the KHS designed and printed 1:250,000 topographic maps for distribution in the various communities. While these maps were very popular and useful for the transmission and continued use of Inuktitut place names, they were not the right vehicle to ensure that all of the oral traditions collected during the project were made available to the public. Another form of cartographic product was needed to deliver on the KHS mandate to actively make available recorded oral traditions to promote their continued use by younger generations.

15.3 METHODOLOGY: A COMMITMENT TO THE ORAL TRADITION

The methodology adopted by the KHS for place name-related information collection during the Kitikmeot Atlas Project was, from the beginning, concerned that place names not be recorded in isolation from their context in Inuit oral traditions. The presence of Inuktitut place names on future official topographic maps for the region was certainly a highly prized objective of the project due to the implications for Inuit identity in their territory, and the guaranteed survival of the names. However, the continued use of the names in the absence of their cultural context – their meaning – would still leave future generations impoverished. While many Inuktitut place names are descriptive of the geographical features named, many others contain a wealth of subsistence, cultural, historical, and mythological information. The place names and their cultural context are the full inheritance of future generations of Inuit. In this light, the KHS defined the objectives of the Kitikmeot Atlas Project area to record:

1. The location and extent of geographical features covered by each place name;
2. The Inuktitut geographical entity or entities represented by the place name;
3. The pronunciation of each place name;
4. The proper representation of each place name in the orthographies used by speakers of the Inuktitut dialects who use the place name;
5. The meaning of each place name; and
6. The oral traditions associated with the name including stories, legends, and songs.

The systematic method for the collection recording of basic data on place names was developed by Ludger Muller-Wille during his extensive place name surveys in Nunavik (Muller-Wille and Weber, 1983).

The methodology applied by the KHS during the Kitikmeot Atlas Project was derived from earlier work associated with Parks Canada National Historic Site research activity. In this work, place names were collected in the field during visits with Elders to the named areas in question (Keith and Scottie, 1995, 1997; Utuq et al., 1995; Utkuhiksalingmiut Elders et al., 1996). Place name locations were recorded on topographic maps and oral traditions video recorded: at the physical location of the named place, within sight of the named place; or, while visiting the general vicinity of the named place. When possible photographs were taken of the named places and formed part of the written field report, this methodology for place name collection – while expensive and time-intensive – maximizes the place name recall of Elders for a given geographical area. It enables them to recognize and recall what one might call micro-level place names around a traditional campsite or in an intensively named area. A good example of this is the Utkuhiksalik area located near the mouth of the Back River where 90 place names were recorded on one 1:50,000 topographic map sheet (Utkuhiksalingmiut Elders et al., 1996). The method is also very successful for the quality of the oral traditions information that can be recorded using video interviews at named sites. This approach allows Elder interviewees to maximize their recall of oral traditions and physically refer to relevant features in the landscape identified in their narratives. The video interview allows place names and their associated oral traditions to be transmitted to young and future generations through the knowledgeable voice and narrative performance of Elders.

As the goal for the Kitikmeot Atlas was to collect place names over a vast geographic area, the limits of financial resources and time did not allow for the superior 'field visit' method of place name collection. Instead, the decision was made to use a method whereby place names were recorded in the context of a meeting of Elders and experienced land-users. These expert meetings were held over several days, as the experts worked their way over topographic maps of their historic land-use areas. As basic location, entity and meaning information was recorded, researchers took note of which place names had significant oral traditions associated with them. These places were revisited at the end of the recording visit and video interviews were recorded for each of the oral traditions. In this way, the Kitikmeot Atlas Project 'village-based' method retained the benefits of video interviews to the transmission of oral traditions, while the group interviews based in villages allowed for the vast areas implicated to be covered in an efficient manner.

Another important part of the 'village-based' method was the recorded of the proper pronunciation of each place name by a fluent Inuktitut or Inuinnaqtun speaker. The purpose of this extra audio-recording step in the method was to ensure that place names were accurately passed on to young and future generations. Place naming is part of Inuktitut oral tradition that has only relatively recently been codified through roman orthography-based and syllabic-based writing systems. These conventions are new and continue to evolve, and the level of literacy in these writing systems is variable in the population. Significantly, as is the case with the Inuinnaqtun dialect of Inuktitut, Elder speakers adhere to an older orthographic form while young people are taught a more recent form. The dynamic state of writing conventions for Inuktitut require that oral pronunciations are preserved so as to be conservatively interpreted in the future.

As already mentioned, the types of information being collected during the Kitikmeot Atlas Project such as audio recordings of pronunciations and video recordings of Elder narrative performances of oral traditions could not be fully communicated using the traditional vehicle of the topographic map. Another medium was needed to transmit the entirety of place name-related knowledge to younger generations.

15.4 CYBER-CARTOGRAPHY: LET ORAL TRADITIONS SPEAK FOR THEMSELVES

In 2005, the KHS was actively looking for a web-based interactive map solution that would allow the use of the multimedia material that had been collected, when Claudio Aporta of Carleton University visited the Kitikmeot Region with place name research interests. Claudio was already working with the Geomatics and Cartographic Research Centre (GCRC) at Carleton, and he put the KHS in contact with Fraser Taylor and Amos Hayes. After preliminary discussions with GCRC, it became evident that the Nunaliit Cybercartographic Atlas Framework could provide an open source solution for the Kitikmeot Atlas Project. The KHS decided to provide financial support to the GCRCs research in return for the development of the Kitikmeot Place Name Atlas that would be hosted by the GCRC. The KHS began working closely with Amos Hayes to create the Atlas (Figure 15.3).

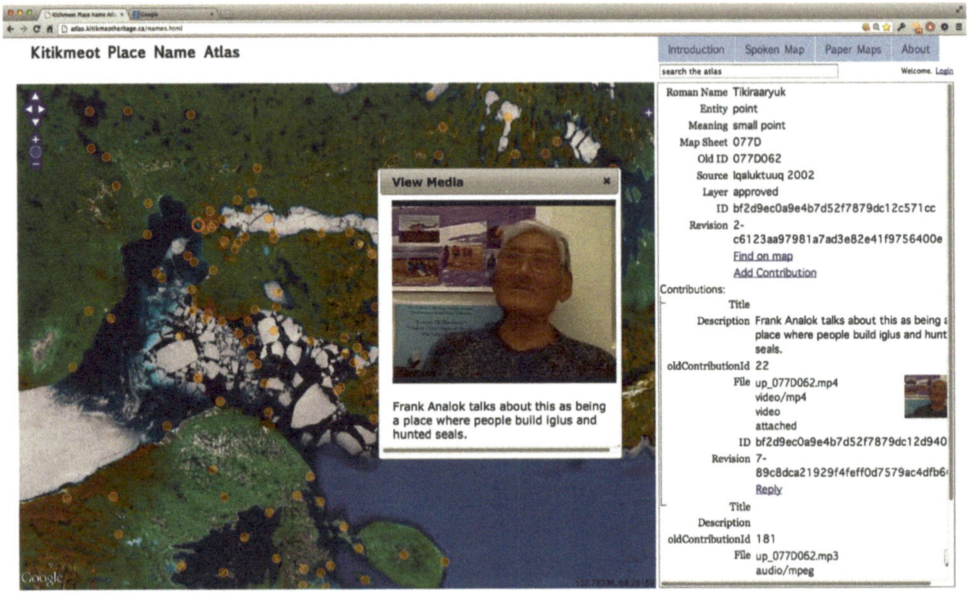

FIGURE 15.3 Kitikmeot Place Name Atlas.

The KHS had the following initial goals for an interactive place name atlas:

1. A continuous, zoomable map that would allow users to navigate freely throughout the entire region and select the place names they want to investigate;
2. The ability for users to hear the pronunciation of each place name on the map as recorded by a fluent Inuktitut speaker – A 'talking map';
3. The ability for users to go beyond the textual information about each name by opening multimedia files with video (and sometimes audio) interviews of Elders sharing oral traditions about the place.
4. The ability to provide a virtual visit to the named place through upload photos or video.

Through creative collaboration with GCRC, additional goals were added to the project with time;

5. Provide a 'wiki'-like function to the atlas that allows community-based users to add place name information and multimedia files to the Atlas directly; and
6. Install a dedicated Atlas kiosk in Cambridge Bay at the May Hakongak Community Library and Cultural Centre to facilitate the contribution of place name data by local residents.

15.4.1 A Continuous 'Zoomable' Map

The initial work that had to be done was to process all the existing place name data collected by the KHS for the Kitikmeot Region and enter it into the Nunaliit Framework. Once all locations and the basic fields of name, location, geographic entity and meaning were entered, we came to a first decision point. In the interest of reducing visual clutter when zoomed out to full extent, it seemed rational to filter or prioritize locations based on a zoom level. However, the decision was made not to do this, as this would hide information about the relative density of names for different geographical areas. Due to the history of land-use in the region, there are areas that formed the core territories of different social or '-miut' groups. These core areas were more frequently used, and therefore also more densely named. This was important information that would be lost by filtering or prioritizing names when zoomed to smaller map scales.

15.4.2 A Talking Map

As explained above, representation of place names using relatively recent and evolving orthography standards is not necessarily conservative of the place name pronunciations. For this reason, it was important to have the most authoritative representation of each name through an audio recording of the proper pronunciation of that name by a fluent Inuktitut or Inuinnaqtun speaker. This issue was elegantly addressed by creating an atlas that 'talks'. When navigating through the atlas, users can hear the pronunciation of each place name when they place their cursor on the place name label. In this way, the atlas stays true to the traditional oral transmission of place names between generations.

15.4.3 A Multimedia Map

Just as it is important to stay true to the oral transmission of place name pronunciations, the cultural context for the name must also be shared in an authentic way. The interview transcript is an important tool in cultural preservation, but the most the ideal format for passing on oral traditions is the narrative performance of these traditions by a knowledgeable and authoritative source. The Atlas accomplishes this by providing users access to video interviews with Elders when they are available.

Atlas users who select a specific place name to investigate further do so by clicking on the name. They then enter another window where basic information about the name is available in a text format (name spelling, identification number, entity named, meaning), as well as links to multimedia if they are available. Clicking on the multimedia permits the user to watch the video or listen to a sound file.

15.4.4 A Virtual Visit

The strength of the 'field visit' method of place name collection mentioned in the methodology section was the ability it gave to Elders to recall more names and oral traditions – but the video and photographic documentation of the named places also aids the student of Inuit place names to understand the named place. The inclusion of multimedia in the Atlas provides an opportunity to add video and photographs taken directly at the named site as these become available. The functionality built into the atlas make it a vehicle for normally community-bound young people to have a 'virtual visit' to named places.

15.4.5 A 'Wiki' Map – Direct Input From the Community

One of the goals of the GCRCs Nunaliit Cybercartographic Atlas Framework project is to have their products be interactive and collaborative with users. The GCRC recommended this approach to the KHS for the Kitikmeot Place Name Atlas. What they have accomplished is an interface that allows community-based users to add their own place names to the map along with all the pertinent textual information, video, sound files, and photographs. With this added functionality, the Atlas has the potential to continue to grow in terms of the number of discrete place names to holds, and also in terms of quality as users add their own multimedia to place names already in the Atlas. This additional multimedia could be in the form of their own contribution to oral traditions, or photos or video taken at the physical site of a place name.

15.4.6 A Dedicated Atlas Kiosk

As powerful as the 'wiki' functionality is in the Kitikmeot Place Name Atlas, there are still some practical limitations to the amount of community input that can be expected. For example, the community members whom are very knowledgeable about place names are not necessarily the same community members who are active and knowledgeable information-technology users. For this reason, it was decided to create a dedicated Kitikmeot Place Name Atlas Kiosk located in the May Hakongak Community Library and Cultural Centre in Cambridge Bay with the idea that:

- The kiosk could be dedicated to the Atlas and therefore interactions with the computer would be limited to the Atlas interface and would not require knowledge of how to use a specific operating system or graphic user interface.
- The kiosk would be in the physical home of the KHS where staff would be available to assist contributors to the atlas with any technical issues they may have.

15.5 DISCUSSION AND FUTURE DIRECTIONS

The Kitikmeot Place Name Atlas is an important example of how community-based heritage groups and academic institutions can collaborate to realize advancements in how technologies can be used by Inuit and First Nations to preserve and promote their languages and cultures. Through its collaboration with the GCRC, the KHS has been able to accomplish one if its fundamental goals, which is to mobilize the products of cultural preservation projects so that they can contribute to the ongoing transmission of Inuit knowledge.

In the future, it is the intent of the KHS to continue to add content to the Atlas and include communities that are not yet involved in the project. The photos and video taken from future cultural projects that visit named sites will be added to the Atlas. The potential is also there to undertake a project to identify historical photographs that have been taken at named places and add digital versions of those photos to the Atlas. The use of photographs in this context is more fully discussed in Chapter 13 on *Mapping Views From the North*.

In collaboration with the GCRC of Carleton University, the KHS has been able to been able to develop an interactive, multimedia cartographic product that allows for the transmission of Inuit traditional knowledge through the voices of Inuit knowledge holders themselves. In the Kitikmeot Place Name Atlas, we see the beginnings of an Inuit adaptation of the technology of cartography as an authentic vehicle for the preservation and transmission of place-related traditional knowledge.

References

Keith, D., Angulalik, M., 2000. Iqaluktuuttiaq Place Name Project. Field Report for the Kitikmeot Heritage Society.

Keith, D., Scottie, J., 1995. Harvaqtuuq: Place Names of the Lower Kazan River. Field Report Parks Canada, Department of Canadian Heritage.

Keith, D., Scottie, J., 1997. Harvaqtuuq 1998: Place Names and Oral Traditions of the Lower Kazan River. Field Report Parks Canada, Department of Canadian Heritage.

Muller-Wille, L., Weber, L., 1983. Inuit place name inventory of northeastern Quebec-Labrador. McGill Subarctic Research Paper 37, 151–222.

Utkuhiksalingmiut Elders, Keith, D., Dwyer, M., 1996. Utkuhiksalik: Place Names of the Lower Back River. Field Report Parks Canada, Department of Canadian Heritage.

Utuq, T., Suluk, L., Keith, D., 1995. Maguse River Place Names Project. Field Report Parks Canada, Department of Canadian Heritage.

CHAPTER

16

The Gwich'in Atlas: Place Names, Maps, and Narratives

Claudio Aporta[1], *Ingrid Kritsch*[2], *Alestine Andre*[2], *Kristi Benson*[3,4], *Sharon Snowshoe*[2], *William Firth*[2], *Del Carry*[4]

[1]Marine Affairs Program, Dalhousie University, Halifax, Nova Scotia, Canada, [2]Gwich'in Social and Cultural Institute, Yellowknife, Fort McPherson and Tsiigehtchic, Northwest Territories, Canada, [3]GIS/Heritage Affiliate, Gwich'in Social and Cultural Institute, Santa Clara, Manitoba, Canada, [4]MDT Communications, Ottawa, Ontario, Canada

OUTLINE

16.1 INTRODUCTION

All peoples and cultures have the need to describe and represent place and space. Spatial frameworks of orientation are both universal and culture-specific. They are universal because all human beings need to relate their locations and trajectories to broader frames of reference in the world. They are culturally specific because such frameworks are intriguingly different across cultures, as people relate them to magnetic phenomena, celestial bodies, winds, coastal and river layouts, mountains, etc. In establishing relationships to specific places, and

Developments in the Theory and Practice of Cybercartography, Second Edition, ISSN 1363-0814
http://dx.doi.org/10.1016/B978-0-444-62713-1.00016-7

to regions, place names are a significant way of articulating people's connections to their surroundings, to features, and to more or less bounded bodies of land or water. They serve to anchor perceptions, memories, and ideas of place in particular locations. They also serve to convey such concepts and information to other people. Place names such as Yellowknife or Halifax possess a functional trait beyond the semantic and symbolic meanings associated to them. They are essential in facilitating people's narratives of place. 'Halifax, Nova Scotia' brings to mind a particular location, thus helping people avoid taxing topographic and geographic descriptions every time the place is referred to.

Place names and maps have been used hand-in-hand by most European and European-descendent cultures since the development of writing, and especially since the production of the first maps and the emergence of cartography (Thrower, 2008; Short, 2009). Maps gradually became essential tools in representing place and space. In the history of exploration, maps were also significant instruments of possession and control in the context of the colonial enterprise.

The exploration and 'discovery' of the Americas by Europeans could be, in a sense, read as a gradual *mapping* of newly found lands. The maps produced by the explorers not only informed Europe about the geographic features of land and water, but also established claims of ownership. Such claims were always established and reinforced through naming. The need to explore, map, name, and take possession were most often part of the same process of 'discovery'. Of course, the aboriginal peoples they encountered had their own approaches to naming and to geographic representation, but the European enterprise of naming after their own cultural frames of reference announced to the rest of the world that they were taking possession of lands they considered to be without previous history and culture.[1]

For a very long time, therefore, Indigenous place names were either ignored or partially adapted to the Colonial naming approach and frames of reference.[2] The native names that 'made it' to our maps (including the name *Canada*) were most often reinterpreted, decontextualized, and deprived of their true cultural meanings and histories.

Beyond instances of map use and map-making in precontact times that have been documented (Lewis, 1998; Mundy, 1996), most Indigenous groups in Canada were characterized by an oral approach to geographic representation. The fact that maps were mostly nonexistent and that Indigenous spatial representations did not accord with European cartographic conventions, made it easier for colonial powers to claim possession of what they intentionally or unintentionally portrayed as wilderness.

In the context of an oral culture, place names are particularly important as spatial and geographic referents, and as crucial elements of spatial narratives (or of any narrative, really, where certain reference to location is required). Place names can also reflect cultural meaning and cosmological views of the peoples who use them. Many explorers, missionaries, anthropologists, geographers, and historians have documented Indigenous place names across North America, from the Apache in the southern United States to the Inuit in the Arctic. In contemporary times,

[1] We are aware that this statement generalizes a phenomenon that was more diverse and complex, across the Americas. However, 'discovery' and colonization of the Americas was a race towards asserting geopolitical claim (Galeano, 1997).

[2] It should be added that fur traders, in particular, used Indigenous place names to identify prominent natural features in the Western Interior, even if the trading posts themselves usually bore European names (Wonders, 1987; See also Freeman (1985)).

and in the context of new geopolitical and cultural realities, an increasing number of Indigenous groups have embarked on the task of documenting and mapping their own place names as part of a variety of projects of cultural revitalization and political activism. In Canada, these systematic efforts started in the 1970s, as part of land-claims processes. This chapter is concerned with the place names of the Gwich'in of the Northwest part of Canada. It focuses on the place names collected and documented by the *Gwich'in Social and Cultural Institute* (GSCI) among the Gwich'in communities now inhabiting the Northwest Territories (NWT), but whose traditional land use included lands in both the NWT and Yukon (Figure 16.1).

The Gwich'in have long used a complex system of place names to refer to a variety of features, events, and processes. Immersed in a context of cultural and language loss, and of political struggle, like most aboriginal groups of Canada, the Gwich'in have been engaged in place names projects since 1992 as part of cultural revitalization initiatives, and as a way of documenting traditional use of the land (Gwich'in Social and Cultural Institute, 2003). This chapter will briefly introduce the Gwich'in, summarize the main characteristics of their place names, describe the place names project undertaken by the GSCI, and, finally, reflect on the effects of mapping oral geographic knowledge and the advantages of the Cybercartographic approach.

16.2 THE GWICH'IN AND THEIR PLACE NAMES

The Gwich'in are one of the most northerly Indigenous peoples in North America, living at the northwestern limits of the boreal forest. Only the Inuit live further north. In anthropological terms, the Gwich'in are part of a larger family of Indigenous peoples known as Athapaskans, which include the Slavey, Tłįchǫ (Dogrib), Tr'ondëk Hwëch'in (Han) and Tutchone. Their language and way of life, however, are distinct.

At the time of contact with Euro-Canadians in the latter part of the eighteenth century, the Gwich'in lived in nine groups, with lands stretching from the interior of present-day Alaska, through the Yukon and into the Mackenzie Valley. Today, the Gwich'in number over 6000 people with 3300 Gwich'in living in the NWT, primarily in the communities of Aklavik, Fort McPherson, Inuvik, and Tsiigehtchic, where they are known respectively by the following names: Ehdiitat Gwich'in, Teetł'it Gwich'in, Nihtat Gwich'in, and Gwichya Gwich'in. The Gwich'in in the NWT still maintain close cultural and family ties with Gwich'in relatives in the Yukon and Alaska, meeting as a Nation on a biennial basis since 1988 at the Gwich'in Gathering–alternating meetings between Alaska, Yukon, and NWT.

The lands of the Gwich'in now living in the NWT traditionally extended from the mountain headwaters of the Peel and Arctic Red rivers in the south to the Mackenzie Delta in the north, and from the Anderson River in the east to the Richardson Mountains in the west. The Gwich'in retain a close relationship with the land and many families living in the present-day settlements also have summer and winter camps outside of their communities. Hunting and fishing remain important, both culturally and economically, with caribou, moose, and whitefish being dietary staples.

In 1992, the Gwich'in of the NWT signed a Comprehensive Land Claim Agreement with the Government of Canada and the Government of the NWT. The land claim outlines

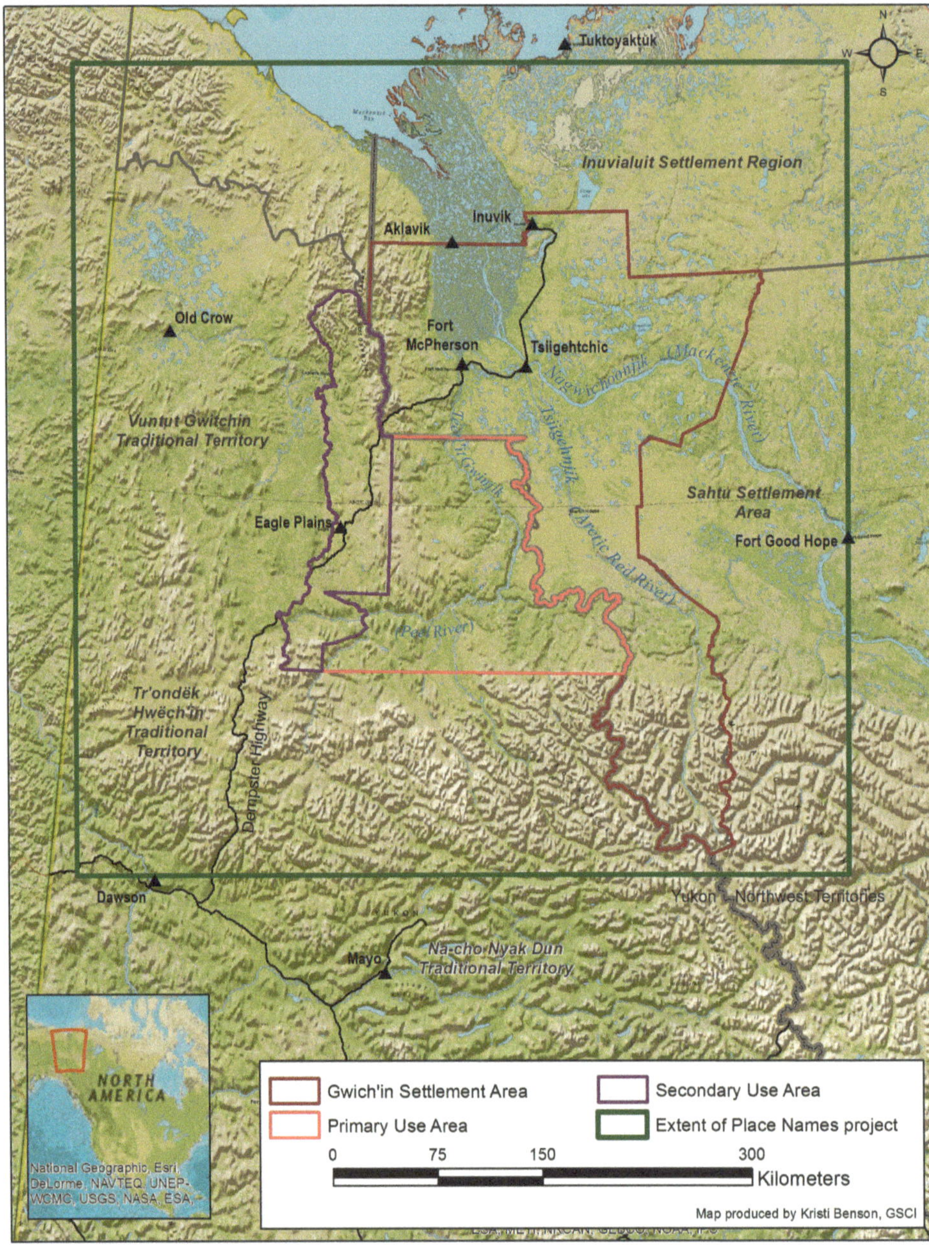

FIGURE 16.1 Map of the Gwich'in Settlement Region (GSR) defined by the 1992 Gwich'in Comprehensive Land Claim Agreement. The GSR is composed of the Gwich'in Settlement Area in the NWT and Primary Use and Secondary Use areas in the Yukon – Trans-boundary areas shared with neighbouring Aboriginal groups. The area within the square indicates the extent of Gwich'in traditional land use and the place names recorded in the Gwich'in Place Names Atlas.

economic, cultural, and political provisions and describes the lands to which the NWT Gwich'in have rights to in both the NWT and Yukon. The claim provides the Gwich'in with increased control and responsibility over their communities, lands, and resources while also recognizing and encouraging the Gwich'in way of life 'which is based on the cultural and economic relationship between the Gwich'in and the land' (Indian and Northern Affairs Canada, 1992: 2).

The Gwich'in Tribal Council (GTC) is the main political body responsible for implementing the land claim, and its mission statement, 'Gwich'in land, culture and economy for a better future', speaks to the need to balance these three entities. To facilitate this, the GTC established a number of Gwich'in organizations following the signing of the land claim. The GSCI is the organization responsible for implementing the heritage resource chapters of the claim, and carrying out cultural and language research and programming. The GSCI's mandate is 'to document, preserve, and promote Gwich'in culture, language, traditional knowledge and values.'

The Gwich'in have a rich oral tradition that is centred on extensive knowledge and use of the land. Travel with the seasons was a necessity to make a living from the seasonal resources that the land provided. The Gwich'in travelled extensively by land and water, naming features and places that were of significance to them, and sharing stories about these places. Names are found everywhere on the landscape, and these names and their associated stories reflect Gwich'in travels, explain the history of the place, and offer insight into Gwich'in culture and knowledge about their lands and about people's relations with the land. Therefore, it could be safely said that Gwich'in culture is tied directly to the land.

Working with a generation of elders who have an in-depth knowledge of the land learned from personal experience and oral tradition, GSCI has been building an inventory of named places within traditionally used lands, both within and beyond the area now known as the Gwich'in Settlement Region (GSR). To date, GSCI has recorded stories and information on approximately 850 named places with many places having more than one name. Most of these names are in the Gwich'in language and the meaning and stories attached to them are informed by many aspects of Gwich'in culture, history, and values. An important concept in the stories that elders tell is that the people and the land are closely linked and that everything and everyone is related. The Gwich'in place names and oral history research carried out by GSCI over the past 20 years has produced a voluminous and precious body of knowledge, and has been the foundation upon which much of the institute's other research and initiatives have been based (Figures 16.2 and 16.3).

Unlike most non-Indigenous place names, Gwich'in named places describe what is important about a location. Gwich'in place names may relate to:

1. the resources available and/or the use of traditional methods for capturing resources (i.e. types of birds, fish or animals for subsistence purposes, stone for making tools, wood to make moose skin boats, plus technology such as fish traps and caribou fences);
2. the physical characteristics of a place (i.e. a distinctive looking round hill or an area where smoke is present from coal-burning beds);

3. connections with a sacred place (i.e. Shiłdii), This place name means 'sitting down' or 'sitting in fear' and relates to a Gwich'in legend about a young girl who broke the rules of her puberty training and caused her three brothers and their dog to turn into stone pillars;
4. connections with legendary or historical events (i.e. how a river was formed, a reference to starvation, or contact with another Indigenous group); and
5. connections with a particular person who is known to have lived or used the location (Kritsch et al., 1994).

FIGURE 16.2 (a) Gwich'in place names map 106M for Teetł'it Zheh (Fort McPherson) from the 1:250,000 series produced by GSCI.

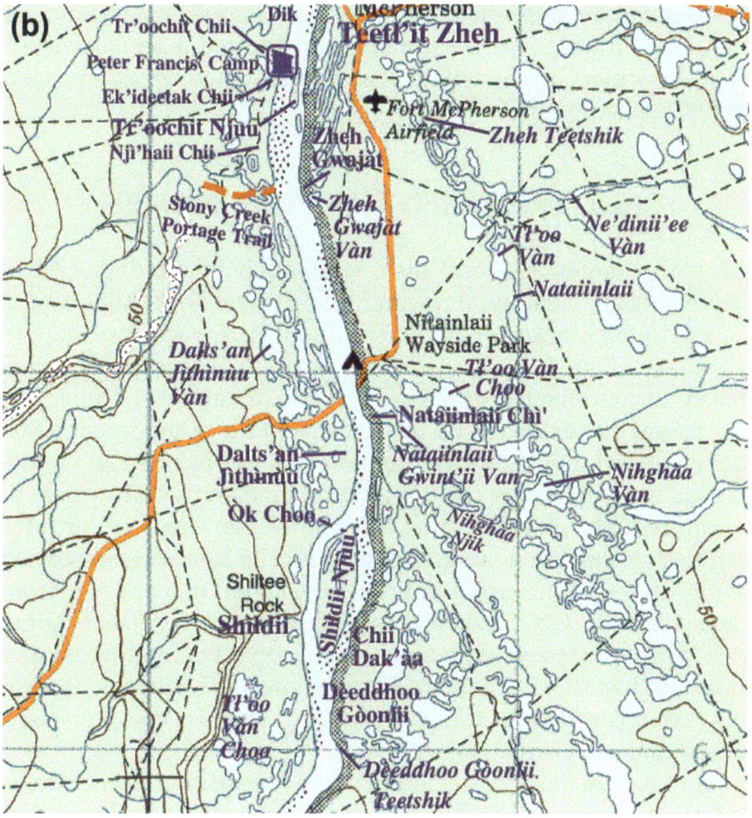

FIGURE 16.2, cont'd (b) Enlarged area of map sheet 106M for Teetł'it Zheh (Fort McPherson) with sample of place names from Fort McPherson to Dèeddhoo Gòonlii Teetshik along the Peel River.

FIGURE 16.3 Travelling along the traditional Teetł'it Gwich'in trail between Fort McPherson and Dawson or Mayo in March 1998. This was the trail that the infamous Royal Northwest Mounted Police 'Lost Patrol' travelled on in 1911 between Fort McPherson and Dawson and lost their lives. The travelers are approaching a range of mountains called Ddhah Dik'ee ('mountain-sharp ridge'), which are located north of the Peel River, across from the Bonnet Plume River. *Source: Ingrid Kritsch, GSCI.*

Most place names can be translated, but a few names are so old, and the language so intricate, that they elude translation. These names are referred to as *Ts'iidaii* (Teetł'it Gwich'in dialect) or *Ts'iidęįį* (Gwichya Gwich'in dialect) referring to their ancestors of long ago – 'stone age people from 500 years or more'.

16.3 DOCUMENTING THE NAMES

The Gwich'in place names and oral history research initiated in 1992 was originally intended to complement archaeological research being carried out at this time in the Gwichya Gwich'in area, north of the Mackenzie River (see Pilon, 1994; Kritsch et al., 1994). In carrying out this research, however, the Gwichya Gwich'in elders in Tsiigehtchic asked the researchers (Andre and Kritsch) to continue the recording and documentation of other areas in their traditional land-use region that extended up the Arctic Red River and into the Mackenzie Delta. The elders were concerned that if this information was not recorded it would be lost as the younger generations were no longer spending much time on the land where this knowledge is traditionally acquired. The elders felt that this knowledge was vital for future generations to understand and appreciate their lands and culture. The researchers soon learned that this view was shared by all the Gwich'in communities consulted in the NWT. From 1993 to 2007, as funding became available, GSCI researchers worked with elders from Tsiigehtchic, Inuvik, Aklavik, and Fort McPherson to record named places and other land-based heritage information, such as trails and historical and archaeological sites. What began as a solitary research project in one Gwich'in community transformed into part of a comprehensive plan to preserve, promote, and manage Gwich'in heritage.

The methodology used for this project included one-on-one interviews with elders and other knowledgeable land-users using 1:50,000 National Topographic System (NTS) topographic map sheets and a tape recorder to record the place names and their meanings, associated stories, trails, traditional camp sites, graves, historic sites, harvesting locales, and sacred or legendary places. Initial interviews were usually carried out in the communities, although in a few cases the interviews took place at people's bush camps. These initial interviews were then followed up with further interviews and verification on-the-land at as many of the named places as possible that could be reached by boat, snowmobile, and helicopter. Each of the named places visited was photographed, and if above ground features were visible, these were also recorded (Figure 16.4).

Verification workshops of the names and associated information were conducted in each community following the mapping, and summary reports were prepared by the researchers. Translation and transcription of all taped interviews were also carried out with the help of local translators. In later years, GSCI held a series of community workshops to verify the names, locations, translations, record the sounding of the place names, and delineate the geographic extents of each name. In order to determine extents, NTS maps were projected onto walls so the elders could indicate the approximate bounds of each place. In some cases, this was quite straightforward (i.e. with lakes), but other features such as rivers, as discussed below, proved more challenging (Figures 16.5 and 16.6).

Determining the extent of the named feature was required for submitting place names for official recognition in the NWT. This requirement follows guidelines set by the Geographical

FIGURE 16.4 Hyacinthe Andre telling a story to his daughter Alestine Andre during the 1993 Gwichya Gwich'in Place Names Project about traditional travel routes up the Arctic Red River while sitting on a hillside of a major fish lake called Nihtavan Diniinlee. The last time Hyacinthe was here was in the early 1950s. *Source: Ingrid Kritsch, GSCI.*

FIGURE 16.5 Annie B. Gordon showing the extents of a named place using a projected map on the wall during the Aklavik community workshop in November 2010. *Source: Alestine Andre, GSCI.*

FIGURE 16.6 Ingrid Kritsch recording the sounding of the names by Walter Alexie during the Fort McPherson community workshop in March 2012 using the Nunaliit application developed by the Geomatics and Cartographic Research Centre, an iPad and snowball microphone. This technology allowed the project to enter the sound files directly into the appropriate place name record – a great time-saving measure. *Source: Alestine Andre, GSCI.*

Names Board of Canada (2007), and is a way to help understand Indigenous naming practices (see Andrews, 2011: 61–63). In Gwich'in, for example, the name for the mouth of a river is often differentiated from the rest of the waterway by a different suffix and this is reflected in the extents. The mouth of the Road River in the Peel River watershed for instance is called Vihtł'òo Tshik whereas the length of the river beyond the mouth is called Vihtł'òo Njik. In most cases, 'tshik' refers to the mouth and 'njik' for the length of the river, although where one begins and the other ends is not always straightforward and speaks to the difficulty of 'bounding' places.

The value of the place name research went beyond recording and documenting this critical land-based knowledge that was at risk of being lost. It also gave GSCI an opportunity to bring elders and youth together on the land, promoting the sharing of the language, knowledge about the land, and Gwich'in culture. The process of knowledge sharing with younger generations that was facilitated in the context of the research project was considered by many in the communities to be as significant as the data collected.

Besides encouraging this intergenerational exchange of knowledge, the place names research continues to be relevant by raising an awareness and appreciation of Gwich'in culture, language, and knowledge of the land. It has assisted GSCI in ensuring that culturally significant heritage sites and places in the GSR are recognized, protected, and/or managed in a manner consistent with Gwich'in values for the benefit of future generations. This work is being carried out through the designation of named places and heritage sites as Territorial or National Historic Sites, and through protective measures in the Gwich'in Land Use Plan.

The place names project has also resulted in the development and publishing of culturally relevant books, exhibits, and educational materials for local schools, museums, and interpretive centres, and for the general public. A prototype 'Talking Map' developed in 1993 by Gwich'in Geographics Ltd. and GSCI to showcase named places within the Gwich'in Territorial Park outside of Inuvik was expanded in 2003 on the GSCI website (http://www.gwichin.ca/Research/placeNameMap.html) as a way to share knowledge gained through this research. The 'Talking Map' provides a virtual tour of the three major rivers in the GSR using place names and associated oral history information, plus photos, and audio sounding of the names. Gwich'in place names are now recognized on highway signage, within Territorial Parks, and on maps of the GSR. The information contained in the names (and the names themselves) also facilitate the review of regulatory land and water permits, providing insight on possible impacts that economic developments could have on Gwich'in culture and resources. The names and knowledge resulting from this research, therefore, are becoming important instruments in regards to heritage and land management issues and are all important measures in the process of Gwich'in nation-building (Kritsch and Andre, 1997).

16.4 MAPPING ORAL GEOGRAPHIC KNOWLEDGE

There are significant challenges in mapping place names and other geographic knowledge pertaining to oral-based Indigenous cultures. The most basic one is epistemological, as Indigenous geographic knowledge is significantly more holistic and connected to other spheres of knowledge and performance than its western counterpart. While, in principle, a western-European place name can be reduced as discrete location, in most cases Indigenous place names only make sense in the context of larger narratives, or in the context of actual travel (as with names that describe features as they appear in the horizon). Therefore, the mapping of a Gwich'in place name involves a process of 'conversion' of an entity that forms part of a wider set of oral narratives and performances – into an artifact developed by western science (a map) designed specifically to place and locate names in reference to a framework of geographic grids and cartographic conventions.

Gwich'in place names, as described above, resist classification, as even the effort of grouping them in different thematic clusters does not reflect the realities of the use of a name in everyday life and in speech. Names referring to land resources, shapes of the land, legendary and historical events, etc. belong to broader narratives in which thematic categories become blurry. The process of mapping or documenting a place name involves, to a degree, isolating it from its broader uses and from its relationships with the land in the context of oral narratives. Place names that refer to, for instance, legendary events are often linked to particular shapes of the land (e.g. a rock), to the presence of animals (e.g. caribou), and to historical events (e.g. a hunting or fishing camp). Place names are in fact multidimensional and they are fully significant not only in the context of broader narratives, but also, and mainly, in the acts through which people interact with the land.

The Gwich'in relationship with their land and environment is intrinsically connected to travel and trails, and place names are fully significant only when they are recognized in familiar horizons. In other words, Gwich'in geographic, historic, and environmental knowledge is embedded in the land. Names are not merely discrete entities fixed in particular places, but they represent connections of people with topographic features and with past events (historic and legendary are not necessarily separate categories in the Gwich'in cosmology). Shiłdii, for instance, has been identified as one of the most important sacred sites to the Gwich'in. The story is told in different ways, but it generally refers to a young woman in puberty, who, in ignoring cultural prescriptions to be followed at this time, sets a series of events in motion resulting in her three brothers (and their dog) transforming into stone. The story is multidimensional, as it not only refers to space, but also to a particular season and to an activity (the brothers are actually returning from a caribou hunt). It is embedded in real space as events unfolded in recognizable places, making the story (and the place name) interconnect with people's lives. As told by William Nerysoo Sr., for instance, the brothers 'traveled to the mountains which you can see over there to the west of Ft. McPherson', and at Dèeddhoo Gòonlii (Scraper Hill) where the family was camping and fishing 'if you look around carefully you will see stones that the women used to bake (a kind of) bannock on,' a reference to another part of the story (Nerysoo, 1976: 98–102). The story, therefore, is anchored in real space, and the place name is part, not only of broader narratives, but also of relationships that people establish with place.

The mapping of the place named Shiłdii, therefore, will in theory separate the name from both the story and the experience of place. Fortunately, reality is less encumbered than theory by such distinctions, and performances (e.g. the act of reading a map by a Gwich'in person) often restore the words to their stories and to their proper place in the land. Semiologist Roland Barthes (1973) made it clear that written text becomes alive with new meaning every time it is read.

The reasons for mapping Indigenous place names are several, but in this paper we will briefly refer to three of them. The emergence and continuous development of mapping projects are a consequence of the need of Indigenous groups to: (1) reclaim their proper place in history, (2) reclaim their traditionally used lands, and (3) take action against a process of culture and language loss. Mapping is now facilitated by the possibilities offered by new technologies, and the new methods of documenting and representing place-based knowledge. These three reasons converge in the pragmatic decision of most First Nations and Inuit groups in Canada to undertake place names projects.

16.4.1 Reclaiming

The first and most obvious reason for mapping is political. Place names mapping projects can certainly be framed as acts of cultural resistance and re-appropriation. Maps have undoubtedly become tools of power, and putting Indigenous place names on the maps are certainly acts of justice. More often than not, places that have been long identified with English (or French) names have their counterparts in the local Indigenous group's language. The names may not refer to the same exact features, as the extent of the features is often conceptualized differently by different cultural approaches, but most English names coincide or overlap with Indigenous names. In Nunavut, for instance, the lake identified as Hall Lake (after a nineteenth century American explorer) is known as Tasiujak ('like a lake') among Inuit from the Igloolik region. At the same time, Inuit also have dozens of names describing shores, bays, hills, camps, and even rocks around the lake. The naming of small features around the lake reveals the intimate knowledge and intensive use of the area by Inuit, while the official English name shows the lack of such cultural context. A Gwich'in example of the same representational conflict is the Peel River, named by John Franklin for Robert Peel on his second overland expedition to the Arctic in 1826. The Gwich'in name for this river is Teetł'it Gwinjik, ('at the head of the waters river'). This river gives the Teetł'it Gwich'in their name as their traditional lands were centred upon this watershed. Many islands, river banks, canyons, hills, and camps along the river have traditional Gwich'in names, revealing their intensive use of the place and showing, again, a significantly different connection to these places than the singular connection implied in the name of the river as the 'Peel'. In the context of Indigenous place names projects, the mapping of place names could be understood as a process through which a western-European tool (the map) is appropriated as a means to empower local groups in their effort to reclaim their history and their rights. In many cases, the documentation of place names is done in the context of land claims negotiations, or to identify places of significance in the context of potential economic development (see, for instance, Milton Freeman Research Limited, 1976).

16.4.2 Language and Cultural Loss

As a result of a long history of contact, land claim negotiations and of formal schooling, maps have become part of people's lives among most Indigenous groups in Canada. While traveling on the land, topographic maps are often used as navigational tools, a practice that is becoming more intensive among younger generations (who are also using Global Positioning

Systems; see Aporta and Higgs, 2005). From a pragmatic point of view, it certainly makes sense for people to use maps with their own place names on them, but the significance of mapping place names goes beyond this, as maps and gazetteers are among the most important instruments to keep Indigenous peoples' histories and knowledge of the land alive. This is especially important in the present context of significant cultural change and language loss common to most Indigenous groups in Canada. Maps are indeed significant educational tools, and efforts to create new artifacts (beyond the traditional map) have been implemented all across Canada. In the Gwich'in context, the GSCI has produced a 1:250,000 map series and a wall map, and has submitted the documented place names for official recognition in both the NWT and Yukon.

16.4.3 New Methods

Finally, the flexibility offered by new mapping technologies and methods (particularly since the introduction of Geographic Information Systems –GIS- and Global Positioning Systems –GPS) has signified a step beyond the former 'static' character of the conventional map. Databases built in these new ways allow for the storage of geographic data in connection to data of different sorts. For instance, the meaning and story associated with a name could be documented and linked, not only to a geographic location, but also to photographs, audio recordings, and video material. Names and other features can also be mapped on location with GPS units, and maps and scales can be changed according to need.

As the technologies of mapping have evolved and merged with other technologies (e.g. multimedia), the possibility of creating a culturally appropriate tool also became clear. The convergence of multimedia, mapping, and web technologies has prompted a number of efforts to document not only the name as an isolated or discrete entity, but also in connection to larger narratives and in accordance with different ontologies of representation, as well as multisensory and interactive experiences of place. It is in this context that the Cybercartographic approach should be placed.

16.5 CYBERCARTOGRAPHIC APPROACHES AND THE GWICH'IN ATLAS

Explanations of the Cybercartographic paradigm, as defined by D. R. Fraser Taylor (2005), can be found in several chapters of this book including chapter one of this book. Of significance to this paper, it suffices to say that a Cybercartographic atlas is an artifact that allows for story-telling or story-building, beyond the representational limits of traditional maps. A Cybercartographic atlas is based on the premise that representations of place and space cannot be circumscribed to the two (or even three) dimensions of a map, but should contemplate different aspects of the concrete perception of place and space as experienced by people. In that sense, the Cybercartographic approach has developed alongside a number of technologies and techniques, allowing for the dynamic combination of data of different sources and different type, including audio, video, and photographs. This type of representation (namely, the combination of different types of data on a map) can be pursued through different approaches, including the use of multimedia products, or web-based mapping of the type offered by Google Earth. Cybercartography, however, offers a different type of approach as it goes beyond representations to more effectively combine the contents of databases. A Cybercartographic atlas consists mainly of a programming framework allowing for communication between datasets, resulting in interactive products that can be built by the user, both through the process of reading and/or while incorporating data.

The fact that this atlas is always in a process of being constructed makes it an ideal instrument for the mapping of Indigenous knowledge. Audio, video, and text recordings of oral histories can be intertwined with photographs and drawings, allowing narratives to emerge through the user's navigation of the atlas, and eventually, through his/her input *into* the atlas.

The types of narrative that emerge through this process are certainly not meant to replace knowledge and experience of place gained on location through traditional frameworks and methods of learning. In fact, a paper map with names and trails is often as useful and as rich in its integration of broader narratives and experiences as a web-based interactive atlas. Rather, a Cybercartographic atlas is useful in two other ways: (1) as a repository of datasets of different kinds, and (2) as a facilitator of learning among younger generations.

From 2010 to 2012, GSCI worked with Gwich'in elders, Carleton University's Geomatics and Cartographic Research Centre, and MDT Communications on the 'Gwich'in Goonanh'kak Googwandak: The Places and Stories of the Gwich'in' project. The goals of the project were to produce an online interactive Gwich'in Place Names Atlas, a series of Gwich'in place name maps for the Gwich'in traditional land-use area, a large wall map, and to submit almost 850 place names for official recognition on future government maps in both the NWT and Yukon.[3] The project was largely funded through the Federal Government's Museums Assistance Program and the GSCI.

The Gwich'in Place Names online atlas showcases almost 850 place names and is based on 20 years of GSCI research with Gwich'in elders living in Aklavik, Fort McPherson, Inuvik, and Tsiigehtchic. The online Atlas uses a Google Earth map image to present all the named places, and contains a database to house all of the information behind the named places. The squiggles, lines, polygons, and circles on the map represent the defined extents or boundaries for each named place. The extents were determined through extensive consultation with Gwich'in elders and other knowledgeable land users.

The Atlas presents a zoom-able map, where Gwich'in place names can be observed in more detail and at different scales. Most of the data collected through the mapping projects have been integrated into the atlas. The combination of geographic data (the place name) with other narratives can be observed, as the clicking of the mouse on the place name will bring up the associated textual oral history, and historical information about the named place and links to its sound files.

Searching the online Atlas can be done using either place names or keywords. For example, entering Chigwaazraii ('Black coloured mountain' – locally known as Black Mountain) in

[3] In a series of community meetings...Canadian government. In December 2011, the GSCI submitted 574 names for official recognition in the NWT. On June 21, 2013, the Government of the Northwest Territories officially recognized 414 of these names, significantly transforming future maps for the northwestern part of the NWT. GSCI also submitted 274 names for official recognition in the Yukon and is awaiting a final decision.

the search field will centre the map on the proper feature and highlight it in brown and blue. The atlas will also show all associated data related to the name and place. Searches through keywords such as 'sheep', 'caribou', 'dog', etc., will bring up all the place name records with these keywords.

Each place name record includes the Gwich'in name for the place, a translation of the name, and alternate names in cases of features that are known by more than one name. Besides the geographic location of the place name, the records include oral history, and also, in some cases, historical and archival information associated with that place. The next stage will involve adding additional material such as photographs and videos of interviews with elders. While this information has been collected over the years, and archived in different datasets, the Atlas allows us to articulate relationships among the data, and to make it available to a broader audience (particularly a Gwich'in audience) in a visually attractive manner.

Another important feature of the atlas is that the user can hear how the name is pronounced in Gwich'in. The pronunciation of the names was recorded with knowledgeable elders, and the audio files represent a valuable record of how the language is spoken before additional transformation and loss takes place. The GSCI hopes that younger Gwich'in will become more familiar with the names, and use them more often instead of the English counterparts. It should be noted that in cases where names are pronounced differently, recordings of multiple pronunciations have been added.

16.6 CONCLUSION

It is clear that the full dimensions of Indigenous place names cannot be captured by any representational technique outside of the original context of use. The limitations presented by traditional cartography, however, have been engaged with and expanded by new mapping and documenting techniques and technologies. While traditional paper maps could only render geographic location and spelling of an orally transmitted place name, new technologies allow for the recording and presentation of the stories and other information connected to the name. The capacity to actually record visual and audio information in connection with the name allows for a more comprehensive type of documentation and representation.

As we have seen with the Gwich'in Atlas, the Cybercartographic approach goes a step further, creating a digital environment for data integration. The importance of this cannot be underestimated, as it sets it apart from other kinds of narrative-constrained multimedia and web-based media. The Gwich'in Cybercartographic Atlas can be understood as a framework within which the user is able to build his/her own narrative through the seamless use and connection of different datasets. Geographic locations, audio recordings of the pronunciations of names, text, and eventually audio and video converge to facilitate the creation of a multidimensional narrative, controlled, to a large degree, by the user. The potential of this approach in the documenting of oral spatial and environmental narratives is beyond doubt, as is the possibilities of this new approach in the context of contemporary education or cultural programs. At the end of the day, however, the full multidimensionality of Gwich'in place names will be only materialized within the context of actual use of the land.

References

Andrews, T.D., 2011. There will be Many Stories. Museum Anthropology, Collaboration, and the Tłı̨chǫ. Ph.D. Thesis, University of Dundee.

Aporta, C., Higgs, E., 2005. Satellite culture: global positioning systems, Inuit wayfinding, and the need for a new account of technology. Current Anthropology 46 (5), 729–753.

Barthes, R., 1973. Le Plaisir du Texte. Editions du Seuil, Paris.

Freeman, M.M.R., 1985. Appeal to tradition: different perspectives on wildlife management. In: Brosted, J., Dahl, J. (Eds.), Native Power: The Quest for Autonomy and Nationhood of Aboriginal Peoples, Universitetsforlaget, Oslo, pp. 265–281.

Galeano, E., 1997. Open Veins of Latin America: Five Centuries of the Pillage of a Continent. Monthly Review Press, New York.

Geographical Names Board of Canada, 2007. Delineation Guidelines, Version 1.0. Geographical Names Board of Canada Secretariat, Ottawa.

Indian and Northern Affairs Canada, 1992. Gwich'in Comprehensive Land Claim Agreement. vol. 1.

Kritsch, I., Andre, A., 1997. Gwich'in traditional knowledge and heritage studies in the Gwich'in settlement area. In: Nicholas, G., Andrews, T. (Eds.), At a Crossroads: Archaeology and First Peoples in Canada, Archaeology Press, Simon Fraser University, Vancouver, BC, pp. 125–144.

Kritsch, I., Andre, A., Kreps, B., 1994. Gwichya Gwich'in oral history project. In: Pilon, J.-L. (Ed.), Bridges across Time: The NOGAP Archaeology Project Victoria, BC, Canadian Archaeological Association Paper No. 2, pp. 5–13.

Lewis, M. (Ed.), 1998. Cartographic Encounters: Perspectives on Native American Mapmaking and Map Use, University of Chicago Press, Chicago.

Milton Freeman Research Limited, 1976. Inuit Land Use and Occupancy Project. Dept. of Indian and Northern Affairs, Ottawa.

Mundy, B.E., 1996. The Mapping of New Spain: Indigenous Cartography and the Maps of the Relaciones Geográphicas. University of Chicago Press, Chicago.

Nerysoo, W., 1976. Shiłdii. In: Ritter, J. (Ed.), Gwich'in (Loucheux) Athapaskan Noun Dictionary, Fort McPherson Dialect Appendix 2. The Queen's Printer for the Yukon, Whitehorse, YT.

Pilon, J.-L. (Ed.), 1994. Bridges across Time: The NOGAP Archaeology Project Victoria, BC, Canadian Archaeological Association Paper No. 2.

Taylor, D.R.F., 2005. Cybercartography: Theory and Practice. Modern Cartography Series, vol. 4. Elsevier, Amsterdam.

Thrower, N.J.W., 2008. Maps & Civilization: Cartography in Culture and Society, third ed. University of Chicago Press, Chicago.

Wonders, W.C., 1987. Native land claims and place names in Canada's Western Arctic. The Canadian Journal of Native Studies 7 (1), 111–120.

CHAPTER

17

The Role of Experience in the Iterative Development of the *Lake Huron Treaty Atlas*

Stephanie Pyne

Geomatics and Cartographic Research Centre (GCRC), Carleton University, Ottawa, Ontario, Canada

OUTLINE

17.1 INTRODUCTION

> [Considering] the possibilities for practices of mapping that lead away from totalisation and open up understandings of place alive to the possibilities of flow, connectivity and transgression; to a logic of movement, desire, and possibility (Painter 2006; Pickles, 2006) [...] frames the problem in more fruitful, performative terms, with the emphasis on practices of mapping, on connectivity, and on other ways of being. It opens up the important question: can we envisage new forms of mapping, for example emergent mapping? ***(Turnbull, 2008: 219).***

The *Lake Huron Treaty Atlas* provides an example of a set of emergent mapping practices that 'open up understandings of place'. A central goal of this cybercartographic atlas-making

Developments in the Theory and Practice of Cybercartography, Second Edition, ISSN 1363-0814
http://dx.doi.org/10.1016/B978-0-444-62713-1.00017-9

project throughout all of its phases has been to gather knowledge and perspectives and present them in decolonizing ways (Pyne and Taylor, 2012; Pyne, in press): building awareness to bridge relationships in a reconciliation context. As an entity that contributes to knowledge sharing, enhanced awareness, and empowerment, the *Atlas* project actively participates in Lake Huron Treaty history. The nonlinear, nonteleological or spatialized approach to the design and development of the *Atlas* involves a revolving series of iterative processes along different dimensions. Iterative processes have to do with change and transformation and are constitutive of the *Atlas*, which goes beyond its online web-based instantiation to include a variety of interactions and experiences that can take varied extents of time and give rise to emergent knowledge (Turnbull, 2007; Kitchin and Dodge, 2007): 'Iterative processes are manifested both within and between atlas projects in the sets of relationships that occur, in information transmission and exchange, in design and development decision-making, and implementation more generally (Brauen et al., 2011: 28).'

This chapter sheds some light on the role of experience in the iterative development of the Survey Journeys maps of the *Lake Huron Treaty Atlas* by recounting the stories of two instances of key decision-making processes that shaped the design and development of the Survey Journeys maps. It further identifies a significant component of the decision-making process: the experience associated with and giving rise to a definite, certain, productive decision that integrates past learning and questions into a focused design solution (most commonly referred to as an epiphany (McDonald, 2008)).

17.2 ITERATIVE PROCESSES IN THE DEVELOPMENT OF THE *LAKE HURON TREATY ATLAS*

The online Atlas is comprised of a series of narrative maps or geonarratives (Caquard et al., 2009; Taylor and Pyne, 2010; Brauen et al., 2011; Pyne and Taylor, 2012) that encourage knowledge sharing and critical reflection along a variety of interrelated dimensions: historical, geographical, legal, political, economic, social, and cultural. The *Atlas* has been evolving in an iterative manner since it began six years ago as a pilot project to create several maps in the Treaties Module of the cybercartographic *Atlas of Indigenous Perspectives and Knowledge*. This initial pilot phase of *Atlas* work was funded by a one-year Inukshuk Wireless grant, and was intended to provide sample content to demonstrate the potential of the cybercartographic approach to tell stories in a multidimensional and collaborative way. Subsequently, funded by a three-year 2009–2012 Social Sciences and Humanities Research Council (SSHRC) grant, the Treaties Module has grown into an *Atlas* with 30 maps-in-progress and more potential maps planned, as the result of collaborations with Anishinaabe and non-Anishinaabe community members interested in sharing their perspectives, contributing historical geographical multimedia content, and exposing assumptions implicit in the Western world-view (Taylor and Pyne, 2010; Pyne and Taylor, 2012). To date, these maps include biography maps; maps related to Anishinaabe language and culture; historical background maps; and maps relating to current affairs. Although it is an ongoing project that is always in the making, the *Atlas* has been publicly accessible via the World Wide Web since May2012[1] as a transdisciplinary

[1] The Atlas can be accessed via the following web page: https://gcrc.carleton.ca/confluence/display/GCRCWEB/The+Lake+Huron+Treaty+Atlas.

research and education tool. The processes involved in knowledge dissemination and facilitating online community-based contributions to the *Atlas* are the focus of a SSHRC Outreach Grant, which began in the summer of 2012. A significant focus for research under this grant has been directed at improving the technological base for the *Atlas* to better enable people to log into the Atlas website, add multimedia content to the existing maps and, in some cases, make maps of their own (Pyne, in press).

The iterative processes comprising map and *Atlas* design and development are linked to a nonlinear, cyclical view of development: incorporating past elements (bringing established elements into the present, or continuing them), bringing new things forward, and leaving certain elements behind, with the possibility of reintegrating them into the development process after any number of project iterations (Cowen and Shenton, 1996; Nederveen Pieterse, 2009). This iterative development is both a function of relationships occurring within the evolving *Lake Huron Treaty Atlas* project, and of relationships between the *Lake Huron Treaty Atlas* and other atlas projects hosted by the Geomatics and Cartographic Research Centre (GCRC), the home of cybercartography. This *Atlas* could not be what it has become *without* the other atlas projects. Several of these other atlases are described in this book. At the same time, it is distinct from them and has generated some unique design solutions, which can in turn be applied to other cybercartographic atlas projects.

In Brauen et al. (2011), we described the iterative *Atlas* development process in the following manner:

> In our experience, digital atlas projects proceed iteratively as fora for the development and exploration of ideas related to a set of topics rather than simply as answers or manifestations of collected knowledge on those topics. This is especially true since these atlases are focused on finding ways to express multiple perspectives, each of which takes time to discover, explore, test, and integrate into an atlas. Iterative processes are manifested both within and between atlas projects in the sets of relationships that occur, in information transmission and exchange, in design and development decision-making, and implementation more generally (28).

Interactions, learning, emergent knowledge, and transformation are central aspects of the types of iterative processes that give rise to *Atlas* maps, and these iterative processes occur along at least four interrelated dimensions: conceptual; financial; technological; and narrative. For example, in the case of the technological dimension, common to the (*Lake Huron Treaty Atlas*) project and other research projects (at the GCRC) is the software base upon which these atlases have been created. This base is developed iteratively in conjunction with this and other atlas developments and is released as 'free and open source software (FOSS)' (Brauen et al., 2011: 29; Taylor and Pyne, 2010). Hayes in Chapter 9 describes the development of this software framework.

Interproject iterative processes involve building on, borrowing from or contributing to other *Atlas* projects. There are also intraproject iterative processes that occur within the life of a cybercartographic atlas project over its successive research and funding phases. The design and development of the Survey Journeys maps, including the evolution of the concept and method of 'geotranscription' (Pyne coined this term in Caquard et al., 2009), provides a good example of this. The iterative development of *Atlas* maps can be roughly divided into the following four stages: conception, prototype creation, prototype review, and prototype revision (Brauen et al., 2011). The genealogy of the Survey Journeys maps primarily in the first two funding phases is provided in order to give the reader a better idea of the iterative processes involved in the design and development of the *Atlas* in general, and of the role of experience in these processes, in particular.

17.3 THE ITERATIVE DESIGN AND DEVELOPMENT OF THE SURVEY JOURNEYS MAPS

> Cybercartographic atlases do not appear in full bloom at a single instant in time, but rather tend to develop over time through an iterative process comprised of design, implementation, and testing phases. Each cycle in this process may involve the production of prototypes (paper and/or digital) around which discussions of both an atlas's conceptual content and cartographic possibilities can occur. A new iteration begins, based on these discussions, through the integration of atlas design and development decisions, feedback from testing, and additional stories and questions that may arise through discussion of information synthesized by already existing versions of the atlas. Story precedes atlas design insofar as it creates a context and purpose for communication but atlas narrative and representational possibilities develop in tandem ***(Brauen et al., 2011: 29).***

A variety of iterative processes contributed to the conception of the Survey Journeys maps. The Treaties Module was not always focused on the Lake Huron Treaty region. At the beginning of the project, the context was a fairly broad definition of the Great Lakes region. Because this was a project to develop sample content, it was clear that after a certain period for preliminary background research, a decision would have to be made with respect to historical and geographical focus. In addition to my prior knowledge and understanding, I began with learning about 'cybercartography' as a theoretical and practical discipline to get a better idea of the conceptual and design framework (Taylor, 1997, 2003, 2005, 2008, 2013). I also tried out mapping ideas suggested in proposal:

> The original mandate [...] was to demonstrate the efficacy of cybercartography in representing information relating to territorial claims and in encouraging indigenous and other community participation. This would involve collecting data and content related to territorial claims in the Great Lakes-St. Lawrence region and transforming it into geonarrative form (with the aid of other GCRC team members) in a manner that would be meaningful to a broad audience, with the added requirement that the content should reflect Indigenous perspectives. In addition to a variety of historical and contemporary research, a polygon-drawing exercise was initiated to gauge the feasibility of mapping out the written descriptions of a selection of historical treaty and surrender documents. For a variety of reasons, it was determined this was not feasible and, as the research and development process began to unfold, the story came to centre on a particular series of relationships surrounding the Robinson Huron Treaty. The decision process involved in focusing on this treaty reflected interactions between atlas-making contributors from several interrelated knowledge communities, and was also the first substantive instance of indigenous community member participation in the project ***(Brauen et al., 2011: 33).***

The first significant Anishinaabe contribution to the map-making process happened about five weeks into the research project in July 2007 when I received a forwarded email from Alan Corbiere, then-Director of the Ojibwe Cultural Foundation, through Heidi Bohaker, a historian and *Atlas*-making team member. Alan had learned about the *Atlas* project through Heidi and Ruth Phillips, a coinvestigator on the project. His email referred to a paper by Michael Marlatt, *The Calamity of the Initial Reserve Surveys under the Robinson Treaty* (2004), which includes a relatively detailed description of the extended survey process for the Lake Huron and Lake Superior region treaties of 1850. After reading the paper, I was sure that the story of the Lake Huron Treaty signing and survey process would serve as an ideal sample content for the Treaties Module: a perfect example for a cybercartographic geonarrative designed to include previously excluded worldviews.[2] Up to this point, in addition to the knowledge and experience that I brought to the project, I had emerging knowledge of: (1) the parameters of the project; (2) cybercartography as a theoretical and methodological framework and the first two atlases produced by the GCRC; and (3) treaty-related history; but I was still not sure exactly 'what' to map, or how. After this point, I was sure. Amazed at how much more there was to the Robinson Huron Treaty than the actual treaty document and impressed by the research in the Marlatt chapter, I now had a working vision for a map and a story to focus on. The experience associated with this significant decision point, which resulted in the initial map conception, can be referred to as a mapping epiphany.

In addition to participating in weekly meetings with an atlas-making team, I attended the 39th Algonkian Conference in Toronto in September 2007 where I met some of the key Robinson Huron Treaty and related issues historical researchers, told them about the idea to map out the Marlatt chapter, and learned about their work. Following the conference, I created a PowerPoint prototype of the proposed map based on the Marlatt chapter with plans to construct a map based on this by February 2008.

Then, in January 2008, another iterative transformation in map design and development occurred. I participated with Alan Corbiere (mentioned above) and Amos Hayes, the GCRC's technical laboratory manager, in mapping the traditional story of Nenboozhoo and the creation of Mindemoya Island for the 'Culture' section of the *Atlas of Indigenous Perspectives and Knowledge*:

> According to this story, Nenboozhoo was in the Lake Erie region playing all kinds of tricks on people in all sorts of ways. Eventually, the people chased him out and he ran away, heading for Manitoulin Island, carrying his grandmother on his back. After travelling across the water from the Bruce Peninsula to the southern shore of Manitoulin Island, Nenboozhoo began to run toward Mindemoya Lake. Finally, he could no longer carry his grandmother; when he reached the lake, he threw her into it. She became what is today called Mindemoya Island: "Enh, mii 'mindemoya' enji-zhinkaadeg. Mii dash [Yes, that is why it is called Mindemoya]" (Johnny Debassige speaking in Anishinaabemowin; English translation by Alan Corbiere). In the atlas, this journey has been rendered through a multimedia map showing the different steps of Nenboozhoo and his grandmother. Audio files of Johnny Debassige telling the story in Anishinaabemowin have been added, along with English translations by Alan Corbiere. The combined Anishinaabemowin/English translated text enables the story to remain in its original language while at the same time making it accessible to a wider audience ***(Caquard et al., 2009: 94).***

Although I already had a map idea with the decision to render the Marlatt chapter into a map, participating in the creation of the Nenboozhoo Mindemoye map gave rise to a second epiphany in the iterative map design and development process. At this point, I had already engaged in a series of phone conversations with Michael Marlatt who had sent me digital copies of archival records including the Dennis survey report, diary and field notes

[2] Marlatt's chapter is particularly instructive for a number of reasons, including the fact that it is based on land claims research conducted by the author on behalf of one of the signatory First Nations and the fact that it provides a critical examination of the survey process from a surveyor's perspective. Marlatt describes the political and economic context of the treaty signing and survey processes and, drawing on the original surveyor's reports, includes a nation-by-nation narrative tabulation of the various errors and omissions in the surveys themselves.

(Dennis, 1851, 1853), and sketch maps. I had also ordered a high-resolution digital copy of the Vidal–Anderson Commission 1849 Sketch Map (Vidal and Anderson, 1849), which Alan had suggested that I do. That evening the idea was born to transcribe and map out – or geotranscribe – the Dennis survey diaries (Caquard et al., 2009):

> The geonarrative focus shifted to a consideration of a more direct "tracking" of the surveyor. Although an initial map idea had already been generated and discussed, the experience of mapping out Nenboozhoo's journey to Mindemoye had inspired a new yet related map idea: To transcribe and map out J.S. Dennis survey diaries, the primary document upon which the Marlatt chapter was based, in addition to mapping out the Marlatt chapter as a critical geonarrative summary of the survey diaries. The diary referred to more than 100 distinct campsites, provided additional details to the Marlatt chapter, and was far easier to follow when translated into geonarrative form. This iterative shift in focus has not only created new possibilities for analysis and critical understanding, but has also helped shape ideas for additional geonarratives for inclusion in the *Atlas* **(Brauen et al., 2011: 34).**

Considering that one of the central goals of the *Atlas* project was to incorporate Indigenous perspectives and knowledge, it is significant that the focus for the Treaties Module was suggested by an Anishinaabe individual and implemented. The decision made in January 2008 to geotranscribe the Robinson Huron Treaty Survey Diaries is an example of intraproject iterative processes, in this case one map benefitting from a design insight inspired by participation in another mapping activity within the same *Atlas* project.

The map creation process began from that evening onward and was followed by approximately three weeks of intensive research using a variety of sources to identify the approximate camp locations indicated in the survey diaries; in addition to transcribing the diary entries themselves. Sources included: the Dennis survey diaries for 1851 and 1852[3]; Google Earth; the online *Atlas of Canada*; Michael Marlatt, who contributed background knowledge regarding survey techniques and instruments of the time and knowledge regarding camp locations; Patricia Kennedy, a senior archivist specializing in file RG-10, the Indian Affairs file for Library and Archives Canada (LAC); and a diverse range of websites from the online Internet community, including First Nations community websites. Since each point on the map could accommodate one place-based image, I was also keeping an eye out for relevant media. I sent emails to the people whose images I had chosen for the maps, describing the mapping project and asking for use permission. Concurrently, I was learning about the history and geography of some of the places being mapped and planned to include some of this contextual information, ideally, as another map layer.

In February 2008, I made a trip to the Lake Huron Treaty region to visit with community representatives. Although I had spent months previous to this visit engaging in telephone conversations with a number of people, this would be the first face-to-face start towards working together on a content development. Much of a February 2008 field trip to a group of Lake Huron Treaty First Nations was devoted to discussing the Treaties Module and the vision for expanding it into an atlas. These initial collaborative investigations contributed

[3] The Dennis diary for 1853 was subsequently transcribed during phase II of the Atlas project. Some camps were located. During the 1853 season, one of the Robinson Huron reserves (Batchewana) was surveyed along with several reserves referred to in the Robinson Superior Treaty, which was signed the day before the Robinson Huron Treaty in the same location, Bawating or Sault Ste. Marie, Ontario.

in a significant way. For example, at the meeting at the Ojibwe Cultural Foundation in M'Chigeeng to review the map prototype, Chief Wilmer Noganosh from Magnetawan First Nation 'expressed concern about the use of "mouth of the river" as a geographical reference demarcating the territorial boundary, noting probable cultural differences in the interpretation of this feature' (Brauen et al., 2011: 37). This comment marked the birth of the idea to create a critical commentaries layer – a layer that would be available to all atlas users, and provide a space for additional discussion within the community.

At this point, the Surveys Journeys maps were partially constructed. By the time the *Atlas* was launched on April 24, 2008, three maps had been created reflecting three phases of the 1851 survey season: en route, in the field, and return. The addition of a video clip with Gimaah Art Petahtegoose, explaining the difference between 'Chief' and 'Gimaah', was actually the first example of a critical comment reflecting the Anishinaabe perspective, even though it was contained in the Background section of the Treaties Module and not in an actual critical layer attached to the Survey Journeys maps.

The prototype Treaties Module includes four sub-modules or *Atlas* parts. The first three sub-modules – Spirit, Welcome, and Background – provide the context for the Survey Journeys Maps in the fourth sub-module. The Spirit sub-module represents in a story form the spirit and intent of the Treaties Module: To put together the various parts of the Lake Huron Treaty process story in a way that enhances awareness of the ethical aspects of its many interrelated dimensions, and to contribute to knowledge of how to engage in healthy and fair treaty relationships today – in short, building awareness to bridge relationships. The Welcome sub-module includes guiding concepts for a critical learning approach. For example, it draws attention to the existence of different perspectives, identifies four important criteria of 'good' treaty relationships (fairness, balance, care, and mutual understanding), and invites viewers to assess the treaty process according to these criteria. The Background sub-module involves a time-line account of some relevant historical background to the Lake Huron Treaty survey journey process. It includes images of Anishinaabe art and artefacts that combine to augment awareness of Anishinaabe understandings and perspectives, and of the historical geopolitical and sociocultural context of the times (Figure 17.1).

In the Survey Journeys sub-module, the 1851 survey diary of J. S. Dennis (Dennis, 1851) was transcribed and mapped with the aid of diary descriptions, survey plans, current and historical maps, and satellite imagery. This represents the first of a two-season field trip by Dennis and his survey party to delimit the boundaries of the reserves referred to in the Treaty. Each camp stop in this journey was overlain on one of two significant historical background maps. The 'En route' and 'Return' maps show a series of camp stops set against a standard settlement map, which includes artistic images denoting 'Indian Territory', and which tells a story in itself, thus illustrating the potential for maps in general as effective narrative vehicles. The 'In the Field' camp stops are set against a map drawn during the 1849 pre-treaty investigative process by Commissioner Alexander Vidal (Vidal and Anderson, 1849). Significant aspects of the reserves survey process are made apparent by overlaying the diary information on this particular historical base map, which depicts the Commissioner's impressions of Anishinaabe territories in the Lake Huron region, in addition to mining lots within the officially demarcated 'Indian Territories' that had been applied for by prospectors, but which could not be officially approved by the Crown until after the completion of

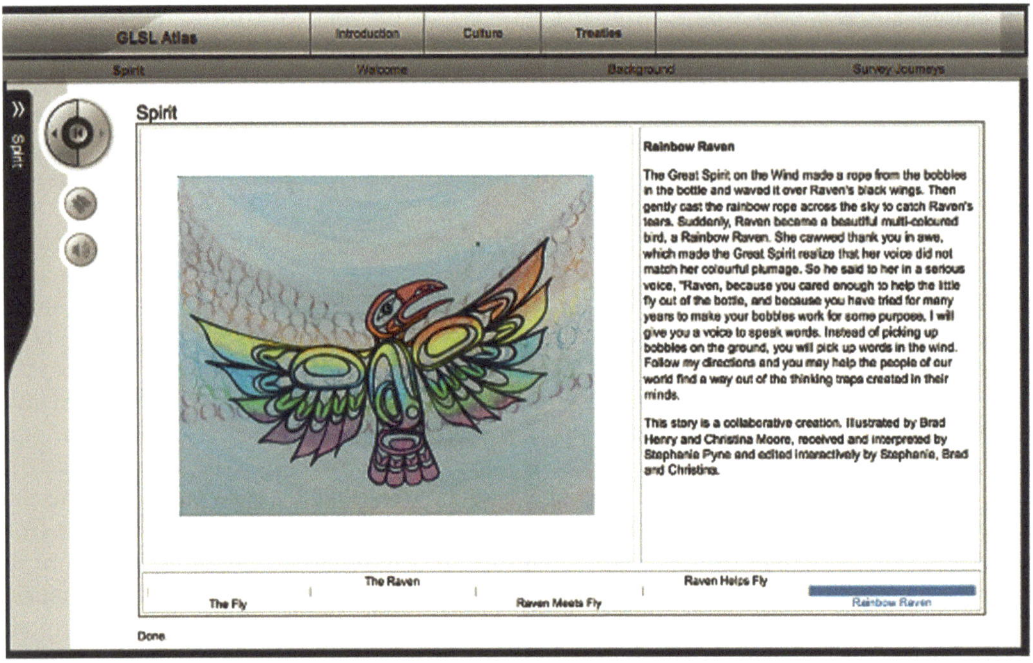

FIGURE 17.1 Screenshot showing final panel of an introductory section to the Survey Journeys Maps in the Treaties Module of the *Atlas of Indigenous Perspectives and Knowledge*.

the reserve surveys (see Figure 17.2). The diary entries are transcribed from Dennis's original handwritten report, and are included to the right of the map together with audio clips of the transcriptions with a voice representing Dennis. The audio clips include sound effects for such things as wind and rain where those phenomena are mentioned in the diary entries, and an effort is made by the voice artist to convey the emotion expressed by the surveyor in his transcriptions (for example, there are entries where the surveyor explicitly mentions dreariness).

During the period following the launch of the *Atlas*, work along the conceptual dimension occurred with the writing of a paper about the project (Caquard et al., 2009) and several conference presentations concerning the Treaties Module. Several months after the launch of the *Atlas*, a decision was made to apply for SSHRC funding to support the expansion of the Treaties Module into an atlas. The preparation for the paper, the conference presentations, and proposal writing included research into critical cartography and provided an opportunity both to situate the *Atlas* project in the ongoing conceptual dialogues, and to draw on insights from those dialogues in the *Atlas* project. Critical cartographers recognize: first, the significant ways in which maps are social constructions (Crampton, 2001); second, how many maps have excluded certain perspectives (Harley, 1989, 2001; Perkins, 2003); and third, how such exclusion has been systemically guided by imperialistic and colonial goals (Craib and Burnett, 1998; Edney, 1997; Given, 2002; Godlewska, 1995). It was proposed that insights from critical cartographic thought and practice would continue to be incorporated into the theoretical grounding of the project

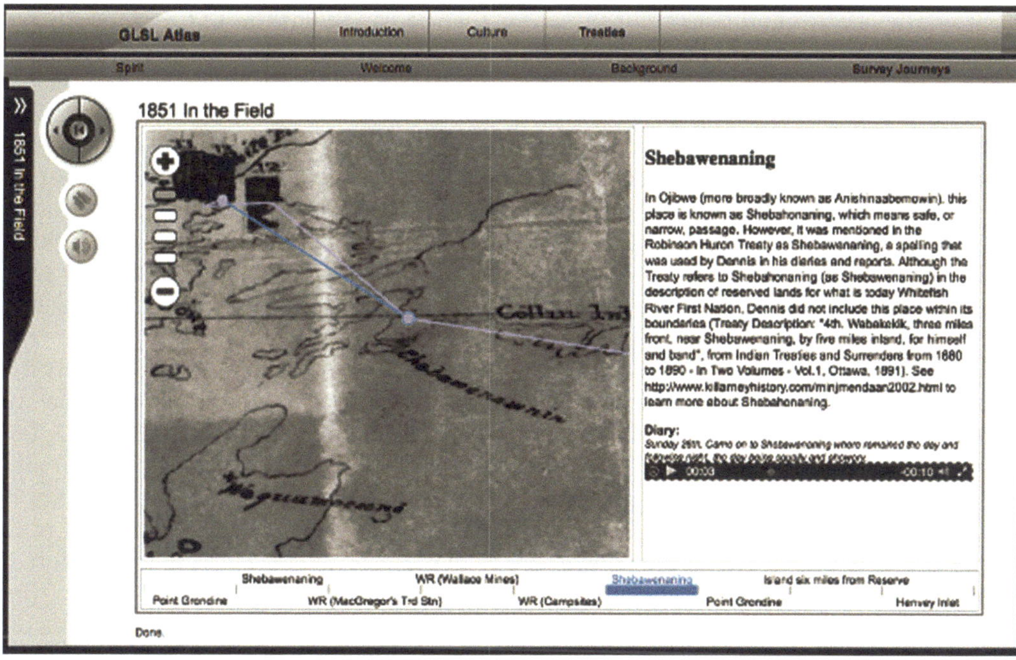

FIGURE 17.2 Screenshot showing J. S. Dennis survey-party camp locations set against the Vidal–Anderson Commission sketch map in the 'In the Field' portion of the Survey Journeys Maps in the Treaties Module of the *Atlas of Indigenous Perspectives and Knowledge*.

with maps being viewed first, as processes (Del Casino and Hanna, 2006) and second, as media of expression that have the potential to rectify previous exclusionary practices (Corner, 1999; Sparke, 1998, 2005) through alternative mapping approaches (Lewis, 1998; Fox et al., 2005).

The funding application was successful and by May 2009, we were ready to begin working on Phase II, *A Cybercartographic Case Study of the Lake Huron Treaty Relationship Process*, funded by a three-year SSHRC Standard Research Grant. The concept for the *Lake Huron Treaty Atlas* was the product of the iterative processes that contributed to the creation of the Treaties Module in phase I of the project. Through the design and development process in phase I, the vision for a more comprehensive series of maps emerged. For example, an important aspect of this vision was the biography map idea. The idea to create biography maps of some of the individuals involved in the treaty signing and survey processes was inspired largely by John Allen's (2004) relational space view, which draws attention to the role of character in institutional power relations.

Along the technology dimension, the conversion process of the Treaties Module from the Nunaliit I cartographic software framework to Nunaliit II represented the main interproject iterative process, since Nunaliit II was developed in the context of other atlas projects underway at the GCRC. Iterative processes occurred along the concept dimension as a function of learning about other approaches in a critical cartography. Iterative processes occurred between the content, technology, and concept dimensions in the further development of survey journeys maps and the extension of the Treaties Module (produced in Phase I) with one three-part map into the *Lake Huron Treaty Atlas*, composed of just over 30 maps.

On May 27, 2009, the first version of the *Lake Huron Treaty Atlas* was uploaded to the server. It was initially comprised six maps, three for season one and three for season two of the Lake Huron Treaty survey journeys. The first iterative transformation to occur in Phase II was along the technology dimension with the decision to convert the Treaties Module content to the Nunaliit II open source atlas framework (described by Hayes, chapter 9). Around this time and following, the technological and content dimensions began to develop iteratively as the technical and content specialists began to have regular meetings concerning the form and functions of the maps envisaged for the *Atlas*. It was decided that the initial priority would be getting these maps up and running in the Nunaliit II framework; then thinking about how to integrate the content of the Spirit, Welcome and Background sub-modules, or whether to integrate them at all. Once the survey journeys maps were working properly, it was thought that attention would turn to the creation of a map template for some of the additional maps planned for the *Atlas*.

The survey journeys maps ended up developing in an unexpected way. Although all of the content for the season two Survey Journeys Maps had been entered into the system, the maps for this season were not published in the Treaties Module that was launched at the end of phase I. It was not until the conversion of the Survey Journeys maps to the Nunaliit II software framework, with its improved usability, that I was able to review the Season Two maps. Through this interaction, I was reminded that at a certain point – contrary to his official instructions – the principal surveyor, J. S. Dennis, delegated some of his duties to a junior surveyor, Arthur Bristow. For several weeks, Dennis was in one place reporting on his own progress in that location, but at the same time appending a summary of the junior surveyor's progress in an entirely separate location (Dennis, 1853). And the decision was born to create a separate map trail for the second surveyor featuring his reports appearing in the vicinity of where he actually was. This addition to one of the Survey Journeys maps was an important innovation and the result of iterative processes along several dimensions: content, conceptual, and technological (Brauen et al., 2011, Figure 17.3).

Once the issues related to mapping the trail of the second surveyor had been sorted out and the new map with two trails had been released, attention could turn to the design of the new maps planned for the *Atlas*. The first idea was to use the data compiled for the first version of the map based on Marlatt (2004) to create the Marlatt Commentaries Map, a map that could also include some of the content from the Background sub-module of the Treaties Module, in addition to new background content.

The work to create a Survey Journeys map with two synchronous trails provided a prototype for mapping multiple trails that can be used in other maps. In addition, the Survey Journeys maps provided the basis for the design of the more generic 'biography map' template, which was subsequently used for most of the later maps added to the *Atlas*. Although it is beyond the scope of this chapter, it is possible to give a genealogical account of the iterative design and development of each of the additional maps that were added to the *Atlas* in phase II by the time the *Atlas* was launched in April 2012.

Closer to the end of phase II, continued discussions about the design of a critical map layer for the Survey Journeys and other maps in the *Atlas* led to the decision to the convert the *Atlas* once more, this time to a newer version of Nunaliit supporting a Couch DB-based document oriented database. The idea of creating a critical layer that can be viewed in conjunction

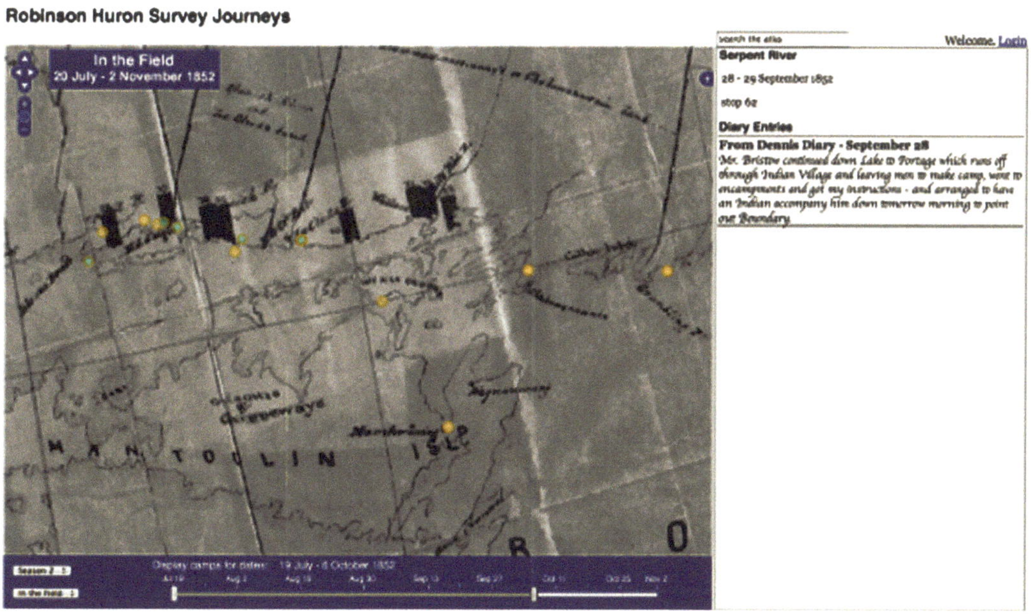

FIGURE 17.3 Screenshot with green dots representing Arthur Bristow's approximate camp spots and yellow dots representing J. S. Dennis' camp spots set against the Vidal–Anderson Commission sketch map in the 'In the Field' portion of the Survey Journeys maps in the *Lake Huron Treaty Atlas*.

with the original comments layer, which was initially proposed in phase I, remained strong throughout phase II as an agenda item for development in phase III.

17.4 EPIPHANIC PROCESSES IN THE ITERATIVE DESIGN AND DEVELOPMENT OF THE SURVEY JOURNEYS MAPS

The description given above provides a general overview or sketch of the iterative development of the Survey Journeys Maps in order to provide the necessary context to consider the role of phenomenological processes involved in the iterative design and development of the *Atlas*. The two instances of significant geonarrative design decision-turning points discussed above, which occurred first in July 2007 and second in January 2008, are of particular interest in this regard. In the context of the iterative processes involved in geonarrative development, the experiences associated with these decisions functioned as map design epiphanies. Although the concept of epiphany has been studied across various disciplines, including sociology, psychology, literary, and religious studies, and has even extended to design research (Verganti, 2011), much remains to be learned (McDonald, 2008):

> Epiphanies [...] are sudden and abrupt insights and/or changes in perspective that transform the individual's concept of self and identity through the creation of new meaning in the individual's life. Epiphanies are momentary experiences of transcendence that are enduring and distinct from other types of developmental change and transformation. Due to this distinction, positive change and transformation that is sudden and abrupt is a relatively under-researched and underdeveloped phenomenon. ***(McDonald, 2008: 90).***

TABLE 17.1 Common Characteristics of Epiphanies (McDonald, 2008) with Brief Descriptions of How the Two Mapping Epiphanies Involved in the Iterative Development of the Survey Journeys Maps Share These Characteristics

Common Characteristics	Epiphany 1 – the Marlatt Chapter Map	Epiphany 2 – the Survey Journeys Maps
Antecedent state Epiphanies are preceded by periods of anxiety, depression, and inner turmoil (Denzin, 1989, 1990; Jarvis, 1997; Jensen, 1998, 1999; Loyttyniemi, 2001; Miller and C'de Baca, 1993, 2001).	A period of uncertainty preceded the mapping epiphany especially in terms of knowing what to map. Dealing with these feelings required that I balance this uncertainty with a confidence that it would present itself.	As I started talking with community members, I started to realize how much more there was to the 'treaty story'. Although the critical review of the survey process provided by Michael Marlatt provided an excellent geonarrative resource, I had a feeling there should be something else.
Suddenness Epiphanies are sudden and abrupt (Beja, 1993; Goud, 1995; Jarvis, 1997; Jensen, 1998, 1999; Miller and C'de Baca, 1993, 2001; Schultz, 2001).	The decision to map the Marlatt chapter came as I was reading the chapter, which I got from the library the day after reading the email.	The decision to map the J.S. Dennis survey diaries came as I was driving home from the day of mapping the Nenboozhoo story.
Personal transformation Epiphanies are an experience of profound change and transformation in self-identity (Denzin, 1989, 1990; Goud, 1995; Jarvis, 1997; Jensen, 1998, 1999; Miller and C'de Baca, 1993, 2001).	Transformation occurred individually in terms of my newfound confidence as a 'mapper', now that I had a story to map; and it occurred in terms of my newfound appreciation of history and geography. The Treaties Module experienced transformation as well, as it now had a geonarrative focus.	I found myself engaged in a diverse array of mapping practices throughout the three to four weeks of tracking the surveyor. I interacted with a variety of places on a variety of maps, and learned a variety of things about history and geography, cartography, and archival knowledge.
Illumination/insight Epiphanies are an acute awareness of something new, something that the individual had previously been blind to (Denzin, 1989, 1990; Goud, 1995; Jarvis, 1997; Jensen, 1998, 1999; Miller and C'de Baca, 1993, 2001; Paris, 1997; Schultz, 2001).	The key insight was the vision for the map, which had crystallized at the moment of decision to map the chapter. A related insight emerged with the first instantiation of the geotranscription method.	The key insight was the vision for the map, which had crystallized at the moment of decision to map the J.S. Dennis field diaries. A related insight emerged with the second instantiation of the geotranscription method.
Meaning making Epiphanies are profound insights because they are deemed significant to the individual's life (Denzin, 1989, 1990; Frick, 2001; Miller and C'de Baca, 1993, 2001).	Significance to my life as a PhD student and to the identity of the treaties module.	Significance to my life as a PhD student and to the identity of the treaties module.
Enduring nature Although the actual epiphany is a momentary experience, the personal transformation that results is permanent and lasting (Denzin, 1989, 1990; Jarvis, 1997; Jensen, 1998, 1999; Miller and C'de Baca, 1993, 2001).	Formed the basis for an ongoing approach to geonarrative that has since been further developed around the original core.	Resulted in a more sophisticated extension of the original map planned and gave rise to the emergence of geotranscription, a unique cybercartographic method.

Closer to the idea of mapping design solutions, the term epiphany is rooted in 'the ancient Greek word "epiphainesthai", which means to "appear" or "to come into view" (Arnold, 2002) and was used to refer to moments of sudden and significant insight (Paris, 1997)' (quoted in McDonald, 2008: 91). Although the narrative details of people's epiphanies vary greatly, certain common characteristics remain not only between these epiphanies, but also in terms of the categories of the characteristics listed by different academic researchers across disciplinary perspectives (this suggests that it is a very transportable concept across disciplines). Mapping epiphanies share the same characteristics with other epiphanies.[4] Table 17.1 lists some common characteristics of epiphanies (McDonald, 2008) with brief descriptions of how the two mapping epiphanies involved in the iterative development of the Survey Journeys maps share these characteristics with other observations of 'epiphany'.

17.5 DISCUSSION AND CONCLUSION

A primary concern for critical cartography is with whether or not cartography is capable of meaningfully conveying such things as experience, Indigenous perspectives and knowledge, and critical academic approaches to the status quo (Johnson et al., 2006; Turnbull, 2007). The iterative nature of the *Atlas* design process, including the phenomenological dimension, begins to meet this challenge. In this chapter, I have designated the experiences surrounding significant map design decisions as mapping epiphanies, which are not only an important part of the design process, but also an important bridging concept between academic and Anishinaabe perspectives with epiphanic decision-making closely approximating the Anishinaabe notion of 'following your heart'. Drawing attention to the phenomenological aspect of mapping highlights the processual nature of mapping and is closely linked to notion of emergent knowledge, a central concept in critical cartographic discussions that emphasize performance (Kitchin and Dodge, 2007; Turnbull, 2007). Through their functional role in the iterative design and development process, mapping epiphanies provide an example of a nonlinear, nonteleological approach to learning and development (Pyne, in press; Pyne and Taylor, 2012) that contrasts with a linear and incremental (rational) approach to learning.

Learning across knowledge spaces is one of the defining characteristics of the transdisciplinary research approach taken in this *Atlas*-making project (Brauen et al., 2011). However, one dimension often missing from transdisciplinary approaches is the phenomenological dimension. The stories of the mapping epiphanies described in this chapter bring together at least two streams of thought in critical cartography: The first is concerned with approaching cartography performatively in terms of mapping processes (Kitchin and Dodge, 2007; Turnbull, 2007); and the second is concerned with a mapping experience (Pearce, 2008; Kwan, 2007). Without the occurrence of the mapping epiphanies in July 2007 and January 2008, there would be no Survey Journeys maps; their role in the map-design decision process was essential. Writing and telling stories about these iterative processes in the making of the *Atlas* maps can be considered as a form of map performance that enriches people's understandings of the maps. Some of these stories are already included

[4] McDonald (2008) identified this list of common characteristics between various multidisciplinary observations.

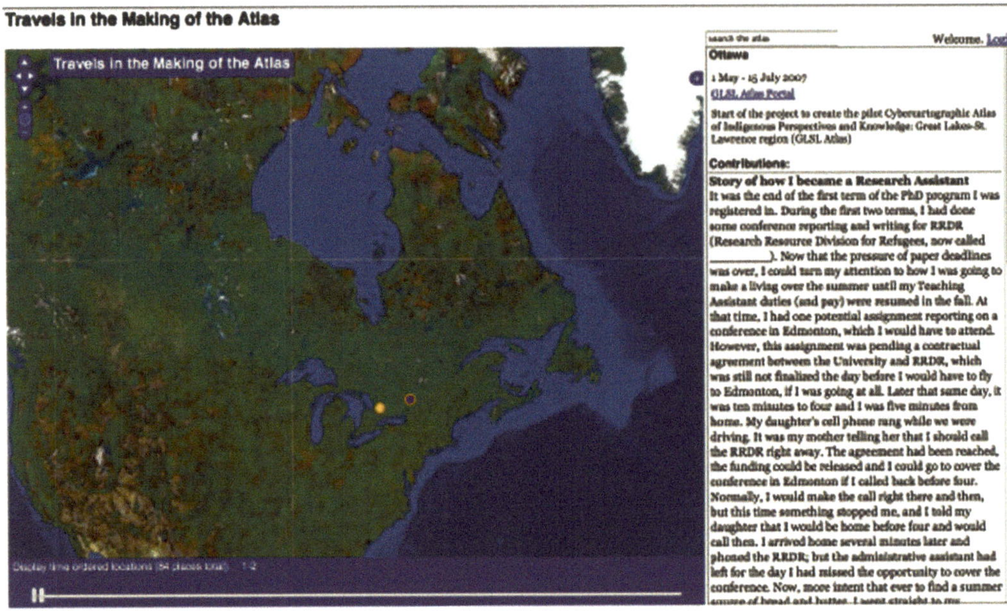

FIGURE 17.4 Screenshot illustrating one background story in the Travels in the Making of the Atlas map of the *Lake Huron Treaty Atlas*.

in the Travels in the Making of the *Atlas* map, which begins to document a diverse array of *Atlas*-making field trips, thus making the experiences not only part of the process of map-making but also a part of the maps themselves (Figure 17.4).

These mapping epiphanies were contingent on mapping interactions involving an Anishinaabe community member and reflected Anishinaabe perspectives and understanding, in addition to being akin to the Anishinaabe advice to follow your heart. There is a growing literature on the role of emotion in decision-making, including neurophysiological literature demonstrating the impossibility of making decisions without emotional processes (Damasio, 1994; Evans, 2002; Hoggett and Thompson, 2002). The multimedia 'geotranscription' method of documenting archival documents emerged through iterative processes incorporating Anishinaabe ways, social interaction, learning, mapping epiphanies, and an iteratively developed technology framework.

The critical cartography literature is full of conceptual transformational thinking. Being on the edge of a novel epistemological and ontological stance with respect to maps and mapping, many of the discussions in this literature are either lacking in exemplary detail or are calling on others to begin mapping from this new epistemological and ontological perspective. The *Lake Huron Treaty Atlas* project participates in this transformation by practising this new stance through acts of mapping. These acts are based on bridging concepts between critical academic and Anishinaabe perspectives that include epiphanies, geonarrative, and an emphasis on performance and the processes involved in the ontogenesis of the *Atlas* project, which is linked to Anishinaabe emphasis on action (verb based language) (Joe LaFord, Anishinaabe Elder, personal communication, December 2012).

These concepts work together in the design and development of the *Atlas*, exemplifying a two-pronged approach (Johnson et al., 2006). This is important for overcoming the incommensurability issue referred to by Turnbull (2007) who recommends that cartography be engaged in bringing together incommensurable knowledges. However, significant commonalities have been revealed between critical perspectives and Anishinaabe understandings in the process of applying them to the design and development of the *Lake Huron Treaty Atlas*. Although these perspectives come from contrasting worldviews to some extent, in various ways they have been questioning some of the basic tenets of the Western world-view – tenets that form the basis for the incommensurability with Indigenous perspectives and knowledge – and replacing these tenets with assumptions that are more in tune with Indigenous ways.

Acknowledgements

I thank all the community members, artists, academics and individuals who have contributed to this project. In addition I would like to thank the members of the GCRC who have been involved in the ongoing Atlas project.

References

Allen, J., 2004. The whereabouts of power: politics, government and space. Geografiska Annaler 86B, 19–32.

Arnold, K., 2002. Anti-epiphany and the Jungian manikin: Toward a theory of prepsychotic perceptual alterations. Journal of Phenomenological Psychology 33 (2), 245–275.

Beja, A., 1993. Epiphany in the Modern Novel. University of Washington Press, Seattle.

Brauen, G., Pyne, S., Hayes, A., Fiset, J.P., Taylor, D.R.F., 2011. Encouraging transdisciplinary participation using an open source cybercartographic toolkit: the atlas of the Lake Huron Treaty relationship process. Geomatica 65 (1), 27–45.

Caquard, S., Pyne, S., Igloliorte, H., Mierins, K., Hayes, A., Taylor, D.R.F., 2009. A 'living' cybercartographic atlas of indigenous perspectives and knowledge of the Great lakes region. Cartographica 44 (2), 83–100.

Corner, J., 1999. The agency of mapping: speculation, critique and invention. In: Cosgrove, D. (Ed.), Mappings, Reaktion Books, London, pp. 213–252.

Cowen, M.P., Shenton, R.W., 1996. Doctrines of Development. Routledge, London; New York.

Craib, R.B., Burnett, D.G., 1998. Insular visions: cartographic imagery and the Spanish–American war. Historian 61 (1), 100–118.

Crampton, J., 2001. Maps as social constructions: power, communication and visualization. Progress in Human Geography 25 (2), 235–252.

Damasio, A., 1994. Descartes' Error: Emotion, Reason, and the Human Brain. G.P. Putnam's Sons, New York.

Del Casino Jr., V.J., Hanna, S.P., 2006. Beyond the 'binaries': a methodological intervention for interrogating maps as representational practices. ACME: An International E-Journal for Critical Geographies 4 (1), 34–56.

Dennis, J.S., 1851. Field Note Book (FNB) 832, Report, Diary and Field Notes, Survey of the Indian Reserves on Lake Huron, vol. 1.

Denzin, N.K., 1989. Interpretative Biography. Sage, Newbury Park, CA.

Denzin, N.K., 1990. Interpretive Interactionism. Sage, Newbury Park, CA.

Edney, M.H., 2005. The Origins and Development of J. B. Harley's Cartographic Theories. Cartographica Monograph, 54; Cartographica 40, nos. 1 & 2. University of Toronto Press, Toronto.

Evans, D., 2002. The search hypothesis of emotions. British Journal for the Philosophy of Science 53, 497–509.

Fox, J., Suryanata, K., Hershock, P., 2005. Mapping Communities: Ethics, Values, Practice. East-West Center, Honolulu, Hawaii.

Frick, W.B., 2001. Symbolic latency: images of transformation across time and space. Journal of Humanistic Psychology 41 (1), 9–30.

Given, M., 2002. Maps, fields and boundary cairns: demarcation and resistance in colonial Cyprus. International Journal of Historical Archaeology 6 (1), 1–22.

Godlewska, A., 1995. Map, text and image – the mentality of enlightened conquerors – a new look at the Description de L'Egypte. Transactions of the Institute of British Geographers 20, 5–28.
Goud, N.H., 1995. Vital moments. Journal of Humanistic Education and Development 34 (1), 24–35.
Harley, J.B., 1989. Deconstructing the map. Cartographica 26, 1–20.
Harley, J.B., 2001. The New Nature of Maps: Essays in the History of Cartography. Johns Hopkins University Press, Baltimore, MD.
Hoggett, P., Thompson, S., 2002. Toward a democracy of the emotions. Constellations 9 (1), 106–126.
Jarvis, N.A., 1997. Taking a Break: Preliminary Investigations into the Psychology of Epiphanies as Discontinuous Change Experiences. Unpublished doctoral thesis, University of Massachusetts, Amherst.
Jensen, K.L., 1998. Lesbian and Bisexual Epiphanies: Identity Deconstruction and Reconstruction. Unpublished doctoral thesis, Union Institute Graduate School, Cincinnati, OH.
Jensen, K.L., 1999. Lesbian Epiphanies: Women Coming Out in Later Life. Harrington Park Press, New York.
Johnson, J., Louis, R., Pramono, H., 2006. Facing the future: encouraging critical cartographic literacies in indigenous communities. ACME: An International E-Journal for Critical Geographies 4 (1), 80–98.
Kitchin, R., Dodge, M., 2007. Rethinking maps. Progress in Human Geography 31 (3), 331–344.
Kwan, M., 2007. Affecting geospatial technologies: toward a feminist politics of emotion. The Professional Geographer 59 (1), 22–34.
Lewis, G.M., 1998. Cartographic Encounters: Perspectives on Native American Mapmaking and Map Use. Kenneth Nebenzahl, Jr., Lectures in the History of Cartography. University of Chicago Press, Chicago.
Loyttyniemi, V., 2001. The setback of a doctor's career. In: McAdams, D.P., Josselson, R., Lieblich., A. (Eds.), Turns in the Road: Narrative Studies of Lives in Transition, American Psychological Association, Washington, DC, pp. 177–202.
Marlatt, M., 2004. The calamity of the initial surveys under the Robinson treaties. In: Wolfart, H.C. (Ed.), Papers of the Thirty-fifth Algonquian Conference, University of Manitoba Press, Winnipeg, pp. 281–335.
McDonald, M.G., 2008. The Nature of Epiphanic Experience. Journal of Humanistic Psychology 48 (1), 89–115.
Miller, W.R., C'de Baca, J., 1993. Quantum change: toward a psychology of transformation. In: Heatherton, T.F., Weinberger, J.L. (Eds.), Can Personality Change?, American Psychological Association, Washington, DC, pp. 253–280.
Miller, W.R., C'de Baca, J., 2001. Quantum Change: When Epiphanies and Sudden Insights Transform Ordinary Lives. Guilford, New York.
Nederveen Pieterse, J., 2009. Development Theory: Deconstructions/Reconstructions, second ed. Sage, London.
Painter, J., 2006. Cartophilias and cartoneuroses. Area 38, 345–347.
Paris, G., 1997. Everyday epiphanies. In: Clarkson, P. (Ed.), On the Sublime: In Psychoanalysis, Archetypal Psychology and Psychotherapy, Whurr, London, pp. 85–95.
Pearce, M.W., 2008. Framing the days: place and narrative in cartography. Cartography and Geographic Information Science 35, 17–32.
Perkins, C., 2003. Cartography: mapping theory. Progress in Human Geography 27 (3), 341–351.
Pickles, J., 2006. On the Social Lives of Maps and the Politics of Diagrams: a story of power, seduction, and disappearance. Area 38, 347–350.
Pyne, S., in press. Spatializing history in the Lake Huron Treaty atlas. In: Hele, K. (Ed.), This Is Indian Land: The Robinson Huron Treaties of 1850.
Pyne, S., Taylor, D.R.F., 2012. Mapping indigenous perspectives in the making of the cybercartographic atlas of the Lake Huron Treaty relationship process. Cartographica, Special Issue on Indigenous Cartography and Counter Mapping 47 (2), 92–104.
Schultz, W.T., 2001. Prison as a turning point in Oscar Wilde's life story. In: McAdams, D.P., Josselson, R., Lieblich, A. (Eds.), Turns in the Road: Narrative Studies of Lives in Transition, American Psychological Association, Washington, DC, pp. 67–90.
Sparke, M., 1998. A map that roared and an original atlas: Canada, cartography, and the narration of nation. Annals of the Association of American Geographers 88 (3), 463–495.
Sparke, M., 2005. In the Space of Theory: Postfoundational Geographies of the Nation-State. University of Minnesota Press, Minneapolis.
Taylor, D.R.F., 1997. Maps and mapping in the information era. Keynote address to the 18th ICA Conference, Stockholm In: Ottoson, L. (Ed.), Proceedings, 1, pp. 1–10.
Taylor, D.R.F., 2003. The concept of cybercartography. In: Peterson, M.P. (Ed.), Maps and the Internet, Elsevier, Amsterdam, pp. 405–420.

Taylor, D.R.F., 2005. Cybercartography: Theory and Practice. Modern Cartography Series. vol. 4. Elsevier, Amsterdam.

Taylor, D.R.F., 2008. Maps, mapping and society: some new directions. In: Proceedings of Global Map Forum Tokyo, Japan.

Taylor, D.R.F., 2013. Cybercartography is charting a new route. Geospatial World, Thought Leaders Edition 3 (6), 50–53.

Taylor, D.R.F., Pyne, S., 2010. The history and development of the theory and practice of cybercartography. International Journal of Digital Earth 3 (1), 1–14.

Turnbull, D., 2007. Maps, narratives, and trails: performativity, hodology, and distributed knowledges in complex adaptive systems: an approach to emergent mapping. Geographical Research 45, 140–149.

Turnbull, D., 2008. What is cartographic rationality? Metascience 17, 215–224.

Verganti, R., 2011. Radical design and technology epiphanies: a new focus for research on design management. Journal of Product Innovation Management 28, 384–388.

Vidal, A., Anderson, T.G., 1849. Report of Commissioners A. Vidal & T.G. Anderson on visit to the Indians on the north shores of lakes Huron and Superior for purposes of investigating their claims to territory bordering on those Lakes. Library and Archives Canada (Ottawa). RG 1, Vol. 266, Reel C-12652, 4.

CHAPTER

18

Considerations for Informed Consent in the Context of Online, Interactive, Atlas Creation

Timothy Di Leo Browne[1], *Gita J. Ljubicic*[2]

[1]School of Canadian Studies, Carleton University, Ottawa, Ontario, Canada, [2]Department of Geography and Environmental Studies, Carleton University, Ottawa, Ontario, Canada

OUTLINE

18.1 INTRODUCTION

The cybercartographic atlases discussed in this volume incorporate personal, social, and cultural knowledge on a wide range of topics and require careful consideration of consent, data storage, data access, and appropriate dissemination. In Canada, many of these ethical concerns are addressed in strict guidelines established by federal funding agencies

Developments in the Theory and Practice of Cybercartography, Second Edition, ISSN 1363-0814
http://dx.doi.org/10.1016/B978-0-444-62713-1.00018-0

(CIHR, NSERC, and SSHRC, 2010). However, even these guidelines do not take into consideration all the ethical issues that may materialize when creating atlases based on community knowledge. Cybercartographic resources, to begin with, are digital and (potentially) accessible through the Internet to anyone in the world, giving rise to new ethical considerations that would not have surfaced prior to the digital era. Because cybercartography is a developing field, researchers and communities alike have had to consider new questions in ways that truly serve the interests of research participants who contribute their knowledge. Informed consent is, then, of central importance: What issues need to be considered before a research participant shares knowledge? What sorts of questions should be asked in a consent form? What specific challenges do digital resources pose, and how can these be addressed early in the knowledge-sharing process? Related to all these questions is the complex issue of Intellectual Property, discussed in Chapter 19. The more effectively these questions are addressed at the outset, the better the participant's and the community's intentions may guide a project for the long term.

In this chapter, some of the challenges surrounding informed consent that Carleton University's Geomatics and Cartographic Research Centre (GCRC) has faced are highlighted. We begin by discussing the genesis of a data legacy project designed to address these issues. We then offer a historical overview of ethics and informed consent, with an emphasis on Arctic research, and we present feedback from a variety of research agencies. In conclusion, some tentative solutions are discussed based on research experiences and community feedback, and many of the issues examined here are also elaborated in the following chapter of this volume. The discussion is necessarily open-ended since the formulation of best-practice procedures for informed consent (especially in the online context) is ongoing. That said, certain themes have surfaced repeatedly, and atlas designers at GCRC are developing a clearer picture of how to approach many of the challenges.

18.2 PROJECT BACKGROUND

In 2009, following the International Polar Year (IPY) of 2007–2008, GCRC was faced with new challenges relating to data legacy – that is, the proper storage, access control, and dissemination of cultural data not just for today but for future generations. This is part of the general challenge of archiving and preservation discussed in Chapter 21, which has been a concern of the GCRC for many years. Carleton researchers had collaborated with Inuit knowledge holders[1] on the Inuit Sea Ice Use and Occupancy Project, funded by the Government of Canada IPY Science Program (see Chapter 14). This project was also part of a larger international IPY project, SIKU (Sea Ice Knowledge and Use: Assessing Arctic Environmental and Social Change, where siku also means 'sea ice' in Inuktitut). Recording Indigenous observations in multimedia cybercartographic atlases contributes both to the transmission of important cultural knowledge within communities and to a better understanding of a Arctic environments among Western scientists in an era of climate change. Additionally, some cybercartographic resources, e.g. the Kitikmeot Place Names Atlas (Chapter 16) and the Atlas of Arctic Bay (Chapter 20), are open to expansion well after a research

[1] In this chapter, we will generally refer to research participants who contribute information to cybercartographic atlases as knowledge holders, to accurately describe the role they play in atlas conception and construction. Participant implies a subordinate role.

project is complete, since local knowledge holders can continue uploading films, pictures, sound files, and texts that are relevant to particular geographic locations in the atlas. But the attributes that make such an atlas an effective means of transmitting knowledge also raise questions that need attention if there is to be an effective consent process. For example, a situation could arise in which a knowledge holder would need to give informed consent after research is complete. How could this be accomplished if the researcher is not present to explain all the ramifications of the project? The consequences of placing information on the Internet may not be clear to Elders who are unaccustomed to computers; and as will be seen, the relationship between researcher and knowledge holder is essential to the trust that underlies the granting of consent. Moreover, how does one determine if a knowledge holder is free to share a particular piece of cultural information broadly to the public, as on the Internet? Some information sharing may require consent of a group of people, even an entire community. Other issues that surfaced in the context of the data legacy project related to refinements of atlas software and the importance of community training. All of these challenges are relevant to the full array of GCRC atlases, whether situated in the Arctic, the Great Lakes, urban environments, or any other setting.

In consideration, then, of concerns arising in atlas developments, the GCRC embarked on an IPY Data Legacy Project, funded by the Government of Canada IPY Federal Program Office from 2009 to 2012. This project focused on three areas: (1) sociocultural concerns; (2) education and training; and (3) technology. Of central importance in the first area were questions surrounding ethics and informed consent, particularly of consent in a purely online context where no researcher is present to explain a project's details to the knowledge holder. The second area, education and training, was to focus on such matters as developing learning resources, increasing equipment availability in schools, and designing courses for Arctic residents (for more on the Education component, see Chapter 20). The technology component would address the further development of tools that would help knowledge holders enter their information directly into an online atlas, as well as improvements to network infrastructure (see Chapter 9).

The important ethical issue of informed consent also became the focus of a technical issue for Amos Hayes, Technical Manager of the GCRC. He was asking, how can a series of consent questions effectively be posed through a sequence of pop-up windows that would appear when knowledge holders are about to upload information to an atlas? What questions should be asked in such a scenario? What suggestions and insights might other Arctic research organizations have regarding consent and ethics issues relevant to digital resources? How might these insights influence product design?

These issues, taken together with the educational concerns of the Data Legacy Project, led to two Arctic research trips in 2011. The first of these, occurring in January of that year, was undertaken by representatives of Inuit Tapiriit Kanatami (ITK), Canada's national Inuit organization. ITK worked in conjunction with GCRC so that an Inuit perspective would inform research into the questions being asked. The organization focused on three communities in the Western Arctic – Sachs Harbour, Paulatuk, and Ulukhaktok, all in the Northwest Territories – that had participated in the Circumpolar Flaw Lead (CFL) project.[2] ITK and CFL representatives asked community members how they felt about the research

[2] The CFL Project, conducted during the IPY out of the University of Manitoba, investigated flaw lead behaviour in the Western Arctic. A flaw lead is an area of thin ice and open water between the solid ice along the land and the moving pack ice further out. Inuit in the Arctic contributed their knowledge to this project.

conducted in their towns, what they liked and did not like, what ethical concerns they had, and what suggestions they could offer for improvement. The second trip, in August 2011, brought GCRC researchers to Iqaluit, Nunavut, during a research methodology workshop (designed especially for Arctic community residents) sponsored by the Nunavut Arctic College (NAC). While the workshop addressed primarily educational and technological concerns, interviews that GCRC researchers held with organizational representatives focused on questions of research ethics and consent. Feedback from these interviews, as well as from ITK's Western Arctic tour, is incorporated into Section 18.4 of this chapter. Many common concerns have surfaced in these discussions, as well as some important differences of opinion.

Although these research trips and follow-up correspondences have helped shape procedures going forward, questions have often led to more questions. We still do not have precise answers to questions such as how best to handle informed consent in a fully online context with no researcher present. What was learned was that a uniform approach may not work, since communities have different needs, and that mutual trust and respect are the key. Some of these issues will be addressed in a forthcoming report by the Canadian Internet Public Policy Interest Clinic, which is working towards the development of a traditional knowledge ethics and consent licence that provides a number of options to select from. That licence and this chapter are informed by the report titled, *Consent and Ethics on Local and Traditional Knowledge in the Internet Era: Some Key Questions* written by Browne as part of the IPY Data Legacy Project. Variation, flexibility, and mutual trust are the attributes that have historically been missing, or not fully developed, in research for Indigenous peoples in the Arctic or otherwise.

18.3 CONSENT/ETHICS HISTORY

Canadian researchers often receive funding from one of three federal organizations, the Canadian Institutes of Health Research (CIHR), the Natural Science and Engineering Research Council of Canada (NSERC), and the Social Sciences and Humanities Research Council of Canada (SSHRC), collectively known as the Tri-Council. In 2010, the three agencies published the second edition of their Tri-Council Policy Statement (TCPS2) addressing ethics in research involving humans (CIHR, NSERC, and SSHRC, 2010). Any researcher funded by these organizations must abide by the TCPS2's regulations. Seventeen pages of the TCPS2 are devoted to informed consent.

Informed consent was defined succinctly by participants at a 2001 seminar in New York focusing on the topic: 'an autonomous action by an individual that authorizes a professional to involve the individual in research' or '[a means] to assure that each person who participates [...] does so willingly and with adequate understanding of the study in which he or she freely agrees to participate' (Wood et al., 2002: 8). This seminar, like many studies and discussions of informed consent, is related to a medical topic (often reproductive health), and it is in the field of medicine that the expectation of informed consent emerged. Following the medical atrocities of World War II, the 1947 Nuremberg Code made a clear articulation of ethical standards in medicine, although similar ethical guidelines can be found as far back as the nineteenth century (Vollman and Winau, 1996). Ethical guidelines, including the requirement of informed consent, extend today beyond health fields to all social sciences, as is clear from the TCPS2. A UN report addresses the particular concerns of Indigenous peoples, who have been vulnerable to unethical research practices, stating that 'Indigenous Peoples' right to free, prior, and informed consent (FPIC) has been recognized by a number of intergovernmental organizations, international bodies, conventions, and international human rights law and increasingly in the laws of State' (Tamang, 2005). The same report also notes that 'there is no internationally agreed upon definition or understanding of the principle [of FPIC] or mechanism for implementation' (Tamang, 2005). There is, therefore, much variation of interpretation.

TCPS2 was released in 2010, and among the changes from its original 1998 version is an extended section relating to research involving First Nations, Inuit and Metis peoples, encompassed in their Chapter 9. This section includes requirements:

- to engage with the community in aboriginal research (110);
- to engage with organizations and other communities of interest (115);
- to recognize diverse interests within communities (116);
- to respect community customs and codes of practice (117);
- to engage in collaborative research (123);
- to assure mutual benefits in research (124);
- to strengthen research capacity[3] (within communities) (125); and
- to interpret and disseminate research results (128), to name a few.

Regarding informed consent, the TCPS2 requires either a signed consent form or 'documentation by the researcher of another appropriate means of consent' (44) – for example, an oral consent.

But consent goes well beyond the signing of a document, as detailed in both the TCPS2 and in Carleton University's own ethics guidelines, which state that consent 'encompasses a process that begins with the initial contact, and carries through to the end of involvement of research subjects in the project' (Carleton University, 2002). Carleton's guidelines go on to say that dialogue, information sharing, and the potential for research 'subjects' to be active participants in the research process are all part of the granting of consent. Informed consent, then, is inseparable from the overall ethics of a project.

These guidelines reflect concerns raised within Indigenous communities. The Nunavut Research Institute (NRI), Nunavut's primary research licensing agency, in conjunction with ITK, the national organization for Inuit throughout Canada, produced a document that details their expectations of Arctic researchers (ITK and NRI, 2007). Along with the related necessity for informed consent and a more active role for knowledge holders, the document highlights ongoing concerns about the way that research has typically been conducted, for example: there is little regard for community needs; funding for locally generated projects is deficient; researchers incorrectly represent local knowledge; knowledge holders are not recognized or compensated while researchers take all the credit; and so on (ITK and NRI, 2007).

[3] Tri-Council defines capacity: 'The ability of prospective or actual participants to understand relevant information presented (e.g. purpose of the research, foreseeable risks, and potential benefits), and to appreciate the potential consequences of any decision they make based upon this information' (190).

These are not new concerns. In 1971, the American Anthropological Society (AAS) adopted ethical guidelines quite like those of TCPS2 and ITK and NRI (2007), albeit with less detail. Spradley (1979) discusses the AAS recommendations:

- 'Consider Informants First' (35) – Spradley emphasizes that ethnographic research must consider the needs of communities first, and incorporate community inquiry.
- 'Safeguard Informants' Rights, Interests, and Sensitivities' (35).
- 'Communicate Research Objectives' (36).
- 'Protect the Privacy of Informants' (37).
- 'Don't Exploit Informants' (38).
- 'Make Reports Available to Informants' (e.g. published papers) (39).

Despite this recognition of the importance of nurturing ethical and respectful relationships, little had changed in research methodologies for years, and power imbalances between researchers and knowledge holders remained. In 1994, Martha Flaherty, then president of the Pauktuutit Inuit Women's Association, expressed concern about the very manner in which research and 'participation' were discussed. In an address to the National Students' Conference of the Association of Canadian Universities for Northern Studies (subsequently published in the Northern Review in 1995), she asserted that: 'The need for guidelines about how to involve the researched suggests to me a presumption that the researched are never truly considered equals and that the "researcher" always retains ultimate control' (Flaherty, 1995: 179). We may wonder, almost 20 years after Flaherty's observations, where even this book chapter on ethics and consent in Arctic research fits into the power dynamic. Here we still are, one might say. And whether atlas contributors are 'research participants' or 'knowledge holders' matters little when these atlases may still seem to be a form of technology imposed from above, however well intentioned. Changing the terminology does not in itself change the relationship; real change occurs only when participants and communities have genuine control over their knowledge and over research conducted in their areas.[4] As Taylor points out in Chapter 1, cybercartography is responding to this by giving as much control as possible to the communities producing the atlases, as evidenced in Chapters 15 and 16. Flaherty further stated, 'As students of northern studies, you will be familiar with the practice where southerners come north, do their field research over a number of months, usually the summer, get to know people in the community, get local help, then go south to write and publish their findings, they are acknowledged as the "experts", more so if they have included Inuit in the data collection and can cite them in their research' (Flaherty, 1995: 184). Flaherty urged researchers to consider community needs and to embrace community-driven ideas and research (Flaherty, 1995: 185). Nadasdy (1999) argued that the supposed integration of Indigenous knowledge into research projects only solidified the concentration of power in non-Indigenous centres of administration, and that Indigenous peoples were compelled to follow state and institutional practices, not their own. Once again, this raises the question: Where are we even now? Is it possible to engage in community research of this nature that is not inherently power imbalanced?

[4] This observation is meant as food for thought. If such thoughts enter the discussion at all, this is a step forward.

There is no easy answer. But in the creation of many of the cybercartographic atlases described in this book, GCRC is addressing these issues. As pointed out in Chapter 20, the processes by which the atlases are produced is as important as the artefacts themselves. There have been extensive community consultations and working relationships developed through long-term involvement and commitments. The atlases have sought to respond to community interests in documenting and sharing Inuit knowledge on local terms (see Chapters 14–17). The continued reflections on these issues – shared within this chapter – highlight ongoing struggles to ensure mutually beneficial and ethical research grounded in Inuit communities and cultural values. Certainly not all of the answers have been provided, but by sharing challenges it is hoped that these contribute to a broader dialogue on these issues and provide some ideas for gradual improvements.

18.4 RESEARCH ETHICS CHALLENGES AND FEEDBACK

In this section, specific challenges that GCRC has faced pertaining to research ethics are discussed, particularly in the context of cybercartographic atlases, with feedback on these issues from organizations involved in conducting and/or facilitating research in the Arctic. Also included are a few examples from other geographic contexts, as relevant. As mentioned above, GCRC has been speaking with representatives of Arctic governmental and research agencies, seeking knowledgeable input that can inform ethics and consent guidelines in Atlas development (Browne, 2012). Several organizations were consulted in this process, including: the NRI, the Government of Nunavut (for information relating specifically to the online cartographic Nunavut Coastal Resource Inventory, or NCRI), and ITK. In summarizing the key issues raised, and sharing organizational feedback, some of the more foundational concerns are discussed with a move towards challenges that arise directly in relation to developing atlases that are shared online and are broadly and publicly available.

18.4.1 Research Licensing and Ethics

NRI, based in Iqaluit, is the main licensing agency for research conducted in Nunavut and is a division of NAC. In principle, all research in Nunavut must be approved by NRI through a formal licensing process.[5] In addition, there may be other permits required for research involving archaeology, fisheries, or wildlife, and this would involve other government departments or regulatory bodies. Researchers are expected to engage in ethical practices for the duration of their work, from the formulation of research questions to the dissemination of research results, with informed consent as a key component of the process. Agencies outside of Nunavut may play a role in the approval process as well. For example, in the development of our *siku Atlas* with different Inuit communities, a Carleton Research Ethics Board (REB) application for approval (entailing rigorous ethical evaluation of our proposed project) had to be completed, as well as securing a research license from NRI. Furthermore, NRI's licensing

[5] The Northwest Territories agency that issues most research licenses is the Aurora Research Institute. In Yukon, researchers go through the Yukon Heritage Resources Unit of the Department of Tourism and Culture.

procedure directly involves communities since the applications are passed on to them and communities have the prerogative to reject proposals or to ask for changes that may better suit their needs.

However, as a representative of NRI explained, not all researchers follow correct procedures, particularly private consultants, corporations, and government employees (Browne, 2012); additionally, NRI's definition of research – 'peer-reviewed academic science' – exempts much investigative inquiry from the licensing requirement (Browne, 2012). Corporations and some governmental agencies may have practices that vary substantially from standards set by NRI. Lack of consistency in the way research is conducted is among NRI's biggest concerns. Inconsistency leads to community mistrust, since community members cannot be sure of a researcher's intentions from the outset. And even those who do follow proper procedures in the beginning do not always continue to do so throughout the research process (Browne, 2012).

Nunavut's brief Scientists Act is the only legal code that specifically addresses research in the territory (Gearheard and Shirley, 2007). This contrasts with, for example, the Navajo Nation's legislation concerning human research, which covers many aspects of the research process, from the requirement of informed consent to the establishment of a Navajo Nation Human Research Review Board (Navajo Nation, 2013).[6] NRI considers the establishment of a comparable review board in Nunavut as a potential means of addressing inconsistencies in research methodology (Browne, 2012). Such a board would also address their concern that many universities outside of North America do not have internal REBs, which are standard in Canada and the United States.

Gaining institutional ethics approval and an NRI license are important foundational elements in engaging in research with Inuit communities, but they are by no means sufficient to address all potential ethical issues that may arise.

18.4.2 Research Duplication

NRI identifies duplication of research as another key problem in the Arctic (Browne, 2012). Communities have found that different researchers will pursue the same or similar research questions, and knowledge holders must repeatedly share the same information. In addition, some communities experience fatigue from too much research. Researchers at the GCRC were acutely aware of these concerns in beginning early research collaborations with Nunavut communities. Through the joint process of developing priorities for research and atlas development, there was an attempt to alleviate these concerns. In addition, because atlas development is a very community-driven process – even if some of the technical experts are responsible for making a community's vision come alive – it helped to minimize any potential duplication with earlier research. Where communities are driving the research agenda, there can be more confidence that results will be of mutual benefit and will not be duplicating earlier work.

Since Nunavut's Scientists Act specifies only that research projects cause no harm, not that they actually benefit a community, legal means do not exist to prevent duplication (Browne, 2012). Once again, this is in contrast to the Navajo Code, which stipulates that research be

[6] Specifically, this is Chapter 25, Navajo Nation Human Research Code, of the Navajo Nation Code.

'beneficial, community-based, and consistent with Navajo Nation priority and concerns' (Navajo Nation, 2013: 295). No formal process currently exists in Nunavut to direct scholars towards areas that need research and away from those that have had too much. However, those who consult with NRI about these issues will receive valuable guidance and can avoid potentially costly and harmful mistakes by heeding their advice. Another possible remedy is to address this issue at the point of consent. According to a NCRI representative, this is a practice that NCRI researchers have used: participants are asked if they have ever shared the information they are offering with another researcher. The NCRI representative argued that this question should be included on the consent form itself (Browne, 2012).

18.4.3 Benefit to Community

In the past, researchers have generated research questions and driven the research process with little thought for how research will benefit communities participating in a project. As mentioned above, nothing in the Nunavut Scientists Act requires that a project be beneficial. Communities are now beginning to generate their own questions and research projects (Browne, 2012), and consideration of community benefit is increasingly an aspect of research projects from the outside as well. Although this issue is broached in the granting of a research license or an REB approval, there is not always a lot of thought given beyond immediate potential benefits (e.g. financial compensation). Some Indigenous nations already require more reflection on this, such as outlined in the Gwich'in Traditional Knowledge Policy (2004). In its discussion of informed consent statements, this policy explicitly asks the researcher to '[i]dentify how this information will benefit the interviewee, their community and future generations of that community' (Gwich'in Tribal Council, 2004: 11). (The Gwich'in Social and Cultural Institute controlled the production of the *Gwich'in Atlas* described in Chapter 16.) Therefore, the point of consent may also be an appropriate place to discuss the benefits of sharing information.

In the processes of atlas development, in each case there was extensive consultation at the outset in order to determine the community priorities in developing a particular atlas. Atlases are seen as a benefit to the community, or else the GCRC would not have had the support or important contributions from knowledge holders to move forward. In the case of the *Kitikmeot Placenames Atlas*, Inuit elders wanted to share the original place names of their homelands, in their language (Inuktitut), so that their descriptive terminology and knowledge of the land may be documented and shared for others to learn from. Having the Inuktitut pronunciations and meanings is important for language preservation, as well as passing on important cultural and geographic knowledge. The online format is easily accessible as a school-based learning resource, and efforts are also being made to integrate a community kiosk locally so that anyone can explore and learn from the *Atlas*.

18.4.4 Varying Levels of Consent

A knowledge holder may wish to share information only under condition that access to the information be restricted in some way. Different levels of consent were considered in the consent form used at Carleton during the IPY project; these were accounted for in data management approaches and in the development of the Inuit *siku Atlas* (Chapter 14). Levels of consent ranged from full inclusion in the online atlas (so the information in question would be accessible to anyone with Internet access) to restrictions limiting access only to within a particular community or only to within the project (and not beyond). Community members who shared their knowledge could also choose to have the information they shared attributed to them (i.e. to be identified by name), or to only use their knowledge in a collective form (i.e. summarized in the context of community-based results). Most knowledge holders wanted to share their knowledge with youth and with the broader public, and so most were happy to openly contribute to the *Atlas*. However, for those who were not so comfortable with this, specific data management safeguards had to be developed (Chapter 9), and in some cases the individual's contributions could simply not be included in the *Atlas* (but may have been made available in other forms locally, only within the community). Similar options were included in NCRI's consent forms, as well as those of other organizations met during field research (Browne, 2012). Such procedures are in keeping with TCPS2, which stipulates that safeguards should protect privacy and confidentiality where these have been requested (CIHR, NSERC and SSHRC, 2010: 127). This could even extend to whole communities, since it may be easy to identify a community just by the information.

18.4.5 Individual vs. Collective Consent

Sometimes community consent is required not just for an overall project (which should be addressed in the research licensing) but for specific aspects of it, or for particular pieces of information. Consider, for example, the case of animal habitats. Information regarding a habitat may not belong to any one individual, and revealing a habitat location may be problematic. First, one must determine which community association has the right to determine whether or not the knowledge can be shared. Should it be the Hunters and Trappers Association? The local Hamlet government? A group of elders? Suppose a nearby Hamlet has access to the same wildlife habitat. Should the other Hamlet be contacted for consent as well? Community consent, in this sense, is a complex notion, one that the TCPS2 barely mentions; reference is made only to the restricted sharing of sacred knowledge in Article 9.8 (CIHR, NSERC and SSHRC, 2010: 117–118).

The Pauktuutit Women of Canada addressed the question of community consent in their 2006 book, The Inuit Way, in a discussion of the traditional means by which consensus might have been reached when group decisions were required: the basic system of making group decisions was consensus. Major decisions affecting the group would be discussed among the adults. People would voice their views, discuss the issues and compromise until the final decision was one everyone could accept. People with special skills, talents or knowledge such as a respected hunter, elder, or a shaman, could be solicited for their opinion on a particular issue but their advice was not binding. Their ability to influence others was limited by the degree to which people chose to follow their advice (Pauktuutit Inuit Women of Canada, 2006: 30).

Reaching consensus among all the adults in a community could, however, prove difficult. The representative of NCRI stated that it might be best not to rely on one knowledge holder for all the information about a given topic. The best means of achieving some community consent is to consult several knowledge holders, typically elders recommended through community consultation (Browne, 2012). Several of the organizations spoken with reported relying on small, representative groups to make decisions that required some form of community consent (Browne, 2012).

Within the context of atlas development, it was this model, of working with a number of knowledgeable individuals in the sharing and representation of knowledge that helped to ensure broad representation and approval of the atlases at the community level. However, this does not effectively constitute collective consent. Within this approach, informed consent is still provided on an individual basis, and the knowledge is attributed individually, even when knowledge holders may emphasize how certain aspects of the knowledge they share were passed on from older generations or their extended family. Within the standard consent form protocols there is no easy way to deal with the issue of collective consent effectively, and indeed, to determine who would need to be involved to provide consent at a family level, much less a community level. Working with various individuals to represent broad community perspectives was the best middle ground to achieve these goals, but the limits of this must still be recognized, in that it can never reflect truly encompassing collective consent. This issue is even more challenging when attempting to visualize what a purely online version of informed consent would involve, and the GCRC has not yet been able to address this problem in a satisfactory manner. Perhaps this can never be truly achieved, and thus the suggested emphasis would be to ensure strong working relationships, with extensive community consultations, and the involvement of key experts as identified by knowledgeable community members, to ensure the more representative results and the closest approximation to community consent.

18.4.6 Authorization to Provide Consent

The question of individual vs. collective consent raises questions about who has the right to share information on behalf of the community, or defining what is sensitive or not.[7] These determinations may vary between individuals. There may even be disagreements between different local government agencies over who has the last say, according to the representatives of both NCRI and NRI (Browne, 2012). One may expect this in any situation where different interests have a stake, but it does make the question of authorization a difficult one in some instances. Building on the example of wildlife habitats mentioned above, consider the particular case of mapping polar bear denning sites. One hunter may be very comfortable sharing his knowledge about polar bear dens, while another may feel this is very sensitive information and would not want this publicly shared. Who would have the authority to include or remove this information from an atlas context? Or would it be the local Hunters and Trappers Organization that should be consulted in this case? Even such authoritative bodies, made up of various knowledgeable community representatives, are not necessarily seen by everyone in a community to be representing their best interests. So, many contentions can arise, and authority is extremely challenging since the response may vary depending on who is asked. Again, this is not an issue that has been effectively addressed as of yet in atlas development. There has been a reliance on individual or small group working relationships (and associated consent from all individuals involved), partnerships with key local organizations, and ongoing communications to seek the feedback necessary to ensure a strong level of confidence regarding what is ultimately included in any given atlas.

[7] Natural Resources Canada has produced the following useful guide on the topic of identifying and disseminating sensitive information *Best practices for sharing sensitive environmental geospatial data* which can be accessed here http://geoscan.ess.nrcan.gc.ca/starweb/geoscan/servlet.starweb?path=geoscan/geoscanfastlink_e.web&search1=R=288863.

18.4.7 Withdrawal of Consent

As part of the consent process for the *siku Atlas*, a clause was included that allowed for withdrawal of consent at any time (which was the recommendation at the time from Carleton's REB). This meshes with TCPS2, which requires the option of withdrawal in Article 3.1:

1. Consent shall be given voluntarily.
2. Consent can be withdrawn at any time.
3. If a participant withdraws consent, the participant can also request the withdrawal of their data or human biological materials.

This stipulation poses some problems for online resources, however, and many organizations spoken with expressed doubt that consent could truly be withdrawn once data have appeared online, since there is no guarantee that the data have not been duplicated in some way. The NCRI representative asks knowledge holders not to share any information that they would not want to be generally disseminated (Browne, 2012), precisely because of concerns over the control of the information once it is placed on the web; NRI, likewise, has reservations about consent rescindment (Browne, 2012). Many knowledge holders themselves are aware that withdrawing consent may be difficult, as expressed in an ITK report from the Northwest Territories community of Sachs Harbour (regarding research conducted during the CFL, described above): 'when asked if they had ever shared information that [they] wanted to withdraw, they said that they did not know how to go about doing this. The researcher should be checking with the interviewee before they put it down on paper. The participants agreed this could be done through transcript verification' (ITK, 2011a). Participants said that if they had signed a consent document and had sufficient opportunity to review the use and interpretation of their knowledge, the need to withdraw consent 'should not be an issue' (ITK, 2011a).

An associated consideration for improvement could be a refinement of the timeframe involved. For example, Carleton researchers involved in the *siku Atlas* have since modified the consent forms for use in other projects, to indicate that the individual is aware that they may change their levels of consent, or withdraw from participating within two months of reviewing their contributions and/or initial results. This was an effort to address the unrealistic nature of saying that an individual can withdraw at any time, but it may still not fully address the challenge of how the participants would go about withdrawing, or how this may apply in the context of content already online. In addition, there are many challenges associated with individuals reviewing transcripts or the interpretation of their contributions (usually in the form of academic writing), based on language (often working through Inuktitut/English translators), literacy levels, time constraints (for the individual and the researcher), and challenges of distance and communication. This is not to say that more efforts should not be made to tackle these challenges, but they are not straightforward issues to address.

18.4.8 Importance of Research Relationships in Consent

The opportunity to discuss and review raises the importance of relationships and trust in the granting of informed consent, a point on which the representatives of NRI, NCRI, ITK, and other organizations all agreed (Browne, 2012). A researcher establishes trust by following through on the initial commitment, revisiting a community multiple times, verifying what is unclear, sharing results, and asking for feedback. One of the roles of the researcher is to inform the knowledge holder about all aspects of the project, so that consent is truly informed. Knowledge holders know where they stand because they have been working with a researcher for an extended period, they know the purpose of the research, they understand what impacts the research may have on their community, and they understand the potential problems that could come up. Particularly in the context of digital research, this trust is critical. As discussed in many chapters in this volume, cybercartographic atlases are iteratively created, as is the underlying framework, and this is done as a collaborative process with knowledge holders in the case of Aboriginal partners or subject matter specialists in the case of the *Risk of Homelessness Atlas* as discussed in Chapter 12.

18.4.9 Online Consent

Much of this emphasis on investigating the ethics of informed consent in working with Inuit communities was driven by the need not just to improve our own methodologies and collaborative research relationships, but also to try to understand how this consent process may be applied in a purely online context. Given the importance of trusting relationships, the granting of online consent is problematic. One of the purposes of the Arctic research trip to Iqaluit in August 2011 was to determine how/if online consent could be accomplished. If an atlas lives on after a researcher's project is complete, and if community members can continue uploading information on their own, how can they grant informed consent? We wanted to determine the best way to do this, and to formulate the questions that should be asked of a knowledge holder who wishes to upload information directly to an already existing cybercartographic atlas through a sequence of pop-up questions.

Potential questions might ask knowledge holders if they would like their names used, if they understand the scope of a project, if they want their information to be generally available or access-restricted in some way, and so on. Many of these are the same questions that knowledge holders were already considering when offering informed consent in more typical one-on-one researcher–participant interviews. GCRC researchers asked research agencies working in the Arctic for recommendations on appropriate questions.

But this returns us to the point raised earlier: all organization representatives spoken with emphasized the importance of relationships in establishing trust between researchers and knowledge holders, and this trust is essential to granting informed consent. How, then, can a knowledge holder be fully informed when no relationship exists? Representatives at NCRI and the Qaujigiartiit Health Research Centre expressed the opinion that this option lacks the all-important trust relationship (Browne, 2012) – in other words, the granting of consent in an online context is not ethically feasible.

Representatives from Nunavut Tunngavik Inc. and ITK thought that the pop-up window option might work if accompanied by a video explaining the project in detail, in both English and other languages (in this case, Inuktitut). Although relationships are hard to establish through a video, there is nonetheless a person (preferably someone familiar) who can explain things in a clear and accessible manner, and thus people could make more informed decisions accordingly. But it is clear that not everyone agrees a relationship really exists in such a scenario.

18.4.10 Levels of Data Access, Control, and Storage

In the subsection on levels of consent, access considerations were discussed. The split between open and restricted use poses some difficult questions. If access and/or control is restricted to the community, what constitutes the community? If information is available for educational purposes, what constitutes education (see Chapter 20)?

Representatives of ITK broached these questions in their CFL follow-up tour of January 2011. The three communities visited did not have the same opinions on all these questions, reflecting the variety of viewpoints existing in Inuit communities. For example, some felt that data should be stored in southern Canadian institutions, such as GCRC or ITK, while others felt that it should stay in the community – which, for Inuvialuit in the Western Arctic, often means at the Inuvialuit Regional Corporation in Inuvik, NWT (ITK, 2011b).

All three communities thought that the community owned its Inuit Knowledge, and that anyone who wanted to use it had to get permission from the appropriate community organization (ITK, 2011b). The legal issues surrounding the data ownership and control are discussed in Chapter 19 and are complex ones. Regarding the question of access, participants were asked to consider different people viewing their knowledge online: a student, a corporate executive, a politician, and so on. Opinions varied once again. Some thought that it would be acceptable (and even necessary) for industry to have access to Inuit Knowledge; others thought that such access could only be gained through a visit to the community and consultation. Opinion also was split over researchers. Some suggested that researchers also owned the research, others that research belonged solely to the community. Some said that future researchers who want to use the data must go to the community. Some suggested that educators should have access to the material.

This wide range of opinions requires a well-organized data management system that can accommodate such diversity and take all possible factors into account. Chapter 9 addresses how the Nunaliit Cybercartographic Atlas Framework deals with this issue.

18.4.11 Secondary use of Data

One aspect of research legacy that has become a growing concern, according to the NRI representative, is unauthorized secondary use of data (Browne, 2012). This is a concern particularly with biological and genetic material, which may have been collected in one context yet used later in another. Studies show that Indigenous people are more sensitive to this issue than the general public because of the ways in which this data may be used to their detriment or because of the sacred status of human remains (see, for example, a study of this pertaining to Hawai'ian Indigenous peoples, Fong et al., 2004). The NRI representative suggests that a clause in the original research contract could address this problem (Browne, 2012).

In the case of the *siku Atlas*, an attempt to work this consideration directly into the consent form was made. For example, it was made clear that project contributions would be used (according to the consented level of attribution) as part of project outputs such as publications and presentations. However, there was an additional section on the sharing of original recordings (i.e. secondary uses for the information recorded). Here, individuals could indicate if they consented to future uses of the original information for: (1) use in an online interactive educational atlas about Nunavut (i.e. broader than the *siku Atlas*); (2) use in general public education and outreach activities; (3) use in this community of ___________ (specify); (4) use in future research projects affiliated with project team members; and (5) unrestricted public, noncommercial use (covers all of the above).

Although the data included in the cybercartographic atlases are not biological, the question arises again of unforeseen uses. Given the accessibility of Internet resources and the problems with consent withdrawal, the importance of elucidating these issues at the outset is clear. The more informed that knowledge holders are, the more informed is the consent that they grant. Nevertheless, it is impossible to anticipate all potential uses of the information, so it must still be made clear that once information is shared online there is really no way to control its use.

18.5 CONCLUSION

As we have suggested throughout, this chapter leaves many questions unanswered. Since digital resources are new and developing media, only future discussion can reach effective solutions to these challenges. Conversations with Northern research organizations are necessarily continuing and evolving.

Procedures going forward will have to take into account a wide range of options and opinions, and data management systems will need to adapt in order to accommodate them. Ethical considerations will vary from community to community, region to region, and nation to nation. A problem in the past has been the imposition of a uniform approach to research and ethics, precisely what researchers should best avoid today. Even so, data managers need to have a sense of what sort of system to construct, what sort of variables and parameters to build in. This has strongly influenced the technical decisions taken in building the Nunaliit Cybercartographic Atlas Framework as described in Chapter 9.

Closely tied to the issues of consent, access, and data management is the question of Intellectual Property raised earlier. Ethics and law are not the same thing, and as we have seen, many of the research protocols that such organizations as the NRI urge upon are not written into legal code. NRI cannot easily enforce its own policies, and does not have the staff to do so at any rate (Browne, 2012). As will be seen in the following Chapter 19, despite the gaps between ethics, consent, and IP law, there are ways in which knowledge holders and communities can protect themselves.

References

Browne, T., 2012. Consent and Ethics on Local and Traditional Knowledge in the Internet Era: Some Key Questions. School of Canadian Studies, Carleton University.

Canadian Institutes of Health Research, Natural Sciences and Engineering Research Council of Canada, and Social Sciences and Humanities Research Council of Canada, December 2010. Tri-Council Policy Statement: Ethical Conduct for Research Involving Humans.

Carleton University, 2002. The Ethical Conduct of Research [Policy Document]. Carleton UniversityAvailable at: http://www6.carleton.ca/secretariat//ccms/wp-content/ccms-files/Ethical-Conduct-of-Research.pdf.

Flaherty, M., 1995. Freedom of expression or freedom of exploitation? The Northern Review 14 (Summer), 178–185.

Fong, M., Braun, K.L., Chang, R.M.-L., 2004. Native Hawaiian preferences for informed consent and disclosure of results from research using stored biological specimens. Pacific Health Dialogue 11 (2).

Gearheard, S., Shirley, J., 2007. Challenges in community-research relationships: learning from natural science in Nunavut. Arctic 60 (1), 62–74.

Gwich'in Tribal Council, 2004. Traditional Knowledge Policy. S.l. Gwich'in Tribal Council. Available at: http://www.gwichin.ca/thegwichin/GTCTKPolicy.pdf.

ITK, 2011a. Draft CFL Team 10 verification tour summary: Sachs Harbour. Unpublished. Ottawa: Inuit Tapiriit Kanatami (ITK).

ITK, 2011b. Draft CFL Team 10 verification tour summary: Community Concerns about Ownership, Control, Access and Possession of Traditional Knowledge Research. Unpublished. Ottawa: Inuit Tapiriit Kanatami (ITK).

ITK and NRI, 2007. In: Nickels, S., Shirley, J., Laidler, G. (Eds.), Negotiating Research Relationships with Inuit Communities: A Guide for Researchers, Inuit Tapiriit Kanatami and Nunavut Research Institute, Ottawa.

Nadasdy, P., 1999. The politics of TEK: power and 'integration' of knowledge. Arctic Anthropology 36 (1/2), 1–18.

Navajo Nation, 2013. Arizona, New Mexico and Utah. Navajo Nation Human Subjects Code. Available at: http://www.nptao.arizona.edu/research/docs/Navajo_Nation_Human_SubjectsCode.pdf.

Pauktuutit Inuit Women of Canada, 2006. The Inuit Way: A Guide to Inuit Culture. Pauktuutit, Kuujjuaq. Available at: http://www.uqar.ca/files/boreas/inuitway_e.pdf.

Spradley, J., 1979. The Ethnographic Interview. HOH, New York.

Tamang, P., 2005. An Overview of the Principle of Free, Prior and Informed Consent and Indigenous Peoples in International and Domestic Law and Practices. United Nations Department of Economic and Social AffairsWorkshop on Free, Prior, and Informed Consent (New York, 17–19 January). Available at: http://www.un.org/esa/socdev/unpfii/documents/workshop_FPIC_tamang.doc.

Vollman, J., Winau, R., December 1996. Informed consent in human experimentation before the Nuremburg code. British Medical Journal 313 (7070), 1445–1449.

Wood, S.Y., Friedland, B.A., McGrory, C.E., 2002. Informed Consent: From Good Intentions to Sound Practices: A Report on a Seminar. The Population Council, New York.

CHAPTER

19

Cybercartography and Traditional Knowledge: Responding to Legal and Ethical Challenges

Teresa Scassa[1], *Tracey P. Lauriault*[2,4], *D. R. Fraser Taylor*[3]

[1]The Centre for Law, Technology and Society, University of Ottawa, Ottawa, Ontario, Canada, [2]National Institute for Regional and Spatial Analysis (NIRSA), National University of Ireland at Maynooth, Maynooth, Co. Kildare, Republic of Ireland, [3]Geomatics and Cartographic Research Centre (GCRC), Carleton University, Ottawa, Ontario, Canada, [4]Member of the Geomatics and Cartographic Research Centre (GCRC), Carleton University, Ottawa, Ontario, Canada

OUTLINE

19.1 INTRODUCTION

Cybercartography has great potential as a vehicle for the representation of traditional knowledge (TK) as outlined in Chapter 1 and several other chapters in this volume. A cybercartographic project to map the TK of a community will rely for its legitimacy on the willing, informed, and engaged consent of that community. The success of any such project will

Developments in the Theory and Practice of Cybercartography, Second Edition, ISSN 1363-0814
http://dx.doi.org/10.1016/B978-0-444-62713-1.00019-2

thus depend upon the development of relationships of mutual trust and respect (Nickels and Knotsch, 2011; Felt and Natcher, 2011; Lyons, 2011). A cybercartography project quite naturally begins with the building of these relationships, and with the collaborative design of the project. Chapter 18 by Di Leo Browne, Chapter 20 by Taylor and others, and Chapter 14 by Ljubicic give excellent examples of this process. Underlying these activities, the law forms an often invisible normative layer that may shape the project and its output (Bell and Shier, 2011). Even without specific legal expertise or reference to law, peoples' assumptions about the normative boundaries of their relationships with each other and their relationships to things implicitly shape how they conduct themselves, how they identify 'relevant' issues, and how they reach agreements. Where people from different cultures and traditions are involved in a project, there may be conflicting normative assumptions at play. That there may not be a complete meeting of minds is a problem that can be exacerbated by the fact that it will be the laws and norms of the dominant culture that will shape how agreements will be interpreted and understood by those outside the particular relationship (Bell and Shier, 2011).

This chapter examines the role of law (defined as explicit and enforceable rules) and normativity (defined as 'ought' statements that guide social relations) in shaping relationships between cartographers and the contributors of TK. This complements the preceding chapter by Di Leo Browne. We consider how those involved in such projects may, by expressly acknowledging the normative framework, use both technological and legal tools to define the terms of their collaboration and to articulate their expectations in sharing their information with a global community.

This chapter begins by defining both TK and cybercartography. We then examine the potential of cybercartography as a means of representing TK. After offering a description of the legal/normative framework for cybercartography projects that incorporate TK, we discuss how, in practical terms, those engaged in mapping TK can take steps to articulate for themselves the normative boundaries of their projects.

19.2 DEFINING TK

TK is best defined in general terms because of the great diversity of TK held by many different peoples, groups and communities around the world (Dutfield, 2001). This diversity is reflected in the way that TK is discussed in several chapters of this book. TK is a central part of Chapters 13 through 20 and is also discussed both in the introductory and concluding chapters of this book. All of the authors recognize its importance but each author describes it in a slightly different way. Nevertheless, certain general characteristics of TK can be derived from definitions expressed in broad terms. The term 'traditional knowledge' is used to describe systems of knowledge that reflect 'a cumulative body of knowledge and beliefs, handed down through generations by cultural transmission, about the relationship of living beings (including humans) with one another and with their environment' (Berkes, 1993: 3). This definition emphasizes the *systemic* nature of TK; it is more than a collection of discrete pieces of knowledge. A system of knowledge implicates how knowledge is both created and disseminated. For example, the Canadian Royal Commission on Aboriginal Peoples (1996: 33) described the Aboriginal tradition as being 'neither linear nor steeped in the same notions of social progress and evolution' as western systems of knowledge. The Commission continued with these observations:

> [Aboriginal tradition is not] usually human-centred in the same way as the western scientific tradition, for it does not assume that human beings are anything more than one – and not necessarily the most

important – element of the natural order of the universe. Moreover, the Aboriginal historical tradition is an oral one, involving legends, stories, and accounts handed down through the generations in oral form. It is less focused on establishing objective truth and assumes that the teller of the story is so much a part of the event being described that it would be arrogant to presume to classify or categorize the event exactly or for all time.

Barsh (1999) argues that TK is distinctive because of the manner in which it is both acquired and used. The focus on the systemic characteristics of TK highlights the important point that TK is not a static catalogue of information; rather it is a body of knowledge that continues to evolve and develop. Article 3 of the World Intellectual Property Organization's (WIPO) revised objectives and principles on the protection of TK specifically acknowledge 'the dynamic and evolving nature of TK and the nature of TK systems as frameworks of ongoing innovation' (WIPO, 2005).

TK is expressed in a wide diversity of ways. It may take the form of: 'stories, songs, folklore, proverbs, cultural values, beliefs, rituals, community laws, local language, and agricultural practices, including the development of plant species and animal breeds' (Convention on Biological Diversity, 2012). Definitions of TK tend to emphasize its oral nature, its transmission from one generation to the next, its collective ownership by a people, and its relationship to a particular geographic area.

19.3 CYBERCARTOGRAPHY

Cybercartography has been defined as: 'the organization, presentation, analysis and communication of spatially referenced information on a wide variety of topics of interest and use to society in an interactive, dynamic, multimedia, multisensory format with the use of multimedia and multimodal interfaces' (Taylor, 1997, 2003; Taylor and Pyne, 2009). Key elements of cybercartography are its interactivity and the breadth of subject matter that it addresses (Taylor and Pyne, 2009, 2012). The most recent description of cybercartography is given in Chapter 1.

The Geomatics and Cartographic Research Centre (GCRC), directed by D. R. Fraser Taylor, has been building cybercartographic atlases incorporating TK for some time as part of the Centre's Indigenous Knowledge and Northern Research themes. The theory and practice of cybercartography includes media and technology, as just discussed, and it is also a practice that is transdisciplinary and involving multisectoral collaborations and partnerships. In addition, the practice of cybercartography is about building meaningful, trusted, and collaborative relationships (Scassa et al., forthcoming). The cybercartographic process implies critically reflecting on knowledge production, cartographic representation, and research methods. In addition, it means examining the legal and policy issues that arise during the process of cross-cultural research and then cyclically embedding that learning into the atlas creation process, such as in computer code (See Chapter 9 by Hayes), user interfaces and volunteered geographic information (VGI) (see Chapter 4), the process of collecting data (see Chapter 18), data licensing (CIPPIC, forthcoming), and attribution.

19.4 HOW CAN TK BE REPRESENTED IN CYBERCARTOGRAPHY?

Cybercartography has great potential as a tool for representing TK. Because of the close ties of indigenous and local communities to particular territories and because of the resultant strong relationship of TK to the land, this type of knowledge may be particularly well suited to representation in a geographic context. Doing so may emphasize in useful ways the interconnectedness of a people, their culture, and the land (Engler et al., forthcoming). This is reflected in several chapters of this volume, especially those dealing with the creation of cybercartographic atlases.

The oral nature of much TK can also be accommodated within the multimedia, multisensory nature of cybercartography to a much greater extent than in traditional cartography as Taylor argues in Chapter 1. Particularly in contexts where oral traditions are made vulnerable by an aging population of elders and the disruption of traditional communities (Krupnik et al., 2010), the ability to integrate audio and video components in digital cartography projects can help build a repository of oral knowledge. The orality of the recorded knowledge may also serve important goals of language preservation, and the resultant atlases can be used as multifaceted teaching tools or community resources (Aporta, 2006). Several of the chapters in this volume that describe the atlases, such as Arctic Bay, Gwich'in and others, testify to the utility and importance of cybercartography.

The interactive and narrative nature of much TK may also be better accommodated through cybercartography than through more traditional forms of mapping activity. There is an iterative nature to the creation and transmission of TK (Pyne and Taylor, 2012). In this sense, as noted by the Royal Commission on Aboriginal Peoples (1996), the person who recounts an event is a part of the event and there are not necessarily 'authoritative' versions or accounts. Omura (2005), writing specifically about Inuit TK, notes as well the tendency against generalization and the importance of experience and perspective in oral accounts. The Royal Commission on Aboriginal Peoples (1996: 33) observed that:

> In the Aboriginal tradition the purpose of repeating oral accounts from the past is broader than the role of written history in western societies. It may be to educate the listener, to communicate aspects of culture, to socialize people into a cultural tradition, or to validate the claims of a particular family to authority and prestige…
>
> Oral accounts of the past include a good deal of subjective experience. They are not simply a detached recounting of factual events but, rather, are "facts enmeshed in the stories of a lifetime". They are also likely to be rooted in particular locations, making reference to particular families and communities. This contributes to a sense that there are many histories, each characterized in part by how a people see themselves, how they define their identity in relation to their environment, and how they express their uniqueness as a people.

Cybercartography is inherently nonlinear (Taylor and Pyne, 2009). By facilitating the mapping of multiple 'stories', audio and audiovisual accounts of place and space, cybercartography may be better suited for the representation of TK than more static and linear accounts. As Taylor and Pyne (2009: 8) note, cybercartography 'allows different stories to be told without privileging one story over another'. It is not limited to one account or perspective, and is not frozen in time as discussed by Taylor in Chapter 1.

The flexibility of the medium of cybercartography can also allow for the challenging or redefinition of basic concepts such as 'map', 'boundaries', and 'territory'. This is particularly important to post-colonial representations of location (Sparke, 1998). Cybercartography can be used to represent alternative conceptions of both territory and land use. Thus, the concept of 'territory' can be reconceived in terms of traditional patterns of use, or layered rights to particular uses (Tobias, 2010). Even the concept of 'land' can be rearticulated, as is the case with the Inuit Siku Sea Ice Atlas described in Chapter 14, which expresses traditional Inuit knowledge of a territory that is not limited to immutable land but also comprises mutable sea ice. Indeed, the mutability of the physical environment is, for Inuit, a fundamental aspect of the context in which their TK, known as *Inuit qaujimajatuqangit* evolves (Felt and Natcher, 2011). This is discussed in several chapters in this volume and in particular, in Chapters 13, 14, and 20.

With its multisensory and multimedia features, cybercartography also diverges from more traditional conceptions of the map. The political nature of maps has long been recognized (Monmonier, 1996) and they carry a certain semantic weight as objective representations of fact in geographic context (Wood et al., 2010; Harley, 1988, 1989). As alternatives to conventional maps, cybercartographic projects may have evolving relevance in land claims disputes, and in claims over indigenous rights to land use (Sparke, 1998; Pyne and Taylor, 2012; Pyne chapter in this volume). By documenting the TK of a people's relationship to the land and patterns of land use, for example, the resultant atlases may form part of an evidentiary record that can be used to support such claims (Sparke, 1998; Frogner, 2010). The multimedia nature of such projects is not necessarily an impediment to their use in this way. The Supreme Court of Canada has noted that '[c]laims to aboriginal title are woven with history, legend, politics, and moral obligations' (*Kruger v. The Queen*, 1977: 109). It has also observed that '[n]otwithstanding the challenges created by the use of oral histories as proof of historical facts, the laws of evidence must be adapted in order that this type of evidence can be accommodated and placed on an equal footing with the types of historical evidence that courts are familiar with, which largely consists of historical documents' (*Delgamuukw v. British Columbia*, 1997: para 87). Cybercartographic atlases may thus have an emerging relevance to sovereignty claims. Cybercartography projects, because of their capacity for worldwide dissemination and broad accessibility can also provide an invaluable means by which a community can present itself, its knowledge, and its values to a broader global audience on its own terms. In addition, some cybercartographic atlases include traditional place names that represent indigenous heritage, land use, and occupancy and these become a kind of geo-narrative (Caquard, 2011) that can surround a treaty and land claims process. These atlases embody the collective memories (Josias, 2011; Nanelli, 2009) of those who have contributed to their creation and have become a means to records the historical, geographical, cultural and scientific facts that have been transmitted orally for centuries. These atlases in some instances are the first official recordings of this aurally transmitted knowledge and elders and communities have authoritatively endorsed each record. The communities who have contributed to and authorized them regard these atlases as living archives.

The challenge lies in building trusted working relationships with communities to design and build projects that can realize this potential. These processes are fully discussed in Chapters 14, 17, and 20 in particular.

19.5 TK IN INTERNATIONAL LAW

Traditional knowledge has received a great deal of attention of late in a diverse range of international fora. It is fair to say that interest in TK as a subject for international law has risen to its current prominence in the context of the exploitation of TK of plant genetic resources by western multinational companies, including those in the pharmaceutical and agricultural industries (Oguamanam, 2011). Thus, although the definitions of TK are much broader than knowledge of plant genetic resources, it is this one lucrative and widely exploited subset of

TK that has dominated the development of international law. Nevertheless, as Oguamanam (2004: 139) cautions, the 'traditional knowledge of non-Western peoples is implicated in virtually every front of human endeavour or in the global arena'. At the international level, those interested in recognizing and protecting TK have sought to have TK acknowledged, to afford it some legal protection, and to develop a regime under which the value flowing from the use of TK can be shared with the source communities. Thus, for example, Article 8(j) of the United Nations Convention on Biological Diversity (CBD) provides:

> 8. Each contracting Party shall, as far as possible and as appropriate: [...]
>
> (j) Subject to national legislation, respect, preserve and maintain knowledge, innovations and practices of indigenous and local communities embodying traditional lifestyles relevant for the conservation and sustainable use of biological diversity and promote their wider application with the approval and involvement of the holders of such knowledge, innovations and practices and encourage the equitable sharing of the benefits arising from the utilization of such knowledge innovations and practices.

While recognizing the goal of protecting TK from undue exploitation, the obligation in article 8(j) is relatively soft. It is 'subject to national laws', meaning that other national laws may limit or trump the obligation. Further, rather than create a property right in TK, it seeks instead to encourage the equitable sharing of benefits that flow from the commercial exploitation of TK (Ni, 2011). The Conference of the Parties and the Working Group on Article 8(j) have since developed a Code of Ethical Conduct (2011) intended to serve as a guide to States Party to the Convention in fulfilling their obligations under the CBD.

Much work has also been done on developing what is known as Access and Benefit Sharing (ABS) regimes to govern the exploitation of knowledge of plant and animal genetic resources. This has been done through the Bonn Guidelines on Access to Genetic Resources and Fair and Equitable Sharing of the Benefits Arising out of their Utilization (Secretariat of the Convention on Biological Diversity, 2002) and the Nagoya Protocol on Access to Genetic Resources and the Fair and Equitable Sharing of Benefits Arising from their Utilization (Secretariat of the Convention on Biological Diversity, 2011a,b). The goal of ABS regimes is to develop a framework to compensate appropriately those communities that are the source of information that leads to lucrative innovation (Oguamanam, 2011).

WIPO has also directed its attention to the protection of TK from an intellectual property (IP) standpoint. In the 1970s and 1980s, its attention was oriented towards protection for folklore, leading up to the Model Provisions for National Laws on the Protection of Expressions of Folklore Against Illicit Exploitation and Other Prejudicial Actions (United Nations Educational, Scientific and Cultural Organization, 1985). The drive to protect folklore did not lead to any binding treaty provisions, and this work was later subsumed into more general work on the broader concept of TK. This was no doubt driven by the growing awareness of the issues around the commercial exploitation of TK linked to plant and animal genetic resources.

Following a lengthy review and study process, WIPO produced a set of revised objectives and principles for the protection of TK (WIPO, 2005). However, unlike its other IP treaties, the Revised Principles do not have as their objective the establishment of positive legal rights in TK. Instead, the goal is to protect TK against undue exploitation. The Revised Principles identify a broad range of goals for the protection of TK. These include recognizing the value of TK, promoting respect for it, promoting its conservation and preservation, empowering holders of TK and supporting TK systems. The Revised Principles also identify the goal of repressing unfair and inequitable uses of TK. While not establishing any obligation on member states to actually legislate to protect TK, the document offers guiding principles for those states that do choose to enact some form of protection. Article 4 of the Revised Principles expressly identifies what TK, at a bare minimum, should be subject to protection:

> Protection should be extended at least to that traditional knowledge which is:
>
> i) generated, preserved and transmitted in a traditional and intergenerational context;
>
> ii) distinctively associated with a traditional or indigenous community or people which preserves and transmits it between generations; and
>
> iii) integral to the cultural identity of a traditional or indigenous community or people, which is recognized as holding the knowledge through a form of custodianship, guardianship, collective ownership, or cultural responsibility. This relationship may be expressed formally or informally by customary or traditional practices, protocols or laws.

The Revised Principles thus offer a starting point for member states who seek to develop their own legislation aimed at protecting TK. However, they are not binding on states, and indeed they have been criticized for giving too much of leeway to states to determine what the appropriate form of protection for TK might be (Ni, 2011).

Much TK is generated by the world's indigenous peoples and the undue exploitation of TK has been linked to the ongoing marginalization of indigenous peoples (Coombe, 2009). The protection of TK is therefore as much a human rights issue as it is one of IP (Helfer and Austin, 2011). The UN Declaration on the Rights of Indigenous Peoples (2007) recognizes their right 'to maintain, control, protect and develop their cultural heritage, TK and traditional cultural expressions' (UN Declaration, 2007: art. 31). Article 31 goes on to affirm that indigenous peoples have 'the right to maintain, control, protect and develop their IP over such cultural heritage, TK, and traditional cultural expressions'. The Declaration, which was adopted by the UN General Assembly in 2007, is not legally binding on states, but nevertheless may have some general normative effect.

19.6 THE LEGAL/NORMATIVE FRAMEWORK

The move towards recognition of TK in international law has not generally manifested itself in domestic law in most countries, although as we will see, the values expressed at international law are reflected in other domestic normative instruments. In this section of the chapter, we consider the normative framework that operates as a backdrop to cybercartographic projects. We begin by outlining how the default principles of IP law determine what is or is not 'intellectual property' and how ownership rights are assigned. We next consider the law of confidential information, with its strong foundation in equity. We then look at how, under the influence of international norms, domestic 'soft law' has emerged to provide principles related to the use of TK.

19.6.1 Intellectual Property

The recognition in international law of the value of TK and the need to offer protection to it has not generally resulted in changes to domestic IP laws. IP legislation is the primary means by which the fruits of innovative and creative activity are rewarded. These systems – predominantly patent, copyright, and trademark law – are based upon entrenched western notions of innovation and creativity (Oguamanam, 2011). These IP norms have achieved global pre-eminence through two centuries of international treaty-making, culminating in the bi- and multilateral international trade treaties that dominate our era. These treaties have led to a global proliferation of IP laws designed to protect the innovation and creativity, which flow from western systems of knowledge. These laws have, for the most part, taken no direct account of TK, offer little or no protection for TK, and often explicitly exclude from protection forms of knowledge production and cultural expression that do not conform to western models of creation and innovation (Dutfield, 2001; Helfer and Austin, 2011; Oguamanam, 2004).

IP laws reward the efforts of identifiable authors and inventors, they insist upon the originality or novelty of works, and they offer a time-limited period of protection after which the work or invention falls into the public domain (Vaver, 2012). It is generally recognized that these and other characteristics of IP regimes are not well suited to protecting TK (Dutfield, 2001; Sunder, 2007; Oguamanam, 2004, 2011). For example, it may be impossible – and in any event, inappropriate – to identify the 'author' of a folktale. In the same vein, knowledge of the use of a particular plant-derived medicine for the treatment of illness that is shared within a community is not an invention attributable to an inventor. IP regimes are premised on individual ownership, whereas TK is a community asset (Oguamanam, 2004). Even if an author of sacred designs or representations could be identified for the purposes of copyright protection, it may not be appropriate for such works to fall into the public domain after the term of copyright protection has expired (Sunder, 2007; Oguamanam, 2004). It may also not be appropriate for an artist – even one from within the particular community – who makes use of traditional designs or representations to acquire a monopoly right to exploit those representations to the exclusion of others within or outside of the community (Bulun Bulun and Anor v. R. & T. Textiles Pty ltd, 1998).

This is not to say that indigenous peoples have not taken advantage of IP protection in some contexts. Certainly artists, writers, and musicians within traditional communities may rely on IP protection in the sale and commercialization of their work. Further, communities have also relied upon certification trademarks as a means of marking their wares as 'authentic' to protect their markets for their works from cheaper imitations (Sunder, 2007). For example, the Cowichan First Nation has registered a collective mark that is used to indicate genuine Cowichan wares made according to traditional methods.[1] Ongoing negotiations at the international level for expanded protection for geographical indications will likely also have benefits for traditional communities linked to defined territories who produce and sell wares that derive from traditional methods of protection and are linked to a particular territory (Sunder, 2007).

[1] For example, the certification mark GENUINE COWICHAN (TMA467837) is owned by the Cowichan Band Council and can be used to designate clothing made according to Cowichan traditions.

Although TK may be difficult to protect using domestic IP laws, a cybercartographic project will attract IP protection – quite possibly in unanticipated ways. This is because a cybercartography project that maps TK operates within a context shaped by the ever-present default rules around the protection of IP. A digital multimedia atlas is inherently a 'work' (in fact, it is a compilation of multiple different 'works') in which copyright subsists. Copyright, which is recognized and protected by an increasingly strong set of legal rules around the globe, arises automatically and without need for any action on the part of the authors of the compilation, or the authors of the individual 'works' that form part of the compilation (Vaver, 2011). Because, as will be explained below, the rules of copyright are a matter of 'hard' law, these rules and their operation may, if left unaddressed, disrupt the expectations of participants in the project with respect to ownership and control over their contributions to the project, and over the resulting atlas itself.

Helfer and Austin (2011) note that when it comes to the assertion of the rights of indigenous peoples with respect to TK linked to cultural expression and innovation, there are 'two distinct and opposing elements – claims *to* IP protection, and claims to be protected *from* IP laws and institutions' (461). Both of these issues can be identified in the context of digital cartography projects that involve TK.

To give an illustration, an Inuit community that shares its TK regarding the characteristics of sea ice and travel routes over sea ice may believe that this knowledge belongs to the community. The prevalence of IP rights discourse in contemporary culture may even lead to the characterization of this information as something that is 'owned' by the community, although, as noted above, copyright law does not go so far as to recognized property rights in information. Copyright law protects only the original expression of facts and not the underlying facts. Facts and information are considered to be in the public domain (Vaver, 2011; Scassa, 2010). Indeed, as Sunder (2007) notes, the concept of the public domain may be a barrier 'to understanding poor people's knowledge as *intellectual property*' (p. 106). Relegating information to the public domain and the characterization of TK as information effectively disqualifies TK from IP protection (Oguamanam, 2004).

Not only does copyright law relegate facts and information to a public domain from which all are free to share, it privileges with ownership rights those who draw upon this public domain to create 'works'. Thus, when knowledge of sea ice characteristics or travel routes is expressed in the context of a digital atlas, copyright law principles would reward with IP ownership not the *source* of the information, but rather the person who has *expressed* the facts or information. The original, fixed, expression of the information becomes the 'work' that is protected in law.

To illustrate, when an elder is interviewed so that his or her account of the characteristics of sea ice can be recorded and added as an audio or video component of the atlas, the elder may well believe that he or she has rights in his/her oral expression. Copyright law, by contrast, would characterize the interviewer who fixes oral expression in some material form (such as an audio-recording) as the author who holds copyright in the resultant work (Scassa et al., forthcoming). From a western IP perspective, it is the interviewer who conceives of the need for the interview, who chooses the topic, and who formulates the questions. These authorial efforts give rise to the interviewer's copyright in the result. The principles of ownership do not apply to the TK that forms the content of the interview, which allows others to exploit it for their own purposes, even converting it into their own IP (Helfer and Austin, 2011).

Although the inequity is more acute with some forms of TK that are exploited in highly lucrative industries such as the pharmaceutical industry, the pattern is the same where TK of other kinds – including songs, narratives, and historical accounts – are transformed into the IP of the person who collects, records, and publishes an account of the TK (Bell and Shier, 2011; Dutfield, 2001).

To reframe these examples in terms of the observation by Helfer and Austin (2011), the community members who participate in the recorded interviews may be unaware that the overarching IP norms not only fail to recognize any rights of the community or of the interviewees in the shared information, they also recognize IP rights in the person who collects, frames, and records the information. The community might well be in need of some form of protection *for* their TK and some form of protection *against* the application of IP norms.

It should be noted that these outcomes are the result of the automatic operation of copyright law. They may run counter to the expectations of both the researchers and community members. They are offered as an illustration of the legal framework for cybercartography, which must be understood and addressed in building a nonexploitative and truly collaborative cybercartography project. Many copyright rules operate as defaults; parties are free to make different agreements as to where rights will lie, and rights holders may also use licence agreements to establish the conditions under which works may be accessed and used. These strategies will be discussed in Section 19.7 below.

19.6.2 Laws around the Sharing of Information

Although copyright law considers information to be part of the public domain that may be drawn upon by creators of works, information nevertheless receives some legal protection in certain circumstances. Bodies of law have developed, for example, around confidential information and trade secrets. To a great extent, these bodies of law protect information in the context of particular relationships. Thus, for example, information shared with a trusted advisor (a fiduciary), or acquired by senior officers of a company in the course of their employment with the company, cannot be used by those individuals for purposes that conflict with the duties of loyalty owed by those individuals (Judge and Gervais, 2011; Sandeen, 2009). In some cases, the law will impose the duty by virtue of the nature of the relationship between the parties; in other contexts, the law will uphold contracts negotiated between parties in which obligations to keep certain information confidential are set out.

The law of confidential information may not be well suited to protect information shared for the purposes of digital cartography projects, since it serves to protect the *confidentiality* of the information, rather than to set parameters for its sharing. Further, the law of confidential information does not create or assign property rights in information (Judge and Gervais, 2011; Sandeen, 2009). As a result, the party who has wrongfully shared or disclosed information may be liable for the harm that they have caused; however, it may not be possible to exercise any further control over information that has slipped into the public domain.

In spite of the limitations of the law of confidential information, concepts of the fiduciary relationship may be relevant in understanding how the law might treat information shared by community members with a researcher. Although much will depend upon the nature of the information and the context in which it is shared, it is conceivable that a court might find that the relationship between the parties constrains the researcher to only those uses of the information that are consistent with the reasons for which it was shared (Helfer and Austin, 2011). However, it is generally not prudent to rely upon an interpretation that a court, at some future date (and at great expense), might give to a relationship. Researchers and communities can use contract law (written or recorded oral agreements) to define the boundaries of their relationship and the terms and conditions under which information is shared.

19.6.3 'Soft' Law around TK

Notwithstanding the fact that IP law may sometimes be of use to indigenous peoples, it does not offer comprehensive protection for TK. Instead, the protection of TK has largely been left to domestic 'soft law'. This involves the emergence of norms, sometimes supported by legal principles that govern how TK should be accessed and used. These soft law norms draw support from the growing international recognition of TK, from the proliferation of guiding principles at the international level, and from national governments' own participation in the development of international law in this regard. These developments at the international level have filtered down to the national level. For the most part, these norms govern relationships. In this sense, then, they establish a model that is closer to the protection of confidential information in equity, although in this case, the special character of the information is not so much that it is confidential, but that it is the TK of a people. Further, the norms that govern its use serve broader goals, including capacity-building, and ABS (Nickels and Knotsch, 2011).

In Canada, there are a number of normative documents and procedures that shape how research in aboriginal communities should be conducted; many of the norms expressed in these documents relate to the gathering and use of TK. Soft law norms can be found in the legislatively mandated review of research proposals. For example, in Nunavut, the government has enacted legislation that requires the review of proposals for most scientific, health and social sciences research (Scassa et al., forthcoming; Felt and Natcher, 2011). Responsibility for such review is delegated to the Nunavut Research Institute (NRI), which has developed a normative guide to conducting research in Nunavut (ITK and NRI, 2007). This guidance document addresses a range of issues including the need for informed consent, the need to address community concerns and to seek local feedback, the importance of reporting research results to the community and of giving appropriate credit, and the obligation to respect local expertise. The guide also speaks to the need to negotiate issues relating to data storage and ownership, and the broader goals of capacity-building within the community.

Obligations have also been imposed upon researchers as part of the ethics review process required by national granting agencies. In Canada, the Tri-Council, consisting of granting agencies for research in the health, science and social science fields, has developed guidelines that specifically address the conduct of research in Inuit, Métis and First Nations communities (Canadian Institutes for Health Research (CIHR) 2010; Nickels and Knotsch 2011). These new guidelines, called the Tri-Council Policy Statement on Ethics in Human Research (TCPS2) emphasize 'engagement with the community [as] an integral part of ethical research involving Aboriginal peoples' (CIHR, 2010: 107). As with the NRI guidance document, the TCPS2 directs researchers to build collaborative relationships, to address data access and ownership issues, and to demonstrate respect for community processes and decision making (CIHR, 2010).

Normative guidelines and research ethics protocols have played an important role in addressing negative and sometimes exploitative research relationships that have 'led to experiences with disproportionately few tangible benefits, fragmented communication of outcomes, and little meaningful involvement in and control over the research itself' (Felt and Natcher, 2011: 111). In addition to granting agency policies such as the TCPS2, Aboriginal and Inuit communities in Canada have also developed their own research ethics protocols (Felt and Natcher, 2011). These include the document on *Negotiating Research Relationships with Inuit Communities* (ITK and NRI, 2006) discussed above, and *OCAP: Ownership, Control, Access and Possession* (National Aboriginal Health Organization, 2007). The OCAP document addresses in its title key principles of ethical research in Aboriginal communities. These include the ownership of TK, the possession of research data, access to research results and a degree of control over the research process.

A common theme across such normative instruments is the need to ensure that research relationships do not perpetuate patterns of exploitation and that they serve the best interests of the communities in which the research takes place. The norms typically require the establishment of mutually respectful relationships with the communities in which research is to be conducted, the involvement of the community in the design of the research study, the return of research results to the community, and the building of capacity within the community. These issues are explored in Chapter 18 by Di Leo Browne. In keeping with the approach taken in international law, the goal is not so much to 'propertize' TK, but rather to ensure that communities exercise some level of control over it, that they are able to benefit from its use, and that its use eschews colonial patterns of exploitation and domination (Felt and Natcher, 2011; Nickels and Knotsch, 2011).

19.7 DISCUSSION

In practical terms, then, there exist normative requirements to deal fairly with communities in contexts where TK is sought and shared and these requirements should shape the inception, design, and execution of any cybercartographic project involving TK. In addition, there are certain legal rules around IP in particular that should be addressed so as to exercise control over both where rights vest and how they are to be exercised. The first step is obviously the building of a relationship of trust and mutual respect. This will be an ongoing process, and in many cases will require appreciable effort to communicate across often significant linguistic and cultural differences (Lyons, 2011; Nickels and Knotsch, 2011). In addition to this relationship, the ethical norms require that agreement on certain key issues be reached. Thus researchers and community members will have to address where the data will be located, how access and control can be exercised by community members, and how the contributions of the community and its members will be recognized and acknowledged. The importance of capacity-building within communities in which research is being conducted is also an important element of a number of the normative documents. Capacity-building means more than simply delivering a finished product. It can be achieved by involving members of the community the design, creation, and maintenance of the atlas (Nickels and Knotsch, 2011).

The nature of cybercartographic projects is such that they may also require different considerations than other forms of research, and they may present different opportunities for addressing concerns as well. The fact that one of the goals of a cybercartographic atlas may be the dissemination of the mapped knowledge in digital format to a worldwide audience means that not only must the researchers and the community reach an understanding about ownership, control and possession of the research materials as between themselves; there must also be attention paid to how this information is to be presented to the world, how control and possession can be understood in the context of global digital dissemination, and ultimately how users of the atlases can be made to understand the significance of the TK that is embedded within the project.

There are certain tools that can assist in the design of cybercartographic projects in a manner that is respectful of TK and that is compliant with the normative framework for protecting such information. There are two main tools; the first is a legal one: the licence or contract. The second is technological. Both will be discussed below.

19.7.1 The Licence

In the context of cybercartographic projects, two different licences or contracts should be contemplated. The first is the agreement between the community and the researchers. Such an agreement should lay out the expectations of the parties as to their respective roles and responsibilities within the project. Existing legal and normative obligations may require these documents in any event; the TCPS2, for example, requires written documentation in the form of consent forms and research ethics protocols (CIHR, 2010), as does the review process undertaken by NRI (Nickels and Knotsch, 2011). However, given that cybercartographic projects have as their goal the creation of a major multimedia work that consists of many smaller 'works' within the meaning of copyright law, the IP issues should also be clearly addressed. It may or may not be feasible or appropriate for the researchers to assign their IP rights in the project to the community, although this should certainly be considered. If assignment is not an option, the parties can agree to limits on how they will exercise those rights vis à vis the community, and they can also confer perpetual exclusive or nonexclusive licences on community members to use the compilation as a whole, or individual parts of it, including photographs and audiovisual content.

The second agreement should be between the researchers and community members on the one hand, and the members of the public who will access and use the project on the other. Such a licence is particularly important in the context of digital cartography projects in contrast to other forms of research output. For example, while such a licence may not be necessary when information is to be published in an academic journal with a very limited circulation, dissemination to a global audience via the World Wide Web may require different considerations. While researchers may be tempted to use readily available licence templates such as Creative Commons,[2] particularly where the goal is to give broad access to the work, these licences address only copyright law. A project licenced under an open licence such as Creative Commons will convey to users very broad rights to use the work and its component parts with relatively few limitations. While this may serve some of the goals of researchers who wish to broadly disseminate their work to the public, it may not adequately address the need to respect the TK that forms a part of the work. In some cases,

[2] Creative Commons, http://www.creativecommons.org.

community members may wish to restrict access to some of the TK of a more sensitive character. In other cases, even if the goal is to share the information broadly, there may still be a desire to draw attention to the nature of the information, and to impose some normative boundaries on its use. It may be, for example, that the community wishes to be consulted or notified with respect to certain kinds of downstream uses of the shared TK. An appropriately drafted licence may be able to incorporate these objectives (Scassa et al., forthcoming; CIPPIC, forthcoming).

19.7.2 The Software Platform

In addition to the use of licences, there are also technological tools that may enable researchers to build the cybercartographic atlases so as to respect the needs and concerns of the community that is a partner in the project. Different layers of access to different types of information can be built into the underlying software such that the atlas is multilayered to serve multiple categories of users. The technological framework used to create cybercartographic atlases is discussed by Hayes in Chapter 9.

Some TK will be of a greater sensitivity to the community, which may wish to limit who can access or view this information. For example, the community may wish to map information about hunting or trapping routes for its own purposes, but may not wish to share it outside of the community. Researchers must therefore work closely with community members to identify what information should be shared more broadly and what information should be limited only to the community or even to certain segments of the community. Cybercartography may be particularly suited to representing information in this way, as digital technologies can facilitate the layering of information with different levels of access. Digitized TK in the form of geospatial data can be managed according to a community's wants and needs at the point of collection. For example, use, management, storage, attribution and dissemination permissions can be built into data collection devices, such as tablets. In this case, licence parameters and normative ethical practices are part of a digitized TK decision-making tree with a user-friendly interface with instructions provided in lay terms rather than legalese. This, combined with project based agreements, provides TK contributors with the power to decide how they want their TK to be seen and shared (Engler et al., forthcoming, Hayes Chapter 9 in this volume).

Of course, while licences define the broad boundaries of a negotiated relationship, technological solutions seek to implement them, and thus may raise many more challenging issues in their application. For example, while technological solutions may permit a period of evaluation before VGI contributions are added to an atlas, they do not assist in making the call between what is or is not TK. Similarly, while technology will permit more sensitive TK to be access-restricted, it may be challenging to identify the authoritative voice or voices on the degree of sensitivity of certain types of TK. In some contexts, it may be difficult to arrive at a consensus with respect to the sharing of some information as there may be divisions within the community (Lyons, 2011). There may also be multiple communities with interests in the same information. The solution to these challenges brings us full circle to the issue of building relationships, which is fundamentally the most important and most challenging aspect of such projects (Lyons, 2011).

The use of VGI raises further issues. VGI has been defined as 'geographical information acquired and made available to others through the voluntary activity of individuals or groups with the intent of providing information about the geographic world' (Elwood et al., 2011: 572). A cybercartographic project can make use of VGI, and it is conceivable that some VGI is also TK. For example, community members may be encouraged to use mobile phones to capture geo-referenced images related to land use, or wildlife patterns of movement, and to upload them to the atlas. In some cases, this may or may not be considered TK; it may not be easy to distinguish individual contributions of general information from information that is TK. Thus, when VGI is integrated into a cybercartographic project involving indigenous communities, it may be important to implement technological filters that allow for a community based vetting or approval of the full or limited sharing of the information. These and other issues related to VGI are more fully discussed in Chapter 4.

While the legal and policy issues related to the cartographic rendering of digitized TK are complex and unresolved, the cybercartographic principles of building collaborative relationships: creation without exploitation and anticipating and mitigating the potential misuse of digitized TK rendered into maps, can provide the necessary platform upon which critical legal and geographical issues can be theoretically informed and then transformed into innovative legal instruments. These can include TK licences, embedding good practices into data collection devices, and then managing TK data accordingly in a cybercartographic framework like Nunaliit and the GCRC data management infrastructure.

References

Aporta, C., 2006. Anijaarniq: Introducing Inuit Landskill and Wayfinding. Multimedia CD-ROM Department of Education, Nunavut Government and the Nunavut Research Institute, Iqaluit.

Barsh, R.L., 1999. Indigenous knowledge and biodiversity, in indigenous peoples, their environments and territories. In: Posey, D.A. (Ed.), Cultural and Spiritual Values of Biodiversity, UNEP, pp. 73–75.

Bell, C., Shier, C., 2011. Control of information originating from Aboriginal communities: legal and ethical contexts. Études Inuit Studies 35, 35–56.

Berkes, F., 1993. Traditional ecological knowledge in perspective. In: Inglis, J.T. (Ed.), Traditional Ecological Knowledge: Concepts and Cases, International Program on Traditional Ecological Knowledge and International Development Research Centre, Ottawa, pp. 1–9.

Bulun Bulun and Anor v. R. & T. Textiles Pty ltd, 1998. FCA 1082 (Federal Court of Australia 1998).

Caquard, S., 7 November 2011. Cartography I: mapping narrative cartography. Progress in Human Geography. http://dx.doi.org/10.1177/0309132511423796.

Royal Commission on Aboriginal Peoples, 1996. Report of the Royal Commission on Aboriginal Peoples, Looking Forward, Looking Back, vol. 1. Royal Commission on Aboriginal Peoples, Ottawa, Canada.

Canadian Institutes of Health Research (CIHR), Natural Sciences and Engineering Research Council of Canada (NSERC) and Social Sciences and Humanities Research Council of Canada (SSHRC), 2010. Tri-council Policy Statement: Ethical Conduct for Research Involving Humans. Canadian Institutes of Health Research, Ottawa, ON.

Canadian Internet and Public Policy Interest Clinic (CIPPIC), forthcoming. Licenses for Local and Traditional Knowledge (LTK) in the Context of Digital Cartography, a CIPPIC and GCRC collaborative report.

Convention on Biological Diversity, 2012. Introduction: Traditional Knowledge and the Convention on Biological Diversity. Available at: http://www.cbd.int/traditional/intro.shtml.

Coombe, R., 2009. First nations cultural heritage concerns: prospects for protection of traditional knowledge and traditional cultural expressions in international law. In: Bell, C., Paterson, R.K. (Eds.), Protection of First Nations Cultural Heritage: Law, Policy and Reform, UBC Press, Vancouver.

Delgamuukw v. British Columbia, 1997. CanLII 302 (SCC), [1997] 3 SCR 1010.

Dutfield, G., 2001. TRIPS-related aspects of traditional knowledge. Case Western Reserve Journal of International Law 33, 233–275.

Elwood, S., Goodchild, M.F., Sui, D.Z., 2011. Researching volunteered geographic information: spatial data, geographic research and new social practice. Annals of the Association of American Geographers 102 (3), 571–590.

Engler, N.J., Scassa, T., Taylor, D.R.F., 2013. Mapping Traditional Knowledge: Digital Cartography in the Canadian North. Cartographica 48 (3), 189–199.

Felt, L.F., Natcher, D., 2011. Ethical foundations and principles for collaborative research with Inuit and their governments. Études Inuit Studies 35, 107–126.

Frogner, R., 2010. Innocent legal fiction: archival convention and the North Saanich Treaty of 1852. Archivaria 70 (Fall), 45–94.

Harley, J.B., 1988. Silences and secrecy: the hidden agenda of cartography in early modern Europe. Imago Mundi 40, 57–76.

Harley, J.B., 1989. Deconstructing the map. Cartographica 26 (2 Spring), 1–20.

Helfer, L., Austin, G.W., 2011. Human Rights and Intellectual Property: Mapping the Global Interface. Cambridge University Press, Cambridge, UK.

Inuit Tapiriit Kanatami (ITK) and Nunavut Research Institute (NRI), 2006. In: Nickels, S., Shirley, J., Laidler, G. (Eds.), Negotiating Research Relationships with Inuit Communities: A Guide for Researchers, Inuit Tapiriit Kanatami, Ottawa, ON, Available at: http://www.nri.nu.ca/pdf/06-068%20ITK%20NRR%20booklet.pdf.

Josias, A., 2011. Toward an understanding of archives as a feature of collective memory. Archival Science 11, 95–112. http://dx.doi.org/10.1007/s10502-011-9136-3.

Judge, E.F., Gervais, D., 2011. Intellectual Property: The Law in Canada, second ed. Carswell, Toronto.

Krupnik, I., Aporta, C., Gearhead, S., Laidler, G.J., Kielsen Holm, L. (Eds.), 2010. SIKU: Knowing Our Ice: Documenting Inuit Sea-Ice Knowledge and Use, Springer, New York.

Kruger v. The Queen, 1977. CanLII 3 (SCC), [1978] 1 S.C.R. 104.

Lyons, N., 2011. Creating space for negotiating the nature and outcomes of collaborative research projects with Aboriginal communities. Études Inuit Studies 35, 83–105.

Monmonier, M., 1996. How to Lie with Maps, second ed. University of Chicago Press, Chicago.

Nannelli, E., 2009. Records of the Commission for Reception, Truth, and Reconciliation in Timor-Leste. Archival Science 9, 29–41. http://dx.doi.org/10.1007/s10502-009-9103-4.

National Aboriginal Health Organization, 2007. OCAP: Ownership, Control, Access and Possession, Sanctioned by the First Nations Information Governance Committee. Assembly of First Nations, Ottawa.

Ni, K.-J., 2011. Traditional knowledge and global lawmaking. Northwestern University Journal of International Human Rights 10, 85–118.

Nickels, S., Knotsch, C., 2011. Inuit perspectives on research ethics: the work of Inuit Nipingit. Études Inuit Studies 35, 57–81.

Oguamanam, C., 2004. Localizing intellectual property in the globalization epoch: the integration of indigenous knowledge. Indiana Journal of Global Legal Studies 11 (2), 135–169.

Oguamanam, C., 2011. Intellectual Property in Global Governance. Routledge, New York, NY.

Omura, K., 2005. Science against modern science: the socio-political construction of otherness in Inuit TEK (traditional ecological knowledge). Senri Ethnological Studies 67, 323–344.

Pyne, S., Taylor, D.R.F., 2012. Mapping indigenous perspectives in the making of the cybercartographic atlas of the Lake Huron Treaty relationship process: a performative approach in a reconciliation context. Cartographica 47 (2), 92–104.

Sandeen, S.K., 2009. The third party problem: assessing the protection of information through tort law. In: Brauneis, R.F. (Ed.), Intellectual Property Protection of Fact-Based Works: Copyright and Its Alternatives, Edward Elgar, Cheltenham, UK, pp. 278–304.

Scassa, T., 2010. Copyright reform and fact-based works. In: Geist, M. (Ed.), From "Radical Extremism" to "Balanced Copyright": Canadian Copyright and the Digital Agenda, Irwin Law, Toronto, ON, pp. 571–597.

Scassa, T., Engler, N.J., Taylor, D.R.F., forthcoming. Legal issues in mapping traditional knowledge: digital cartography in the Canadian North. Cartographic Journal.

Secretariat of the Convention on Biological Diversity, 2002. Bonn Guidelines on Access to Genetic Resources and Fair and Equitable Sharing of the Benefits Arising Out of Their Utilization. Available at: http://www.cbd.int/doc/publications/cbd-bonn-gdls-en.pdf.

Secretariat of the Convention on Biological Diversity, 2011a. Nagoya Protocol on Access to Genetic Resources and the Fair and Equitable Sharing of Benefits Arising from Their Utilization to the Convention on Biological Diversity. Secretariat of the Convention on Biological Diversity, Montreal, QC.

Secretariat of the Convention on Biological Diversity, 2011b. The Tkarihwaié:ri Code of Ethical Conduct to Ensure Respect for the Cultural and Intellectual Heritage of Indigenous and Local Communities Relevant to the Conservation and Sustainable Use of Biological Diversity. Available at: http://www.cbd.int/traditional/code/ethical-conduct-brochure-en.pdf.

Sparke, M., 1998. A map that roared and an original atlas: Canada, cartography, and the narration of nation. Annals of the Association of American Geographers 88 (3), 463–495.

Sunder, M., 2007. The invention of traditional knowledge. Law and Contemporary Problems 70, 97–124.

Taylor, D.R.F., 1997. The concept of cybercartography. Keynote address at the18th International Cartographic Conference in Stockholm (Sweden).

Taylor, D.R.F., 2003. The concept of cybercartography. In: Peterson, M. (Ed.), Maps and the Internet, Elsevier Science, Oxfod, pp. 405–420.

Taylor, D.R.F., Pyne, S., 2009. The history and development of the theory and practice of cybercartography. International Journal of Digital Earth 3 (1), 1–14.

Pyne, S., Taylor, D.R.F., 2012. Mapping indigenous perspectives in the making of the cybercartographic atlas of the Lake Huron treaty relationship process: A performance approach in a reconcialiation context. Cartographica 47 (2), 92–104.

Tobias, T.N., 2010. Living Proof: the essential data-collection guide for indigenous use-and-occupancy map surveys. Ecotrust and the Union of BC Indian Chiefs.

United Nations Declaration on the Rights of Indigenous Peoples, 2007.GA Res. 61/295, UN GAOR, 61st Sess., (Suppl. 49) vol. III, UN Doc. A/61/49 (2007).

United Nations Educational, Scientific and Cultural Organization, 1985. Model Provisions for National Laws on the Protection of Expressions of Folklore against Illicit Exploitation and Other Prejudicial Actions. Available at: http://www.wipo.int/wipolex/en/text.jsp?file_id=184668.

United Nations, 1992. Convention on Biological Diversity. 5 June, 1760 U.N.T.S. 79, 31 I.L.M. 818.

Vaver, D., 2011. Intellectual Property Law. 2nd ed. Toronto: Irwin Law.

Wood, D., Fels, J., Krygier, J., 2010. Rethinking the Power of Maps. Guildford Press, New York.

World Intellectual Property Organization (WIPO), 6–10 June 2005. Intergovernmental Committee on Intellectual Property and Genetic Resources, Traditional Knowledge and Folklore. The Protection of Traditional Knowledge: Revised Objectives and Principles. Eighth Session, Geneva.

CHAPTER

20

Cybercartography for Education: The Application of Cybercartography to Teaching and Learning in Nunavut, Canada

D. R. Fraser Taylor[1], *Cindy Cowan*[2], *Gita J. Ljubicic*[3], *Carmelle Sullivan*[3]

[1]Geomatics and Cartographic Research Centre (GCRC), Carleton University, Ottawa, Ontario, Canada, [2]Nunavut Arctic College, Iqaluit, Nunavut, Canada, [3]Department of Geography and Environmental Studies, Carleton University, Ottawa, Ontario, Canada

OUTLINE

Developments in the Theory and Practice of Cybercartography, Second Edition, ISSN 1363-0814
http://dx.doi.org/10.1016/B978-0-444-62713-1.00020-9

20.1 CYBERCARTOGRAPHY AND EDUCATION: SOME THEORETICAL CONSIDERATIONS

In the first edition of this book (Taylor, 2005) Chapter 21 (Baulch and others) outlined the potential of using the *Cybercartographic Atlas of Antarctica* as an educational resource, especially for students and teachers involved in Students on Ice, an organization that takes students of high school age from all over the world to the Polar regions (Students on Ice, 2013). The relationship with Students on Ice was important in the decisions about the design and content of that Atlas, which is still being used today. This chapter takes the educational applications of cybercartography in new directions by looking at its application to both high school and community college education in Canada's most recently created Territory Nunavut. The involvement of educators, students, and community members in Nunavut has been central to the ongoing development of the theory and practice of cybercartography and the technical aspects of atlas creation are outlined elsewhere in this volume (Hayes and others, Chapter 9). This chapter concentrates on the processes used to create atlas content, and associated educational potential.

20.2 CYBERCARTOGRAPHY AND EDUCATION

In Chapter 1, six ideas central to cybercartographic theory and practice are outlined. A number of these are of particular importance to the teaching and learning process. Teaching and learning are most effective when individuals are actively involved, as has been argued by educational theorists for decades since Dewey's (1938) pioneering work. In the Nunavut context, active involvement is not confined to using cybercartographic atlases but also to creating atlas content and this is a key factor that will be described later. Cybercartographic atlases are highly interactive and engage users in new ways. People learn in different ways and prefer teaching and learning materials in different formats. Cybercartographic atlases provide information in many different formats and users can choose the format best suited to their own needs. People use all of their senses in learning and cybercartography creates multisensorial representations (Taylor, Chapter 1). Inuit society is an oral society and story-telling is of great importance in learning. People need ways to create their own online narratives and cybercartography provides a way of doing this. In Nunavut, community involvement in creating atlas content is of great educational importance, as will be outlined later. Cybercartographic atlases make full use of a variety of media formats that again increase user involvement and enrich the information content.

Many topics of interest to society are very complex and there are multiple ontologies or ways of knowing to consider. There is no simple 'right' or 'wrong' answer to the causes and effects of issues such as global warming and climate change. Cybercartographic atlases allow the representation of a variety of different opinions or narratives on such issues rather than privileging just one definitive answer. Of particular importance in Nunavut is the incorporation of traditional knowledge (TK), especially regarding peoples' relationships with the environment. Inuit 'ways of knowing' must be part of the atlas creation process to increase the utility of these atlases in an educational context.

20.2.1 Cybercartography and Howard Gardner's Multiple Intelligence Theory

One of the most widely applied cognitive learning theories is Howard Gardners's MI Theory. In the early 1980s, Gardner questioned not only the typical uniform style of teaching but also the concept of a singular intelligence. He argued that students learn in very different ways and outlined seven intelligences to describe these differences. These intelligences were presented in his widely read book Frames of Mind (1983). Gardner argued that we all exhibit these seven intelligences. However, we have strengths and weaknesses in different ones. The seven intelligences are:

1. *Linguistic intelligence*: 'involves the sensitivity to spoken and written language, the ability to learn languages, and the capacity to use language to accomplish certain goals;'
2. *Logical-mathematical intelligence*: 'involves the capacity to analyse problems logically, carry out mathematical operations, and investigate issues scientifically;'
3. *Musical Intelligence*: 'entails skill in performances, composition, and appreciation of musical patterns;'
4. *Bodily kinesthetic intelligence*: 'entails the potential of using one's whole body or parts of the body (like the hand or mouth) to solve problems or fashion products;'
5. *Spatial Intelligence*: 'features the potential to recognize and manipulate patterns of wide space as well as the patterns of more confined spaces;'
6. *Interpersonal Intelligence*: 'denotes a person's capacity to understand the intentions, motivations, and desires of other people and, consequently, to work effectively with others;' and
7. *Intrapersonal Intelligence*: which 'involves the capacity to understand oneself, to have an effective working model of oneself – including one's own desires, fears, and capacities – and to use information effectively in regulating one's own life' (Gardner, 1999: 41–43).

Gardner later added an eighth intelligence, Naturalist intelligence: a person's ability to identify plants and animals in the surrounding environment (Gardner, 1999, 2000a) and Gardner is now arguing for nine intelligences (Gardner, 2008), adding spiritual and existential intelligence.

Gardner claims that educators tend to focus on teaching to the first two intelligences (Linguistic Intelligence and Logical-mathematical intelligence), often neglecting students who exhibit other learning forms. Therefore, he sees MI theory as a basis for education reform that will encourage educators to look for ways to reach these other intelligences. Gardner continues to research MI theory and its use in education together with the Education Research Group at Harvard. His latest thinking can be found in the 2011 edition of Frames of Mind and in the electronic publications on his web site (http://HowardGardner.com/). MI theory is less popular with theoretical psychologists such as those who argue that the empirical evidence for MI is weak and that there is really only one intelligence (Schaler, 2006). However, the principles behind MI theory closely emulate many of the central concepts of cybercartography. Cybercartographic theory and practice developed completely independently of the theory of MI until 2004 when the link with Gardner's work was made (Baulch et al., 2005). Gardner now specifically looks at the applications of MI theory for education and in the preface Gardner (2011) argues for the 'individualization' and 'pluralization' of education. By 'individualization' he means tailoring teaching and learning as far as possible to the abilities of the individual leaner:

> put concretely we can approach topics in a number of different ways (often termed 'entry points' to the same topic). We can make use of some of the analogies drawn from a range of domains and we can express the key notions or concepts in a number of different symbolic forms...By pluralizing I mean that an educator should decide on which topics, concepts, or ideas are of greatest importance and should then present them in a variety of ways. Pluralization achieves two important goals: when a topic is taught in multiple ways, one reaches more students. Additionally, the multiple modes of delivery convey what it means to understand something well. When one has a thorough understanding of a topic one can typically think of it in several ways. Conversely, if one is restricted to a single mode of conceptualization and presentation, one's understanding (whether teacher or student) is likely to be tenuous" ***(Gardner, 2011: xvi).***

In an earlier paper, Gardner argues 'as human beings, we are the kind of creatures who can learn in many ways: through exploration with our hands, the use of our several senses, the silent observation of other persons, conversation and argument, and the development of many kinds of symbols – ranging from painting or graphs to semaphore or dance notations' (Gardner, 2000a: 32) and in 2008 while looking at four new directions for MI Theory he argues that one of the four new directions should be 'devising of computer software and virtual realities that present or teach the same topic via the activation of different intelligences' (Gardner, 2008: 1).

This supplements his earlier arguments on the potential of multimedia: '...the new technologies make the material vivid, easy to access, and to play well – and they readily address the multiple ways of knowing that humans possess' (Veenma and Gardner, 1996). Gardner is, however, aware that indiscriminate use of technology can be disruptive (Gardner, 2012) and he argues for the careful use of technology to meet defined teaching and learning goals.

Gardner's ideas fit well with the theory and practice of cybercartography in relation to education, especially in the Nunavut context. All of his intelligences have relevance to education in Nunavut but his eighth intelligence – naturalist intelligence – is particularly appealing in the context of developing a curriculum that incorporates Inuit experience and ways of knowing as will be outlined later in this chapter. Whether this is a unique intelligence, as Gardner asserts, or whether it is part of a more general intelligence as his critics argue, there is no doubt that it must play a large part in developing new approaches to teaching and learning in the Nunavut context.

There is a limited literature on the application of Gardner's theories in geography and cartography (Baulch et al., 2005) and none that we are aware of that are specifically related to Nunavut. Many of the assertions made on theoretical grounds will have to be tested in the realities of the educational contexts outlined later in this chapter where Gardner's theoretical constructs provide a general basis for the development of cybercartographic atlases for educational purposes.

20.2.2 The Importance of Content and Process

The most important concern for creating dynamic learning tools such as cybercartographic atlases is not the interactive technology but the quality and relevance of the content of the atlases as well as the processes by which that content is created. A well designed atlas does not equate to a well-designed teaching and learning tool if the content has not been thoroughly thought out in terms of its relationship to the curriculum and teaching and learning goals in both formal and informal education. In Nunavut, the involvement of Inuit communities in providing and selecting that content is of great importance and the following section will describe the educational context in which the atlases will be used.

20.3 THE EDUCATIONAL CONTEXT IN NUNAVUT

Prior to elaborating on the educational potential of atlases, specific to the geographic and cultural context of Nunavut, it is important to provide some context around the history and evolution of northern education in Canada. While a full history and discussion of the many nuances and complexities of Inuit experiences with formal education is beyond the scope of this chapter, however it is important to feature some examples. Helpful overviews are provided in McGregor (2010b). The following provides a brief overview of some important periods related to Inuit education, and gives the context and rationale for incorporating cybercartographic atlas development into Nunavut educational programming at both the high school and college levels.

20.3.1 Evolution of Inuit Education

Traditional Inuit education was facilitated by family members, especially grandparents and Elders. These individuals drew from their own experiences and knowledge passed on to them by previous generations to teach their youth. This particular approach to education centred on tasks and responsibilities that were part of everyday life, as well as traditions of food sharing, spirituality, games, and narratives, where youth learned through observation and experience (Bonesteel, 2006; Friesen and Friesen, 2005). The environment played a major role in Inuit education due to the harsh climate and nomadic lifestyle, which influenced many of their daily activities (McGregor, 2010b). The relationship of Inuit with the renewable resource base underlies almost every aspect of their lives, from songs and family customs to political organization and the settlement of disputes (McGregor, 2010b). John Amagoalik (2001: 9) describes this relationship with the environment by explaining how: 'the land shaped our mind and language, our culture, our legends, our philosophy and our view of life.' Each region had a unique ecology and therefore the hunting techniques that were developed varied accordingly to respond to these differences (Bennett and Rowley, 2004). This began to change with the increasing influence from outsiders (often referred to as Qallunaat, or non-Inuit).

Although Inuit had contact with early explorers and traders, major outside educational influences came with the arrival of missionaries. In the early 1900s, missionaries established churches, hospitals, and schools, usually near Hudson's Bay Company posts, where they provided spiritual, health, and educational services for Inuit families when they came to trade or visit (Bonesteel, 2006; Laugrand and Oosten, 2009; McGregor, 2010b). Missionaries developed the Inuktitut syllabics system still used today to translate gospel into Inuktitut and to teach syllabic literacy to Inuit (Bonesteel, 2006; McGregor, 2010b). Once the use of Inuktitut syllabics spread, missionaries would offer basic instruction in Inuktitut literacy, Western hygiene, and arithmetic (McGregor, 2010b). McGregor (2010b) states that at that time, the goal for education was not explicit assimilation but rather to teach the Inuktitut syllabics to aid in reading the Bible and prayer books, for communication with others via written text, and to provide some knowledge of global issues. Arithmetic and Western hygiene were taught to strengthen and aid in trading relationships. However, missionary schools disrupted the transmission of culture and language from one generation to the next (Fogwill, 1994).

> Across much of the Eastern Arctic [...] Inuit children were exposed to the paternalistic, assimilative approach of Qallunaat educators. Inuit parents had no say as to how or what their children learned in school [...] Indeed, the purpose of early schools was to change Inuit children into knowing, being and doing like Qallunaat, to prepare them for employment on Qallunaat terms. It appeared that the Inuit system of experiential, learner-centred, environment-based education could not co-exist with that of the formal school system. Many Qallunaat and Inuit felt that this period marked the passing of the relevance of traditional Inuit knowledge, culture and education ***(McGregor, 2010: 83).***

The federal government had limited interest in the North for many years. However, with increased military and economic interest in the region, combined with political pressure to reinforce territorial claims, the federal government began taking more responsibility and control. Control over education had passed to the Canadian government by the late 1950s, with small schools being built across much of the Eastern Arctic as well as the first residential schools in the North, in addition to those previously established by missionaries (Laugrand and Oosten, 2009; McGregor, 2010b). Once this power shift occurred the education system became more a means of assimilation than a tool for personal and communal strengthening and enhancement (Castellano et al., 2000; Laugrand and Oosten, 2009).

> Residential schools were also intentionally designed to impose Western standards in every realm of their lives, manifested by the disallowance of Inuktitut and Inuit clothing and food, and the devaluing of Inuit social norms, beliefs or practices. This imbedded in many Inuit youth a sense of shame about their own culture and heritage ***(McGregor, 2010: 88).***

While some families saw residential schools as a means to new opportunities (Laugrand and Oosten, 2009), many parents felt pressured to send their children to school (e.g. being threatened with removal of social assistance support). Residential schools also had a number of long-lasting consequences on Inuit culture, education, and family networks:

- a loss of land-based survival and cultural skills;
- limited ability to maintain family connections or land-based lifestyle;
- increased dependency on government provisions;
- decreased self-confidence or sense of Inuit identity;
- loss of Inuktitut language and literacy (only allowed to speak English); and
- interruption of Inuit values (Berger, 1977; Bonesteel, 2006; McGregor, 2010b; Castellano, 2006b).

As part of the devolution process that was occurring in most government sectors in the late 1970s, the responsibility for education was transferred from the federal government to the

Northwest Territories (McGregor, 2010b). This shift of responsibility was a significant attempt to break with the past and aimed to produce a 'made-in-North West Territories (NWT)' curriculum that was more Inuit-centred (McGregor, 2010b).[1] The goal was to create a northern focused curriculum through:

- the provision of schooling in Native languages during the early elementary years;
- the development of programmes and classroom materials that reflect northern content;
- the increased involvement of community members and parents in the education process; and,
- expectations that teachers would support students in various career choices (McGregor, 2010b: 109–110).

One key factor in accomplishing these goals was the training of Inuit teachers; initially to bridge the language barrier between students and Qallunaat teachers, and later, after the NWT Teacher Education Program was developed, to certify teachers to manage their own classrooms (Bonesteel, 2006; McGregor, 2010b). Classroom assistants improved communication with students who spoke English as a second language (ESL), and improved communication with parents and the community. There was also more continuity in teaching staff as Inuit were less likely to be transient compared with Qallunaat educators. These teachers also contributed to increased cultural content in the curriculum (McGregor, 2010b). However, despite the substantial rhetoric about change, the reality and implementation of change was much less impressive. One of the main factors holding back the education system from achieving its goals was that many Qallunaat continued to believe that they knew what was best for Inuit education, as demonstrated through the fact that no Inuit were involved in any educational decision-making under the NWT (McGregor, 2010b). Limited and negative experiences for Inuit in these educational institutions, including long distances to travel to schools, contributed to low attendance and high dropout rates (Bonesteel, 2006; Berger et al., 2006). A move towards more Inuit-driven approaches was fostered with the establishment of the Nunavut Land Claims Agreement (NLCA).

The NLCA was signed in 1993 forming the basis for the creation of the territory of Nunavut in 1999, and bringing with it the control of institutional education:

> Nunavut's creation as a territory enabled and empowered Inuit leaders to attempt to address the local, educational, social, and economic needs of its isolated northern communities...This was the first time that the Inuit had the opportunity to define and implement their vision of what constituted a quality education *(Tyler, 2008: 13)*.

Berger (2006) explains that within the territorial government a proportional percentage of Inuit representation is assured for Inuit employment; however, this goal has not been realized to date. Although many Inuit are employed in support positions, such as administrative or secretarial area, there is a need for Inuit with post-secondary education to fill more of the leadership roles in government, including managerial positions (Berger, 2006). Education was not specifically identified in the NLCA and therefore neither funding policy nor legislation was

[1] There are multiple First Nations and Métis groups in NWT to consider, but for our purposes we are focusing on Inuit; it is not representative of all groups.

secured. However, Berger (2006)[2] explains that in order for these rights and privileges to be realized, bilingual (Inuktitut and English) education needs to become a focal point (Section 4.1.2). There are a number of challenges that remain in order to create a culturally appropriate and effective educational system in Nunavut, which are associated with the history of Inuit education described above.

20.3.2 Current Challenges

Continuous efforts have been made to: improve the education system in Nunavut; make learning more relevant for Inuit students; increase graduation rates; and to help prepare students for their future endeavours. These efforts include various Nunavut-created resources for principals and teachers, as well as curricula created by the NWT to integrate relevant learning. However, Berger (2009) argues that the government education system has not succeeded in graduating students who, for the most part, are more connected to their Inuit heritage, nor are they fully equipped to engage in the wage economy (Berger, 2009). There are a number of current challenges that have hindered the current system from reaching its full potential:

- schooling in Nunavut remains primarily based on Euro-Canadian values, curricula, and pedagogy (e.g. Alberta curriculum is still the broad foundation, majority of teachers are Qallunaat);
- English is privileged over Inuit languages;
- Inuit knowledge, values, and approaches to education are not well integrated;
- the history and trauma associated with residential schooling still influences families today (which can affect attitudes/support for formal education);
- lack of teacher training and preparation in relation to the northern cultural context; and
- socioeconomic barriers discouraging school attendance (including, but not limited to, high rates of overcrowded housing making it hard to study), and high rates of unemployment.

These can contribute to negative attitudes about the importance of education when potential economic returns are limited locally (Berger, 2006, 2009; Bonesteel, 2006; Berger et al., 2006; Laugrand and Oosten, 2009; McGregor, 2010b; Simon, 2008).

As a result of these challenges, the rate of student graduation from secondary schools in Nunavut in 2008 (26.9%) was quite low compared with the national average (71.3%) (Brockington, 2009). Of those who do graduate, many are often not prepared to attend post-secondary institutions and are further discouraged by the long-distance travel involved, most often to southern Canada (Bonesteel, 2006).

20.3.3 Integrating Cultural and Academic Learning

The current challenges highlight the need for culturally relevant educational curricula that will contribute to higher retention rates of Inuit students throughout primary and secondary

[2] The Conciliator's report was requested as part of a lawsuit for $1 billion launched by NTI against the Government of Canada for breach of contract regarding implementation of the NLCA (AADNC, 2011).

grades (Bonesteel, 2006). Here is where cybercartographic atlases have a role to play as will be discussed shortly. Various actors, such as Inuit organizations (e.g. Inuit Tapiriit Kanatami (ITK)), government bodies (e.g. the Government of Nunavut (GN)), and researchers (e.g. Berger et al., 2006), have identified the convergence of Indigenous and Western knowledge systems as an important approach to the education of Inuit youth (Inuit Circumpolar Conference, 1992; RCAP, 1996; Government of Nunavut, 2008; ITK, 2011). Champagne and Abu-Saad (2006) argue that this convergence must result in the education system being designed to support the continuity and future development of Indigenous communities but should also facilitate access to mainstream institutions. To achieve successful educational outcomes, Berger et al. (2006) recommend the implementation of true Inuit control over the education of youth, along with the meaningful consultation of community members, development of a curriculum that reflects Inuit pedagogies, and extensive cross-cultural and ESL training for non-Inuit educators working in Nunavut. Increased teaching and learning in Inuktitut would also be a positive step forward.

GN has undertaken the task of revising the mainstream education curriculum in a way that encompasses traditional Inuit culture and skills while simultaneously focusing on the development of confident, responsible, and capable individuals who can contribute to Nunavut society and at the same time prepare them for enrollment in post-secondary institutions (GN, 2008). There are also several government policies that complement this goal including the following:

- *Inuuqatigiit* – This is curriculum written from the Inuit perspective, as first conceptualized and developed in 1996 by the NWT Department of Education, Culture and Employment. This curriculum is based upon Inuit philosophy and provides Inuit knowledge to students from Kindergarten to Grade 12. Inuuqatigiit means: 'Inuit to Inuit, people to people, living together, or family to family. It implies togetherness and family unity between people. This is the foundation of the curriculum: a unity of Inuit philosophy for the benefit of children, teachers, schools, and communities.' (DECE, 1996: 3). Inuuqatigiit focuses on the 'enhancement and enrichment' of the language and culture of Inuit students. It also promotes the integration of Inuit knowledge with the mainstream curriculum.
- *Inuit Qaujimajatuqanit (IQ) principles* are Inuit ways of knowing, being, and doing. These IQ principles were documented as a GN initiative to promote the integration of Inuit knowledge and values into government functions. There are eight IQ principles defined by the Department of Education (2007), and all government departments in Nunavut are required to integrate IQ principles into their operations and services according to the Bathurst Mandate of 1999. IQ principles have become a key resource for the development of local school programmes, expectations for students and educators, curriculum development, and the establishment of educational policies (Aylward, 2007). The GN Department of Education is currently developing a new environmental science curriculum that integrates IQ principles with current scientific information about land, oceans, and weather systems, and is working to revamp curriculum at all levels based on a newly developed IQ framework (GN Department of Education, 2007).
- *Nunavut Education Act* – The Nunavut Education Act (2008) outlines the new educational policy for Nunavut, and IQ principles are the foundation for supporting student

intellectual, physical, emotional, and spiritual development. The importance of involving Elders, parents, and other community members in the education system is recognized, as well as the value of bilingual education and experiential (land-based) learning opportunities.

Inuit education has experienced many changes over the last century. Despite the negative influences from missionary and especially earlier federal government control, education is still seen as a key means to promote engaged, successful Inuit youth who are confident in themselves and their cultural identity. Education efforts are centred on the increase of cultural learning and integration with academic learning, including the Inuuqatigiit curriculum, IQ principles, the Nunavut Education Act, and local cultural programmes. Cybercartographic atlases are being built with educational communities in Nunavut and are designed to contribute to this process. Among several challenges that northern educational institutions face is the challenge of creating culturally relevant educational materials and learning experiences for Inuit youth. Two examples discussed next describe efforts to integrate atlas development (including the research that goes into creating them, and atlases as a vehicle for sharing research results), one in the context of Nunavut high school education followed by Nunavut Arctic College (NAC) level education.

20.4 THE *INUIT SIKU (SEA ICE) ATLAS* AND HIGH SCHOOL EDUCATION IN NUNAVUT

The *Inuit siku (sea ice) Atlas* (http://sikuatlas.ca) was developed as an educational resource to share the knowledge, stories, maps, language, and lessons shared through years of interviews, focus groups, sea ice trips, and workshops with Inuit sea ice experts in Cape Dorset, Igloolik, Pangnirtung, and Clyde River, Nunavut (see Chapter 14). This resource is available online, using interactive and multimedia features to enable northern students, and the general public, to explore and learn about various sea ice topics from Inuit perspectives. The main purpose of developing this Atlas was to respond to Inuit elders' and hunters' expressions of interest to:

- share their knowledge with youth;
- see more Inuit knowledge and northern content in the northern education system; and
- share their knowledge more broadly with scientists and the general public.

This educational resource is a compilation of Inuit sea ice knowledge and use, as documented by university and community researchers between 2004 and 2008. It can be used to learn about:

- Inuit knowledge of sea ice features, uses, hazards, and changes;
- Nunavut community elders, hunters, and researchers involved in the project;
- maps of Inuit sea ice knowledge and use;
- Inuktitut sea ice terminology in different seasons;
- the Floe Edge Service – satellite imagery available for communities across the Canadian North; and
- background on the larger sea ice project and researchers involved.

The following outlines some important feedback received from educators, students, and northern community members during informal community consultations and workshops associated with *Atlas* launch events, as well as efforts to integrate the siku Atlas into the Nunavut educational curriculum.

20.4.1 Community Consultations

Prior to the public launch of the siku Atlas northern community members, researchers, and anyone else interested in these topics, were invited to take a few minutes to fill out a short online survey (using Fluid Surveys) to provide their feedback on the utility of the Atlas. This consultation ran from mid-January to the end of February, 2011. Over 30 completed surveys, and over 60 surveys with partial feedback, were received, representing locations across Canada (North and South), including a few from the United States and even one from Germany. All age ranges were represented, from under 18 to over 65, with the majority falling between 25 and 64 years of age. Occupational categories were also broadly represented, including among others students, teachers, Inuit organization representatives, hunters/harvesters/fishers, government employees, and researchers. Overall, the feedback was very positive, with 57% indicating that they liked it very much, felt it was highly relevant and useful, and 31% identifying that they liked it, felt it was a good overview of key topics. The main positive comments we received included the fact that respondents:

- liked having this knowledge available and accessible for educational purposes;
- appreciated the public sharing of research results;
- could see the contributions to language preservation;
- found this a valuable way to engage students (i.e. they are very into digital media);
- liked the how elder contributions and background were highlighted;
- were glad that it enabled Inuit voices to be heard directly (e.g. through audio or video clips); and,
- were happy about the compilation of sea ice maps and terminology.

However, some concerns were also raised, many of which could not adequately be addressed prior to the launch of the Atlas, but which work towards improving these over the longer term is in progress, and these are:

- ongoing issues of limited bandwidth across the North;
- accessibility/use of this online tool by Elders, hunters, or those without computers or who are unilingual in Inuktitut;
- the need for more visual appeal (which was initially kept simple though to help deal with low bandwidth issues);
- interest in having more communities represented;
- the need for even simpler language;
- the need to include more interviews and historical information;
- the ability to turn the site into a book or CD-ROM;
- more clarification of plug-ins needed to run all aspects of the *Atlas*;
- an interest in having the site bilingual throughout;

- accessibility to regularly updated satellite imagery; and
- the ability for community members to update information themselves.

Additional feedback was also sought specifically from Inuit Elder and youth perspectives in each of the key participating communities in association with Atlas launch events.

20.4.2 Elder/Youth Workshops

The siku Atlas was officially launched on March 17, 2011, and was celebrated with simultaneous community events in Cape Dorset, Igloolik, and Pangnirtung, Nunavut, as well as at Carleton University in Ottawa, Ontario (see Chapter 14). Along with launch events, several local Elder/youth workshops were hosted at each community high school. These provided an opportunity for Elders who contributed to the project, and students who might use the siku Atlas in the future, to learn more about it – but most importantly to learn more from each other.

20.4.2.1 The Atlas Facilitating Elder/Youth Interactions

In all three communities, the workshops were hosted by the local high schools, with which ongoing contact was made throughout the life of the research project. In all cases, the students would navigate through the *Atlas* to show the Elders different sections, while the Elders would elaborate on the various topics presented and shared their more detailed and practical knowledge about sea ice conditions with the students.

The Pangnirtung Elder/youth workshop was held at Attagoyuk Ilisavik (high school), a partner over the life of the sea ice project to document Inuit knowledge of sea ice processes, uses, and changes. This school also collaborated earlier to host multimedia training and editing workshops to help students learn about making direct contributions to the *Atlas*. For this particular event, the three Elders Jamesie Mike, Michael Kisa, and Peterosie Qappik who provided major contributions to the Atlas in earlier interviews, worked with three classes of Grade 10, 11, and 12 students, with a total of 25 students and 4 teachers. The session was also facilitated by Amos Hayes (who helped develop the technical aspects of the Atlas discussed in chapter 9) and Rebecca Mearns (Carleton Undergraduate Student working with Gita Ljubicic). Particular attention was paid to the Pangnirtung mapping section (http://sikuatlas.ca/sea_ice_map.html?module=3) with different features drawn by various local sea ice experts, as well as showing Elders the audio and video clips where they were featured. Session participants voiced their hope that the Pangnirtung terminology section could be expanded to be more like the detail and interactivity of the Cape Dorset terminology diagram (http://sikuatlas.ca/cape_dorset_terminology.html).

The Cape Dorset Elder/youth workshop was held at Peter Pitseolak High School with which was closely connected to the sea ice project. For this event, the five Elders who had provided major contributions to the Atlas in earlier interviews, were Mary Kellypalik, Mangitak Kellypalik, Etulu Etidlouie, Adamie Nuna, and Atsiaq Alasuaq participated. They worked with one Grade 10 class of 10 students and 1 teacher for 2 periods (80 min), while a few other teachers and the principal dropped in. This session was also facilitated by Pootoogoo Elee (long-time interpreter and local research assistant for the project) and Karen Kelley (who had done her MA research in Cape Dorset as part of the larger sea ice project). Most of the focus was on the Inuktitut sea ice terminology diagram for Cape Dorset (http://sikuatlas.ca/cape_dorset_terminology.html), with students clicking on various terms and Elders elaborating with stories and demonstrations through hand gestures. The students were engaged for the full length of the session, and the teacher commented that it almost brought tears to her eyes to see such concentrated student engagement and to watch the interactions between Elders and youth.

The Igloolik Elder/youth workshop was held at Ataguttaaluk High School, which was also closely connected to the sea ice project. For this event, Abraham Ulayuruluk, a knowledgeable local Elder who had provided major contributions to the Atlas in earlier interviews, participated in all the workshops, while Levi Qaunaq assisted with a few. These sessions were also facilitated by Theo Ikummaq (long-time interpreter and local research assistant for the project) and Chase Morrison (who had done his MA research in Igloolik as part of the larger sea ice project). They conducted 5 sessions with students ranging from Grades 8 to 12, and a total of 92 students. The students, especially the boys, were very interested in the map pages (http://sikuatlas.ca/sea_ice_map.html?module=2), and picking out places that they knew. Abraham also encouraged the girls to get involved, saying it was very important for them to learn about sea ice as well (and that this online learning forum provided them more access to this, since they had less exposure to travelling on the sea ice). It was also hoped that the Igloolik terminology section could be further expanded to be similar to the detail and interactivity of Cape Dorset (http://sikuatlas.ca/cape_dorset_terminology.html). The students also further discussed the atlas in an evening community radio show in Inuktitut.

Overall, the workshops exhibited common traits:

- there was great interest in sea ice map pages, terminology pages, and audio/video clips;
- Elders were happy to see video and audio clips included, and see/hear themselves telling their own stories;
- the terminology pages would be better if they had Inuktitut 'months' or yearly cycles included; and
- some students and/or Elders wondered about being able to continue to update/expand content, or access the content in book format (since not everyone uses computers).

In terms of the siku Atlas' educational potential, the workshops yielded the following key observations:

- the *Atlas* is a great tool to begin dialogue with Elders and students;
- students navigated the site, and Elders told different stories related to what they saw, and explained areas on the map or sea ice terms in more detail than what was included in the *Atlas* – which led to more questions. The *Atlas* therefore seems to be a helpful tool to get Elders and youth connected, enabling direct learning between each other (i.e. the Atlas is more of a facilitator than a stand-alone educational resource);
- students' interest in the Atlas was based on the combination of mixed media, local context, and the inclusion of relatives or familiar people;

- it sometimes took a little time for students to learn how to navigate map or terminology pages, and concurrently students and Elders were eager to view areas they were familiar with, and describe them further; and
- Elders quickly took on instructional roles, referring to visuals or terms as needed, but mainly sharing their own expertise on topics related to what they were viewing.

Through these interactions, it became clear that the *Atlas* by no means replaces the importance of learning through experience, or sea ice in context, and in Inuktitut from Elders who are most knowledgeable about ice conditions as that type of learning ensures that the necessary navigational and survival skills for safely travelling and hunting on the sea ice is done. However, the *Atlas* does provide an important starting point for discussion and questions. The *Atlas* was developed based on Elders' interests and concerns to share their knowledge with youth, and it helped bring people together to learn from each other, which is an important outcome in itself.

20.4.3 Integration into Tariuq (Oceans) Curriculum in Nunavut

Part of the reason for developing the siku Atlas was to respond to Elders' interest to have their knowledge documented in written and other media formats, in order to have more Inuit knowledge and northern content available for incorporation into the northern education system. Furthermore, in the process of undertaking several community-based research projects, as well as witnessing the many efforts researchers are currently undertaking to develop educational materials as an important part of their research project outputs, it became clear that these sorts of materials are rarely effectively utilized in schools if they are not part of the formal curriculum. Teachers are already overwhelmed, and it takes time to search out these resources and develop related connections to teaching responsibilities, and even though they may be very valuable, they are often overlooked. Therefore, to formalized educational connections, GN's, Department of Education, was contacted to see how/if the siku Atlas could be incorporated into their curriculum development efforts.

20.4.4 GN Curriculum Development

The GN is currently in the process of revamping the entire territorial curriculum to develop a made-in-Nunavut curriculum that would replace that current reliance on Alberta and Northwest Territories curriculum resources as described earlier in this chapter. The shift is being made towards a Nunavut curriculum centred on:

- Inuit Qaujimajatuqangit principles;
- northern context (language, examples, activities);
- experiential, project-based, flexible, integrative, cumulative learning;
- engagement of Elders and community members; and
- meeting national educational standards while also being more culturally appropriate.

The siku Atlas was developed with a northern high school student target audience of in mind, making the high school curriculum highly relevant in the project.

The high school science curriculum in Nunavut is now being strategically developed based on thematic foci, which include interdisciplinary learning of relevance to: Nuna (Land) in Grade 10; Tariuq (Oceans) in Grade 11; and, Sila (Weather/Climate) in Grade 12. This project was mostly involved with Tariuq because sea ice is such a prominent and important part of Arctic marine environments. The content of the Atlas was designed to be in alignment with the Tariuq portion of the curriculum for from 2009 to 2011, which incorporated various aspects of the siku Atlas (along with some results from previous sea ice projects) into teaching resources, activities, assessments, and examples.

20.4.5 Contributions to Tariuq

The objectives for formally incorporating the *siku Atlas* into Tariuq curriculum activities and learning modules were to contribute to better engaging students in the classroom setting (by having more relevant examples, project-based learning, and more balanced perspectives from Inuit and scientific knowledge), as well as encouraging more student-Elder interactions inside and outside the classroom.

Specifcally, *Atlas* contributions have been made to Module 5 of Tariuq, which focuses on 'Navigating the Sea Ice'. This new curriculum is organized according to four strands based on IQ principles, and Tariuq is part of Nunavusiutit, an integrated core curriculum that focuses on heritage, culture, history, geography, environmental science, civics, economics, current events, and world views, as well as linking Nunavut concepts to national, circumpolar, and global issues (with delivery involving educators, Elders, and community experts). Each learning unit is also organized according to various northern-centred icons that represent language development, materials, background, opener, connector, activity, reflection (content, collaboration, personal), follow-up, classroom reinforcement, accommodating diversity, assessment (student and teacher), activities and teaching/learning support materials for use in delivering the lessons.

Specifically, the *siku Atlas* (as well as some results from previous sea ice projects) has been incorporated into learning activities and Blackline Masters (key background information for teachers) related to six learning units within Module 5, including:

- Unit 2 – Siku and Sikulirijiit (worldviews and relationships with sea ice according to Inuit and Western scientific perspectives);
- Unit 3 – Inuktitut Sea Ice Terminology (local Inuktitut terminology used to describe sea ice and processes of melting and freezing);
- Unit 5 – Sea Ice Monitoring (sea ice monitoring using instruments and satellite imagery, as well as developing traditional observational skills);
- Unit 6 – Safety on Sea Ice (environmental indicators and survival skills for safe sea ice travel);
- Unit 7 – Navigation and Mapping Skills (identify important travel routes, place names, seasonal ice conditions, use traditional wayfinding indicators, and use of GPS); and
- Unit 8 – The Ocean, Ice and Climate Change (understand connections and feedbacks between sea ice and climate change).

One area where the siku Atlas is most directly incorporated is into Unit 3 on Inuktitut sea ice terminology. For example, learning competencies for this unit include:

- learning sea ice terminology in the local dialect of Inuktitut by making a sea ice dictionary;
- being able to identify major ice types with each type of terminology during field observation, and creating a sea ice key;

- learning about the seasonal freezing and melting stages of sea ice by creating models for a class sea ice cycle;
- being able to describe influential factors in freezing and melting processes through local community research;
- being able to describe why it is important to communicate about ice, and understand the different types of terminology; and,
- being able to communicate about sea ice conditions in Inuktitut with other students, Elders, and hunters through practice, making daily observations, and field experiences with community members.

Specific activities in this learning unit make reference to the siku Atlas introduction for teachers (http://sikuatlas.ca/teachers.html) and the background on Inuktitut sea ice terminology (http://sikuatlas.ca/sea_ice_terminology.html). In addition, specific activities were designed to encourage students (and Elders) to explore the Cape Dorset interactive dictionary (http://sikuatlas.ca/cape_dorset_terminology.html) and then create locally relevant terminology resources that reflect the local dialect of whichever community school is doing the activities. Furthermore, pictures and Inuktitut terminology contributed to pictoword and matching activities, along with sea ice monitoring sheets (which also links to Unit 5).

At the time of writing this chapter, the Tariuq curriculum was still being finalized, and was just moving into the piloting phase. Therefore, feedback from teachers or students as to the effectiveness of this more integrated approach to learning and benefits of including the Atlas in the process is not yet available.

20.5 THE ARCTIC BAY ATLAS AND COMMUNITY COLLEGE EDUCATION IN NUNAVUT

NAC serves the vast and beautiful Nunavut Territory, which became Canada's third territory on April 1, 1999, as the final result of the land claims described in Section 20.3. Nunavut includes about 20% of the landmass of Canada – essentially everything above the tree line. About 34,000 people live in the new territory, more than 85% of whom are Inuit. Today, many Inuit continue to live off the land as hunters and fishers, though this activity is usually combined with a role in the newer wage economy. The changing economy and the newly established public territorial government offer the residents of Nunavut enormous opportunities. It is the College's task to ensure that these individuals have the training and education they need to take full advantage of this situation. Its mission is to strengthen the people and communities of Nunavut by advancing the life-long learning of Northern adults through high-quality career programmes appropriately delivered with our partners throughout the Arctic, and by making the benefits of traditional and southern science more available (www.arcticcollege.ca).

The Arctic Bay interactive mapping project has its roots in the early 2000s as a literacy research project in regional place names undertaken by Nunavut Youth Consulting (NYC). NYC is a not for profit youth organization whose mandate is to engage youth in the community in positive undertakings, to address issues such as suicide and substance abuse (Inuit Heritage Trust, 2007). The NYC youth decided that to develop themselves and the community they would concentrate on education and health. Over the years, as part of that effort, they have continued to initiate projects and obtain funding to operate literacy projects involving

Inuit elders and youth. NYC's activities are supported and championed by Ron Elliott. Elliott was the local adult educator in Arctic Bay and on faculty with NAC until he was elected to the Nunavut Legislative Assembly in 2011.

20.5.1 Early Visions for the Arctic Bay Atlas: Documenting Traditional Inuit Placenames

Recognizing the learning potential of intergenerational research, the College supported Elliott in initiating and guiding the project and facilitating with NYC the collection and processing of data with elders and community members, as well as providing the use of the Community Learning Centre (CLC) classroom and computer lab. From 2003 to 2006, NYC worked in partnership with NAC, the Canada/Nunavut Geoscience Centre, and Inuit Heritage Trust (IHT) on a literacy research project called Traditional Name Placing Project. Funded by GN, Culture, Language, Elders, and Youth (CLEY), the learning project involved youth working with Elders. Based upon interviews youth were able to complete 12 maps with 350 traditional place names recorded, on 12 – 1:250,000 scale maps covering a vast land area (IHT, 2007). In addition to the mapping of traditional geographical place names, some of the traditional trails that were used to travel between Arctic Bay and Igloolik were also documented.

Through this work, Elliot and the youth team became aware of Dr Claudio Aporta's work with IHT, and at the Geomatics and Cartographic Research Centre (GCRC) at Carleton University. They sought to develop a similar – but humbler – version of his Anijaarniq (Aporta, See Chapter 16) interactive CD-ROM documenting traditional Inuit land skills, wayfinding, and related environmental knowledge in Igloolik, Nunavut.

Elliot approached Dr Aporta to assist NYC in digitizing the place names, and so in February, 2006, Dr Aporta (also a member of the GCRC) became involved in the Arctic Bay mapping project. Dr Aporta assisted the youth in gathering more complex information from local elders about place names in the area. In addition to demonstrating the process of mapping place names, a series of by-products were also created, including a database of place names as GPS waypoints and digital recordings containing valuable information about the meanings and stories of each name.

20.5.1.1 Map-Making as the Facilitator of Intergenerational Interaction and Learning

In March 2006, NYC was in their fifth year of existence, and they secured CLEY funding to extend their place naming documentation efforts with a project entitled 'Reclaiming the Land'. This youth-driven project was again primarily focused on Inuktitut Language, but was designed to build and continue the intergenerational exploration, dialogue, and documentation of geospatial locations of cultural significance. An essential goal of the extension project was to begin to use current technology familiar to youth in everyday life. Youth were keen to engage with global society through social media, but also wished to maintain and foster valuable cultural ties. This process of active learning through identifying and mapping places of cultural importance not only engaged youth, but soon community members, such as hunters, dropped in to share stories of their hunting experiences in different locations. Elliott observes: 'the maps are like magnets. Each time they are taken out and put on display people are drawn to them. Hours go by as Elders talk about places they have not visited for years.

The Elders say that their spirits are lifted by doing these projects' (NYC, 2006: 3). For this reason NYC proposed to take land trips with elders out to the various traditional camps to live, learn, and document (through photos, and audio or video clips) the stories and experiences that Elders shared in relation to important places.

In addition to the goal of preserving historical and traditional cultural land use, the project sought to create a dialogical learning experience with youth sharing their knowledge of contemporary technology and the Elders their traditional geospatial knowledge. It was hoped that visiting locations of cultural significance would create a form of spiritual healing for Elders who could revisit and remember these long forgotten locations (NYC, 2006). Additionally, it was intended that the final outcome would be the production of local educational resources that could be shared with the schools and the college.

Students spent 25 days participating in land trips over two seasons – spring and fall. During the quiet winter months, they reviewed and transcribed interviews, maintained language glossaries of new terms and words, and used video footage of Elder interviews as the basis of more consultations with Elders. At this point the mapping had been conventional – locations and names were placed onto enlarged paper maps. Elders were also invited to determine the next steps for the ongoing community-based project. From community dialogues, two things became obvious: (1) the mapping process was, in itself, a remarkable mutual learning experience; and (2) the data obtained could be used in other projects beyond the production of paper maps. This became the impetus for creating a cybercartographic *Atlas of Arctic Bay* (Figure 20.1) (http://arcticbayatlas.ca), and became the foundation for a more formal collaboration between NYC, NAC, and the GCRC.

FIGURE 20.1 Screen capture of Arctic Bay atlas.

20.5.2 Integration of Atlas Development into Arctic Bay College Learning and Research Activities

Since the software framework (Hayes and others, Chapter 9) ideally does not require advanced technical skills, it provides an opportunity for anyone with an Internet connection, individuals and communities such as Arctic Bay, to document and represent Inuit identity in relation to space, time, land, and spiritual forces in new and exciting ways. This is done on their own terms and in their own language so that the community-defined knowledge is privileged and the community controls the way their environment is portrayed (Taylor, 2007). For Inuit, this is critical. Research exhaustion and disappointment with the results of research has led Inuit, and even entire communities including Arctic Bay, to withdraw from research engagement, thereby silencing the Inuit voice in research methods, outcomes, and purposes. Cybercartography reverses the traditional research relationship between the researcher and the research community by facilitating reciprocal knowledge sharing. From the perspective of Inuit, this makes it a potentially powerful decolonizing research tool as is argued in several of the other chapters in this book.

In the case of Arctic Bay, the potential for Inuit control over the interaction, accumulation, production, and dissemination of knowledge was an essential criterion for working on the cybercartographic mapping project. This particular characteristic of the GCRC work and philosophy is consistent with NAC values of learner-centred programming that integrates Inuit culture and knowledge (research ethics and methods are discussed in chapters 18 and 19). The director responsible for community and adult learning, Cindy Cowan (co-author), also shared an academic interest in exploring new ways to engage adult learners in research as a motivational factor in student success. Of special interest is the idea, new to the College, that atlases can be seen as 'narrative forms that tell stories and stimulate imagination' (Wood, 1987). This is because Inuit culture is an oral culture and storytelling is central to the traditional transmission of knowledge critical to the sustaining of the Inuit. As noted by Taylor (Chapter 1) paper and text produced by non-Inuit researchers does not adequately capture and communicate the multiple dimensions of this knowledge – a point that is also made by Bonny and Berkes (2008).

The *Arctic Bay Atlas* was thus proposed as a pilot project that would link academic research with Inuit communities in an innovative way and increase the understanding of Indigenous knowledge as well as the theory of community-based research. It also involved a new partnership among academia (college and university) and community (Arctic Bay, a small Inuit community on Baffin Island) to facilitate new knowledge construction networks. Here was a solution for the college. The transmission of Inuit knowledge is difficult within the framework of a formal post-secondary institution that does not lend itself to the acquisition of predominantly experiential and informal knowledge.

Because the project strove to facilitate many outcomes of value to adult learners, the college, community and distance learning division formally became a full partner by assisting in efforts to fund and develop the *Arctic Bay Atlas*. Elliot was supported to work with adult learners on a class project involving ongoing research, developing interactive maps, and documenting routes and place names. They would explore the use of methods such as utilizing voice input, storytelling, video, photographs, art, and text in Inuktitut. The GCRC and the College left it open to Elliot and participants in the project to define the project's overarching concept, main topics, objectives, timing and costs. Once defined, the GCRC agreed to provide the technical support to compile and incorporate them into the framework of the *Nunavut Atlas* being constructed by GCRC, as well as research training regarding data collection and formatting (using GPS, drawing, digital photo, video, and audio recordings). Using the Nunaliit Framework (Hayes and others, Chapter 9) also enabled the community members involved to contribute additional materials as they were acquired through an ongoing local research.

With funding from a Social Sciences and Humanities Research Council of Canada (SSHRC) Initiative on the New Economy Public Outreach Grant, a three-day training workshop on community atlas building was hosted in Iqaluit, Nunavut in September, 2007. This was a collaborative effort between the NAC, Arctic Bay, and the GCRC, with an emphasis on: (1) identifying community needs, interests, and capacities to create an atlas; (2) allowing the modification and adaptation of cybercartographic theory and to respond to the local realities; and (3) identifying a concrete plan for the atlas production. Initially, the workshop was to be held in Arctic Bay, but Iqaluit (the capital of Nunavut) was decided upon in order to reach a broader audience and potential partners (e.g. with the Department of Education), and to capitalize on the facilities at the main College campus. The workshop was successful and demonstrated Carleton's ongoing 'commitment to provide communities with the tools, knowledge, and ongoing self-capacity to make their heritage accessible to new generations' (Hayes, 2007).

20.5.3 Conceptualizing and Creating the Arctic Bay Atlas

Building on earlier iterations of *Atlas* visions, funding from the Heritage Canada Gateway Fund (Canadian Culture On-line) was secured through a joint effort of IHT, GCRC, NAC, and NYC in January of 2007. These funds could be used directly to support the development of the *Arctic Bay Atlas* as a community-based interactive, multimedia website that presents Inuit culture, geographical locations/trails of significance, place names, oral history, and geographic representation to the Canadian public. It was designed as a community-based research project to engage youth and Elders of Arctic Bay in researching, documenting, and representing their multifaceted special knowledge. This project also sought to develop the local capacity in Arctic Bay to digitally capture and disseminate via the Internet the importance of Inuit knowledge and how this community 'historically conceptualized and represented its identity in relation to space, time, land, and spiritual forces' (Taylor, 2007). This meant that NYC could further engage Elders in this process, and were able to undertake an additional 50 days of data collection on the land (in 20–30 km day trips) to support the development of online content. Three additional research trips were conducted to document seasonal activities: (1) spring hunting; (2) fishing and camping locations; and (3) fall hunting on sea ice. The goal was not only to share the cumulative place name mapping in an interactive form, but also to digitally document the 2007 Arctic Quest, a spring dog-team race that goes from Arctic Bay to Repulse Bay (1000 km) along a traditional trail. This journey documentation was altered due to a full season delay, but instead the participants followed the 2008 Pangagujuinq Nunavut Quest dog-team race from Iglulik to Arctic Bay (See Figure 20.2).

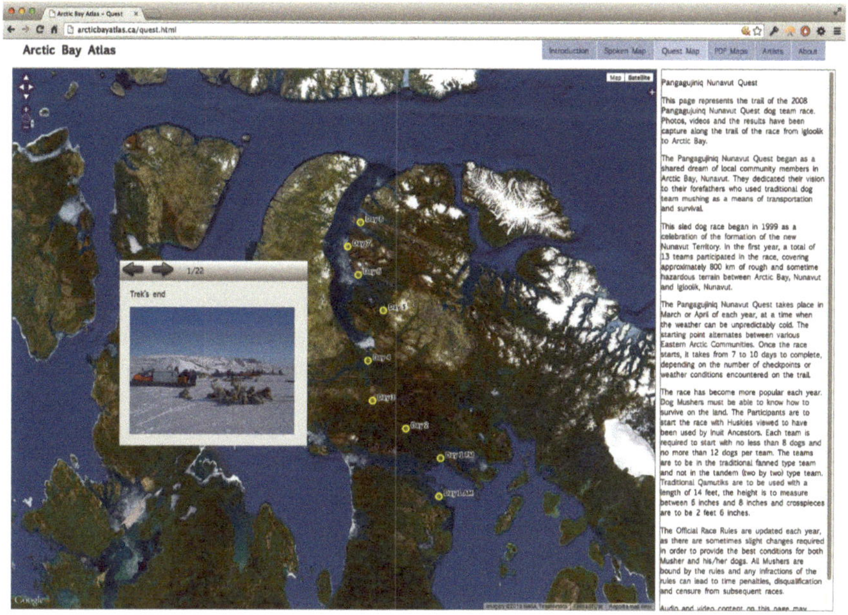

FIGURE 20.2 Arctic Bay atlas quest map.

20.5.4 Summary of Challenges Faced

While project outcomes were successful and the *Arctic Bay Atlas* has been successfully launched, some of the following challenges have been encountered:

- slow Internet connectivity of Nunavut;
- the vagaries of local networks;
- the ever-changing population of adult learners;
- distance to technical support and data management knowledge (challenges of access to GCRC technical staff based in Ottawa);
- automated file formatting and size reduction software still under development (at the time);
- the need for more community-based research training (workshops and manuals) as there is little local research and technical supports;
- issues of accounting for missing documents; and finally
- reliance on a few individual community educators and dedicated youth (i.e. local champions).

20.5.5 Results

Through *Atlas* development, the resulting maps were not necessarily as rich and complex as hoped, but there is great value in the collaborative process. The *Arctic Bay Atlas* is multi-authored, involving the cross-generational sharing of knowledge, as well as a partnership between formal university research practice and the community. Many important leaning objectives were realized as follows:

1. the mentoring of youth in gathering and reflecting upon traditional Inuit knowledge;
2. skills development in working with elders and collecting oral narratives;
3. stronger survival skills though knowledge of geography;
4. enhanced language acquisition from learning traditional place names; and
5. unique geographic and multimedia technical skills (including data management).

The resultant *Atlas* provided a mechanism for Inuit to represent and disseminate their geospatial knowledge in new ways, as well as being an important advance in cybercartographic theory and practice. On March 30, 2009, the *Atlas* was launched at the Tunnganusarvik Campus of NACin Iqaluit and is available online at http://arcticbayatlas.ca. The project was well received with excellent coverage from Aboriginal Peoples Television Network (APTN) and Canadian broadcasting Company (CBC), and by NYC youth who continued to be actively involved in launch activities and *Atlas* promotion in the media.

The goal for NAC is that ultimately this kind of research, learning, and atlas-building experience would be integrated into educational programming in the 25 CLCs across Nunavut. This effort was facilitated through ongoing collaborations between the NAC and GCRC, especially with additional funding secured from the Government of Canada Program for International Polar Year (IPY) Training, Communications and Outreach, for a project entitled Nunavummi Nunarjuattigut Illiniarniq – Learning in Nunavut Through Our Earth.

20.5.6 Integrating Research into the Educational Programming of NAC

The Gateway project demonstrated that the learning between elders and youth through intergenerational research could be applied to the adult learning environment of the CLCs as a means of engaging and motivating adult learners in a creating new and interactive educational learning experiences and resources using cybercartographic technology. Being part of this larger team and receiving additional funds through IPY provided critical support to continue to develop the *Atlas* as envisioned, as well as more in-depth community-based training in data management in Arctic Bay, and finally the development of associated college level teaching and learning resources to be used throughout the NAC programmes across Nunavut.

The development of a post-secondary level credit-based NAC course in research methods was facilitated by ongoing NAC and GCRC collaborations, as well as contributions from ITK (the national Inuit organization) and IPY funds. The course was completed in August, 2011, and was presented jointly by GCRC and NAC to college faculty during their annual orientation. It was subsequently piloted in September, 2011. It is now an elective in the College Foundation, college access programme. The goal of this course is to remove the distance between people studying the North, and those who feel they are 'being studied'. The course focuses on understanding the research process, so that participants can develop the skills to help shape and participate in studies taking place in their community, and can begin to form research projects of their own. There is also an option for the course to be delivered as a general non-credit workshop to interested community members who wish to acquire the skills to become researchers themselves. For example, in March 2013, the course was offered in Repulse Bay to better prepare community members to participate in a three-year longitudinal study pertaining to fish stock monitoring.

20.5.6.1 Why a Research Course at NAC CLCs?

Like most colleges the educational programmes at NAC is a focus on providing career training for employment within Nunavut. There is little encouragement within faculty for engaging in research or presenting at academic conferences. However, the College does have a unique relationship within the world of academic research. The Nunavut Research Institute (NRI) is affiliated with NAC and indeed the Environmental Technology Program is delivered through NRI bringing a closer relationship between academic research and applied practice. However, NRI primarily functions as a research licensing body as well as providing support to scientific research projects.

The Language and Culture program at NAC, which was established in 1996, is founded on the belief that Inuit students need a grasp of both the past and the present to adequately be prepared for the future. The programme's learning outcomes focus on exposing the student to both the traditional Inuit worldview and to southern academic thought. It also seeks to utilize a decolonizing research methodology to support student-research projects. Most of the anthropological literature published on Inuit is done from a Western Judeo-Christian perspective. Oral histories and information on TK collected by Inuit tend to be included in the academic research presented from this particular frame as verbatim transcriptions. The approach in the Language and Culture programme was to develop a collaborative research model whereby anthropologists and Inuit college students worked together to meld a methodology that integrates Inuit analysis and interpretation with an academic anthropological perspectives (Sammons and Kakkik, 2003). Southern academics and Inuit elders are brought together to form a learning project that would assist students in their research. The subject areas are selected by the students (student-driven research) as topics they want to learn more about and which the students believed would contribute to preservation of local and TK. Participation in a research project helps build confidence in their research/academic skills working as research informants and assistants, helping them gain skills and experience. Importantly, this type of collaborative research-learning project creates pride in Inuit culture and heritage. Elders reinforce this positive self-image by acknowledging the important work the students are doing in learning from their elders and preserving this knowledge in college publications. Utilizing this methodology, the students have benefitted greatly, and the college has produced a series of books documenting traditional Inuit knowledge.

Aside from the two examples cited above, which occur in programme niches, there are few community-based projects truly embedded in, and directed by, the communities where the research is taking place. Research questions, especially those that address social problems, are not driven by community interests or needs but framed by southern Eurocentric perspectives (Sammons and Kakkik, 2003). There is also research fatigue, where Elders in many communities are exhausted as a result of participating in multiple projects (often being asked similar questions), and where there is limited awareness from outside researchers of Inuit protocols for societal engagement. Overall, there is a need for more community-driven research in the areas of education, parent-teacher engagement, decolonizing curricula, and student motivation. Elders' knowledge and life stories are essential narratives that are being lost, and new programmes and policies must be developed if GN is going to successfully address these issues using an Inuit ideology and value system as outlined earlier in the chapter. This is where NAC has an important role to play. More work needs to be done at the college in extending research experiences to learners in communities to prepare them for leadership roles within their communities – Nunavut and more broadly in Canada and abroad. This could contribute towards advances in building the capacity of Inuit to participate in shaping their communities and to improving their quality of life. Basic education should do more than provide upgrading in local CLCs, and it should provide new learning opportunities that contribute to leadership, engaged citizens, and decolonizing policies and practice. This is the rationale behind the development of the research course, which arose from experiences in developing the *Arctic Bay Atlas*.

20.5.6.2 Working in Research: A Guide to Understanding and Participating in Research Programmes Course at NAC

Within the College Foundation access year the Working in Research: A Guide to Understanding and Participating in Research Programmes course is offered as a 45-h credit course. The foundation programme functions as an access year for all college programmes but is specifically designed to provide access to nursing and teacher education degree programmes. The course aims to build research understanding and capacity, better prepare students for post-secondary education, foster critical enquiry, and provide the skills to preserve TK and apply these to contemporary issues of concern to communities and leaders.

The course is designed to be offered in the context of a community-based workshop, non-credit, and flexibly delivered to an audience of adults who wish to learn more about the research process and to develop projects of their own. Course materials include an instructors' resource so that the adult educator in collaboration with visiting researchers, graduate students, and Inuit who have engaged in research at some level personally can deliver the course. An example of how this collaborative model might work is the proposed delivery in Arviat with a local group of youth, Elders, community members, and academics within a community-university affiliation examining the role tourism might have in their region. They have applied for an SSHRC grant to kick off their fieldwork, and interested community members will take the course.

The Council of the Ministers of Canada includes in their description of the many functions of post-secondary education the function to 'advance, preserve, and disseminate knowledge and understanding' (Stevenson, 2010: 12). The question is of course whose knowledge will be disseminated. Scholarly research and copious government and client-based research proliferate in Nunavut. However, community-based policy research conducted in collaboration with an academic faculty committed to supporting an Inuit worldview and research agenda responsive to the needs of communities is absent. Inuit with strong research skills employing traditional methodologies are the key to realizing this vision.

NAC campuses, CLCs, as well as the NRI provide opportunities to support increased awareness and understanding of the important role research plays in defining policies that speak to the real needs and interest of Inuit through education and provision of courses such as Working in Research. The practical application of research skills can be realized in collaborations with institutions such as the GCRC. The Arctic Bay Cybercartographic mapping project aimed to increase the capacity of Nunavut research and education communities to address their priorities from multiple perspectives. It compelled us to look more closely at developing community-based research capacity that supports appropriate and effective means of cross-cultural knowledge representation and sharing. The course is meant to provide a preliminary foundation for the understanding and development of linkages between Inuit and scientific modes of learning.

20.5.7 Some Additional Observations

Communities throughout the world now have new opportunities to collect, share, and integrate geographic information. However, as experienced by the Arctic Bay projects, the complexities of the geographic information technology domain can make it difficult for communities to effectively engage with these technologies. Communities using geographic information technology not only need to manage the technology; they also need to fully consider the social implications of technology use and this process is best addressed by understanding the role research plays their life-world. The ability to understand and engage in research is essential for improved knowledge sharing and integration within academia.

The Arctic Bay Cybercartographic Atlas project aimed to increase the capacity of Nunavut research and education communities to address their priorities from multiple perspectives. Developing community-based research capacity supports appropriate and effective means of cross-cultural knowledge representation and sharing. It is the foundation for the understanding and development of linkages between Inuit and scientific modes of learning. Central to the research process was the vision of Arctic Bay youth in NYC who sought to engage with Elders in understanding traditional worldviews and knowledge pertaining to land and language. McGrath (2011: 255) builds a model for knowledge renewal from her conversations with respected elder Mariano Aupilardjuk that 'knowledge is relational and therefore knowledge renewal is relationship renewal'.

To support this objective, NAC continues to expand its learning programme to support community-based knowledge building through research. New curriculum on research theory and practice, data management, and digital mapping research, combined with the experiences of local experts in conducting projects, will assist communities to share and preserve their knowledge with generations to come.

20.6 CONCLUSION

The creation of the two atlases discussed in this chapter, the *Arctic Bay Atlas* and the *Inuit siku (sea ice) Atlas*, have made important contributions to both the theory and practice of cybercartography and its application to both formal and informal education. Cybercartography has, from the outset, had considerable theoretical potential to make a positive contribution to education (Baulch et al., 2005) but rarely has this been demonstrated so well as in the cases discussed here. The technological potential of a multimedia and multisensory teaching and learning tool is, of course, great and in theoretical terms and the link between Gardner's MI and cybercartography is interesting. However, the major contribution of the work in Nunavut comes from an increased understanding of the importance of processes rather than artefacts. In both cases, the process by which the atlases were created – and the processes of the learning they facilitated – may be more important than the atlases themselves. In both cases, the atlases were created with significant input by the Inuit people themselves who decided on content and determined what was relevant to them. As the chapter illustrates, considerable efforts were made in the creation of both atlases in particular to ensure community involvement, and control and ownership of the atlases in ways that have not been presented before. The creation of the atlases was in itself an educational activity for the Inuit involved and in both content and process terms these are truly indigenous atlases. The atlases will, of course, continue to be useful for the formal and informal education both of Inuit communities and for a much wider audience interested in Canada's North but the processes, which the atlas creation engendered are also of great importance. The Working in Research course in NAC, which the atlas creation helped to set in motion, for example, will give Inuit educators a new and more purposeful role and help Inuit take greater control of research in their own right rather than being the objects of research by outsiders. The incorporation of the *Siku Atlas* into the formal high school curriculum will contribute to the creation of more relevant teaching and learning content and processes. We hope that our work on the application of cybercartography to social, economic, and especially educational issues of importance to the people of Nunavut will continue in a cooperative fashion.

References

AADNC., 2011. Annual Report on the Implementation of the Nunavut Land Claims Agreement (NLCA) 2006–2008. Ottawa: Aboriginal Affairs and Northern Development Canada (AADNC).

Amagoalik, J., 2001. What is this land? In: Blohm, H. (Ed.), The Voices of the Natives: The Canadian North and Alaska, Penumbra Press, Toronto, pp. 9–10.

Armstrong, T., 2009. Multiple Intelligences in the Classroom, third ed. Basic Books, New York.

Aylward, M.L., 2007. Discourses of cultural relevance in Nunavut schooling. Journal of Research in Rural Education 22 (7), 1–9.

Baulch, S., Macdonald, R., Pulsifer, P.L., Taylor, D.R.F., 2005. Cybercartography for education: the case of the cybercartographic atlas of Antarctica. In: Taylor, D.R.F. (Ed.), Cybercartography: Theory and Practice, Elsevier, Amsterdam, pp. 491–515.

Bennett, J.R., Rowley, S., 2004. Uqalurait: An Oral History of Nunavut. McGill-Queen's University Press, Montreal and Kingston.

Berger, T. R., 1977. Northern frontier, northern homeland: the report of the Mackenzie Valley Pipeline Inquiry. Government of Canada, Minister of Supply and Services Canada, 1 (Social, economic and environmental impact).

Berger, P., 2009. Eurocentric roadblocks to school change in Nunavut. Études/Inuit/Studies 33 (1–2), 55–76.

Berger, P., Epp, J.R., Møller, H., 2006. The predictable influences of culture clash, current practice, and colonialism on punctuality, attendance, and achievement in Nunavut schools. Canadian Journal of Native Education 29 (2), 182–205.

Bonesteel, S., 2006. In: Anderson, E. (Ed.), Canada's Relationship with Inuit: A History of Policy and Program Development, Government of Canada, Ottawa Indian and Northern Affairs Canada (June).

Bonny, E., Berkes, F., 2008. Community traditional environment knowledge: addressing the diversity of knowledge, audience and media types. Polar Record 44 (250), 243–253.

Brockington, R., 2009. Summary Public School Indicators for the Provinces and Territories, 2000/2001 to 2006/2007. Government of Canada, Statistics Canada, Ottawa.

Castellano, M.B., Davis, L., Lahache, L., 2000. Aboriginal languages and communications: voicing the promise. In: Castellano, M.B., Davis, L., Lahache, L. (Eds.), Aboriginal Education: Fulfilling the Promise, UBC Press, Vancouver, pp. 25–28.

Champagne, D., Abu-Saad, I., 2006. Introduction: seeking common ground through education. In: Abu-Saad, I., Champagne, D. (Eds.), Indigenous Education and Empowerment: International Perspectives, Altamira Press, Lanham, pp. 1–11.

Coleman, C., 2009. Just the Tip of the Iceberg: Report on the Arctic Bay Interactive Online Atlas Project (Cybercartographic Atlas). Prepared for NAC, Geomatics and Cartographic Research Centre, Carleton University, Government of Canada Program for the International Polar Year. Unpublished.

Department of Education, 2006. Qalattuq: 10 Year Educator Training Strategy 2006–2016. Department of Education (DE) and Nunavut Arctic College.

Department of Education, 2007. Inuit Qaujimajatuqangit Education Framework. Government of Nunavut, Department of Education (DE), Curriculum and School Services.

Department of Education, Culture and Employment, August 1996. Inuuqatigiit: The Curriculum from the Inuit Perspective. Government of the Northwest Territories, Department of Education, Culture and Employment (DECE), Yellowknife.

Dewey, J., 1938. Experiences and Education. Kappa Delta Pi, Indianapolis, USA.

Fogwill, L., 1994. Literacy: A critical element in the survival of aboriginal languages. Chapter 16. Retrieved from , http://search.proquest.com/docview/62628405?accountid=9894.

Friesen, J.W., Friesen, V.A.L., 2005. First Nations in the Twenty-first Century: Contemporary Educational Frontiers. Detselig Enterprises, Calgary.

Gardner, H., 1983. Frames of Mind: The Theory of Multiple Intelligences. Basic Books, New York.

Gardner, H., 1999. Are there additional intelligences? The Case for Natural, Spiritual and Existential Intelligences, Transformation. Prentice Hall, Saddle River pp. 111–131.

Gardner, H., 2000a. Can Technology Exploit Our Ways of Knowing in the Digital Classroom. Basic Books, New York pp. 32–35.

Gardner, H., 2000b. Intelligence Reframed: Multiple Intelligences for the 21st Century. Basic Books, New York.

Gardner, H., 2005. Development and Education of the Mind: The Selected Works of Howard Gardner. Taylor and Routledge.

Gardner, H., 2008. The 25th Anniversary of Howard Gardner's Frames of Mind: The Theory of Multiple Intelligences. Basic Books, New York.

Gardner, H., 2011. Frames of Mind: The Theory of Multiple Intelligences, sixth ed. Basic Books, New York.

Gardner, H., 2012. Truth, Beauty and Goodness Reframed: Education for the Virtues in the Twenty-first Century. Basic Books, New York.

GN., 2008. Nunavut Education Act 2008: A Reference Guide for the Commission Scolaire Francophone and School Principals. Government of Nunavut (GN).

Hayes, A., 2007. E-mail Message to Cowan, January 11. .

Inuit Heritage Trust, 2007. Proposal to Gateway Fund, Canadian Culture Online, Department of Canadian Heritage for 2007–2008.

ITK., 2011. First Canadians, Canadians First: National Strategy on Inuit Education. Inuit Tapiriit Kanatami (ITK).

Laidler, G., 2007. Report on Hands-On Workshop: Creating Interactive Atlases for Education. Unpublished.

Laugrand, F., Oosten, J., 2009. Transfer of Inuit Qaujimajatuqangit in modern Inuit society. Études/Inuit/Studies 33 (1–2), 115–152.

McGrath, J., 2011. Isumaksaqsiurutigijakka: Conversations with Aupilaarjuk towards a Theory of Inuktitut Knowledge Renewal. (Ph.D. dissertation). Carleton University.

McGregor, C., 2010a. Nunavut to Integrate Inuit Knowledge in Science Class. The Canadian Broadcasting Corporation (CBC).

McGregor, H., 2010b. Inuit Education and Schools in the Eastern Arctic. UBC Press, Vancouver.

McGregor, C., 2012. *Aulajaaqtut*. Curriculum and School Services, Department of Education, Government of Nunavut. Retrieved November 28, 2012, from Council of Ministers of Education, Canada Web site , http://aboriginal.cmec.ca/documents/NU-Aulajaaqtut.en.pdf.

Nicklen, P., August 2007. Arctic Ivory. National Geographic Magazine, Washington. Available at: http://ngm.nationalgeographic.com/2007/08/hunting-narwhals/hunting-narwhals-text.

Nunavut Arctic College, 2008. Corporate Plan 2008–2013. Available at: http://www.arcticcollege.ca/publications/corporate/CorporatePlan2008-final.pdf.

Nunavut Tunngavik Inc, 2004. Tukisittiarniqsaujumaviit: A Plain Language Guide to the Nunavut Land Claims Agreement. Nunavut Tunngavik Inc (NTI).

Nunavut Youth Consulting, 24 March 2006. Reclaiming the Land. Proposal to Culture, Language, Elders and Youth. Government of Nunavut.

Report of the Royal Commission on Aboriginal Peoples (RCAP)., 1996. Government of Canada, Indian and Northern Affairs.

Sammons, S., Kakkik, M., 2003. Building capacity in Arctic societies: dynamics and shifting perspectives. Proceedings of the Second IPSSAS Seminar, Iqaluit, Nunavut. Québec, CIÉRA, Université Laval, pp. 145–153, Available at: http://www.ciera.ulaval.ca/publications/parutionsAnterieures.html.

Schaler, J.A., 2006. Geocentric Theory: A Valid Alternative to Gardner's Theory of Intelligences. Northern Bridge, New York.

Simon, M., 2008. Education Summit is a step in developing a National Strategy on Inuit Education 'Toward a National Strategy on Inuit Education'. Remarks to Delegates at the Summit on Inuit Education by Mary Simon, President, Inuit Tapiriit Kanatami April 15. Inuvik, NWT.

Stevenson, B., 2010. Dialogue towards a University in Canada's Far North: Environmental Scan. Walter and Duncan Gordon Foundation.

Students on Ice, 2013. Available at: http://www.studentsonice.com.
Sullivan, C., 2013. Integrating Culturally Relevant Learning in Nunavut High Schools: Student and Educator Perspectives from Pangnirtung, Nunavut and Ottawa. Ontario, (Unpublished Master's thesis). Carleton University, Ottawa.
Tagalik, S., 2010. Inuit Qaujimajatuqangit: The Role of Indigenous Knowledge in Supporting Wellness in Inuit Communities in Nunavut. National Collaborating Centre for Aboriginal Health.
Taylor, D.R.F., 2007. A Cybercartographic Atlas of Arctic Bay: Enabling Community Based Atlas Production. Grant Application to Social Sciences and Humanities Research Council of Canada.
Taylor, D.R.F., 2005. Cybercartography: Theory and Practice. Elsevier, Amsterdam.
Tester, F.J., Irniq, P., 2008. Inuit Qaujimajatuqangit: social history, politics and the practice of resistance. Arctic Institute of North America 61 (5), 48–61.
Tyler, K., 2008. Explaining why Students Stay in School. Inuit Perceptions of Modern Guidelines that will help students stay in high school.
Veenma, S., Gardner, H., November/December 1996. Multimedia and multiple intelligences. The American Prospect, 69–79.
Watt-Cloutier, S., 2009. Returning Canada to a Path of Principle: An Arctic and Inuit Perspective. Speech Notes for: the 9th LaFontaine-Baldwin Lecture Institute for Canadian Citizenship. Iqaluit, Nunavut.
Wenzel, G.W., 2004. From TEK to IQ: Inuit Qaujimajatuqangit and Inuit cultural ecology. Arctic Anthropology 41 (2), 238–250.
Williams, W.M., Blythe, T., White, N., Li, J., Sternberg, R.J., Gardner, H., 1996. Practical Intelligence for School. Harper Collins College, New York.
Wood, D., 1992. The Power or Maps. Guildford, New York.

CHAPTER

21

The Preservation and Archiving of Geospatial Data and Cybercartography as a Proactive Preservation Process

Tracey P. Lauriault[1,3], *D. R. Fraser Taylor*[2]

[1]National Institute for Regional and Spatial Analysis (NIRSA), National University of Ireland at Maynooth, Maynooth, Co. Kildare, Republic of Ireland, [2]Geomatics and Cartographic Research Centre (GCRC), Carleton University, Ottawa, Ontario, Canada, [3]Member of the Geomatics and Cartographic Research Centre (GCRC), Carleton University, Ottawa, Ontario, Canada

OUTLINE

Developments in the Theory and Practice of Cybercartography, Second Edition, ISSN 1363-0814
http://dx.doi.org/10.1016/B978-0-444-62713-1.00019-2

21.1 INTRODUCTION

Maps, atlases, and geospatial data are repositories of cultural and scientific knowledge and are a key part of our collective geographical memories. In Canada, all levels of government create maps. Natural Resources Canada (NRCan) produces, acquires and maintains the largest collection of maps, satellite and radar imagery, air photos, and geospatial data sets. These are disseminated via geospatial data portals, map servers, and special programmes such as *GeoGratis*. Numerous other federal departments also produce geospatial data and the largest collection of freely accessible data sets being disseminated in the Treasury Board Secretariat's Open Data Pilot portal are geospatial (IASSIST, 2011). Provincial and territorial governments also produce maps, are responsible for cadastres, and have databases to help manage transportation networks, watersheds, and administer programmes such as health and education. At the level of the city geographic information system (GIS) units manage infrastructures and other municipal responsibilities.

Academic institutions create maps and atlases that are normally funded by government granting agencies and non governmental organizations produce maps and on a variety of topics such as food security (e.g. Community Data Portal). While the private sector, through companies such as MDA Detwiler or Environics, form part of Canada's multibillion-dollar geomatics industry (StatCan, 2010). Finally, the public and communities are engaged in map-making endeavours by: contributing volunteered geographic information (Scassa, 2013; Chapter 4) into Mashups (RAPLIQ, 2013); building local infrastructures (CCSD, 2013; SPCO, 2013); creating mobile device apps[1]; and by partaking in participatory mapping (Chapter 14) or citizen science endeavours (e.g. WeHub, Digital Fishers). Digital maps and geospatial data are ubiquitous artefacts in the twenty-first century even though the digital era in map making is a relatively recent phenomenon of the last 50 years. It has accelerated very rapidly over the last decade as software is easier to use and less expensive, geospatial data are more accessible, organizations have opened their application programming interfaces and Internet social networking services have proliferated. This source of knowledge is, however, not being systematically preserved. For example, Brauen (Chapter 10) and Lauriault (Chapter 12) were unable to access most of the maps they surveyed within the last 10 years.

This chapter begins by discussing the rescue and salvage of the Canada Land Inventory (CLI). As data management (including preservation) and access to data have become key policy issues, partially as a result of a number of public consultations that will also be discussed. This is accompanied by a brief examination of national legislation, policies, and directives, and four geospatial data preservation examples will be introduced, as will be three preservation frameworks. The preservation of memory comes in many forms and the Geomatics and Cartographic Research Centre (GCRC) has been involved in the mapping local and traditional knowledge by applying participatory research techniques in the creation of atlases with indigenous communities. A few are discussed as 'archives' of local, cultural, scientific, and historical knowledge. The chapter concludes by suggesting strategies for the preservation of geospatial cultural and scientific assets.

[1] Open-data initiatives make data accessible to developers who create GeoApps such as Restonet (http://resto-net.ca/), which maps food inspection results; Edmonton's Historic Buildings Android App (http://contest.apps4edmonton.ca/apps/21) or the BC Apps4ClimateChange contest, which produced Waterly (http://www.waterly.ca/) that tracks rainfall and combines these with watering restriction schedules.

21.2 THE RESCUE AND SALVAGE OF THE CLI

The CLI is an example of geospatial data that were nearly lost in the late 1980s. It was initiated by the Department of Forestry and Rural Development in 1960 'to guide the development of policy on the control and management of land-based resources' (Ahlgren and McDonald, 1981: 61). The CLI mapped 2.6 million square kilometres of Canada and 'the original cost of the programme was in the order of 100's of millions of dollars in the 1970s' (Wilson and O'Neil, 2009). In 1963 the Canadian Geographic Information System (CGIS) was developed to automate its data management and mapping. The CGIS was the world's first GIS, it was a milestone in the history of geographic and government computerization and it revolutionized mapping. It was both a set of electronic maps and the 'computer programs that allowed users to input, manipulate, analyse, and output those maps' (Schut, 2000). It ultimately grew to contain thousands of maps and unknowingly became a technology that 'spawned an industry that today is worth billions of dollars' (Schut, 2000).

By the late 1980s, priorities changed, people retired, and institutional memory faded. Numerous boxes of tapes and racks of documentation were left behind and only a few computers were capable of reading nine track tapes let alone run the program. In 1995, an informal trans-organizational group of individuals from Statistics Canada, the *National Atlas of Canada*, Archives Canada and the private sector joined forces to restore it. Fortunately, members of this salvage team had either formerly worked on the CGIS, which meant they had the requisite skills, knowledge, and more importantly they recognized the value of these data. They also understood that the CLI was part of Canada's technological and scientific heritage. On June 18, 1998, the agriculturally relevant portions of the CLI were handed over on one CD and it worked flawlessly on the analytical tools built in anticipation of the new format and eventually the CLI was distributed in the *GeoGratis* portal for free along with metadata. The CLI 'rapidly became [GeoGratis'] most popular product' (Schut, 2000) with 45,373 downloads in 2011–2012 (Trottier, 2012).

The 1960s CLI was a massive, expensive, national scale project that was technologically, scientifically and procedurally innovative marking a turning point in the history of computerization and science (Morris, 1991), which informed the Canadian economy and society (Canada Encyclopedia, 2012). Not all of the CLI data were saved and the cost of the effort was substantial. Its salvage illustrates the importance of considering long-term preservation at the point of creation and provides an argument for the adoption of open specifications, standards and interoperability or for 'proactive archiving' (Doorn and Tjalsma, 2007; Lauriault and Hackett, 2005).

21.3 CANADIAN DIGITAL DATA CONSULTATIONS AND STUDIES

A number of public consultations have taken place between 2002 and 2013 about digital data, information strategies, preservation of research and geospatial data and open data as seen in the timeline below (Figure 21.1).

The preservation of geospatial data has been the topic of interest by GeoConnections since 2005. The *Archiving, Management, and Preservation of Geospatial Data Study* (GeoConnections,

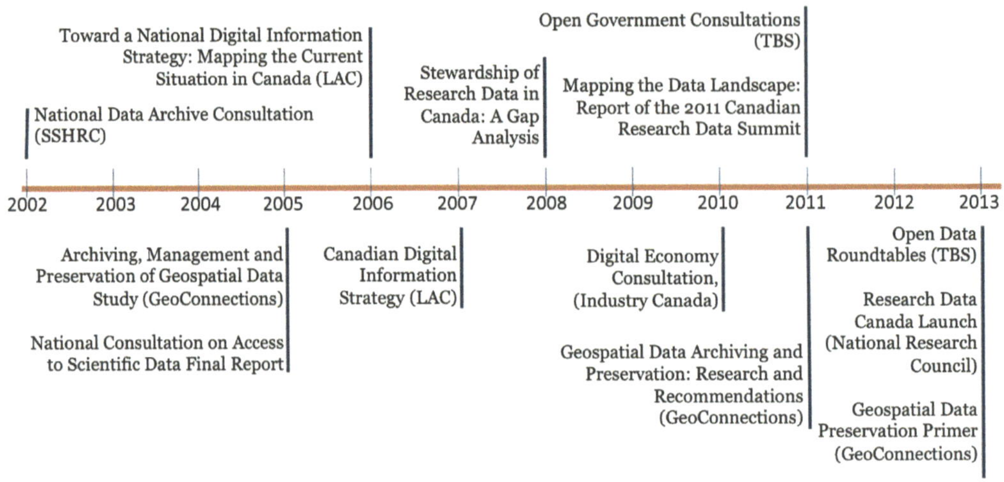

FIGURE 21.1 Canadian digital data and information initiatives.

2005) provided a well-rounded analysis of geospatial data preservation issues, which included technological obsolescence, formats, storage technologies, temporal management; and metadata. The *Study* provided a list of technological preservation solutions and stated that 'data producers must adopt an information life-cycle management approach, which will ensure that their data will be managed proactively from creation to disposition' (2005: i). The *Study* also recommended:

- defining and implementing policies and practices for the creation, use, retention, dissemination, preservation, and disposition of geodata and building a business case for their preservation;
- establishing authoritative responsibility centres to define and apply information management principles to ensure the integrity of an organization's geospatial data holdings with custodianship built into the geospatial data infrastructure;
- making geospatial data preservation part of a national information and data management policy and strategy in collaboration with creators and preservers; and
- the Canadian Council on Geomatics should create a task force with stakeholders to develop policy priorities.

GeoConnections later commissioned a research report entitled *Geospatial Data Archiving and Preservation* (Lauriault and Kennedy, 2011), which discusses Canadian preservation and management policies, directives, legislation, and research and provides a set of recommendations. An outcome of these recommendations is the *Geospatial Data Preservation and Archiving Primer* (Lauriault et al., 2013), which includes preservation case studies and guidelines for geospatial data creators and preservers. Like the 2005 GeoConnections *Study*, the recommendations call for cross-departmental collaboration, records management as part of a dataset's life-cycle, and making preservation of geospatial data part of national digital preservation strategies. The *Primer* also includes tools such as business plans, records retention schedules, policies, standards, models, frameworks, and checklists.

The launch in 2013 of *Research Data Strategy* (RDC, 2013) was part of the *National Consultation on Access to Scientific Data* and the *Mapping the Data Landscape: Report of the 2011 Canadian Research Data Summit* process, which started in 2002. It is the culmination of collaborative and multidisciplinary effort aiming to address the challenges and issues surrounding access to and preservation of data created by Canadian researchers. While the focus is on research data, the ideas are transferable to the preservation and management of geospatial data. The approach and principles of data stewardship recommended by *RDC* have been echoed in many of the other initiatives and consultations discussed here.

The *Open Data Roundtable Report* (2013a) is an outcome of the Treasury Board Secretariat's (TBS) Open Government (2011a,b) plan released at the Open Government Partnership (OGP) in Brazil in the spring of 2012 and the TBS Open Data Consultations (2011) informed it. It includes better record keeping and the lifting of restrictions to archived material at Library and Archives Canada (LAC). The TBS Consultation report and the Plan however do not discuss preservation (Lauriault, 2012) nor does the Advisory Panel on Open Government (TBS, 2011a,b) include a representative from libraries, archives or science. Interestingly, the most popularly requested data sets in the TBS Consultation were geospatial and geospatial datasets predominate in terms of number of data sets in the *Open Data Pilot Project* portal (IASSIST, 2011). Participants at the 2013 Roundtable focused strongly on geospatial data, metadata, and access to longitudinal data while also emphasizing the need to archive the data in the portal.

The Industry Canada (IC) *Digital Economy Consultation* (2010a) did not result in any actions. The IC *Consultation Paper* (2010 b) however did mention improving access to research data, data discovery, and open-access publishing, while the geomatics sector and related data were not mentioned at all. As discussed, geospatial initiatives need to be positioned as information strategies if they want traction. Submissions and the idea forum called for the creation of a digital data and information archive and a network of trusted digital repositories. The summary of key findings (2010c) failed to mention these.

The *Stewardship of Research Data in Canada: A Gap Analysis* (2008) was conducted by members of *Research Data Canada*. A life-cycle model was used as the framework to analyse the current state of research data. The *Gap Analysis* reviewed: policies; funding; roles and responsibilities; repositories; standards; skills and training; rewards and recognition systems; R&D; access; and preservation. Not surprisingly, the analysis concluded that most research data created in Canada are greatly underutilized and are at a high risk of being lost. While the report was about research data, geospatial data share similar characteristics.

The 2007 LAC *Canadian Digital Information Strategy* report reflects the broad consensus that was reached at a *National Summit* held in 2006, which included contributions from more than 200 specialists some of whom are involved in geomatics. There are numerous actions listed, none of which are specific to geospatial data although preservation assumptions mirror contemporary geospatial data productions practices and issues. For example, technological change is constant; a distributed, interoperable, standards based, open and accessible strategy is key; stakeholders and communities of practice input are essential.

The LAC 2006 *Toward a National Digital Information Strategy: Mapping the Current Situation in Canada* states that 'the stewardship of digital information produced in Canada is disparate and uncoordinated' and 'the volume, diversity and complexity of digital information is growing exponentially' while the 'technologies, standards and practices that will better ensure the ongoing accessibility and integrity of digital information are not yet consistently applied'. The report includes a brief description of the GeoConnections programme and lists a number of other geospatial data initiatives at different levels of government in Canada in its Annexes. The report does not provide specific recommendations on how to manage these resources but it does suggest that the:

> preservation of digital information requires a management response, not just a technological response. The response must be active and sustained through time and it must address the preservation of digital information from both the strategic and tactical perspectives and within the context of the management frameworks that govern the businesses of organizations themselves. It must also be comprehensive and address all facets of the required preservation infrastructure: the policies that assign accountability, the standards, practices, procedures and technologies that enable the implementation of preservation strategies, and the people that make it all happen (2006: 50).

MacDonald and Shearer also provide a number of next steps towards the creation of a national digital information strategy.

The 2005 *National Consultation on Access to Scientific Data Final Report (NCASRD),* developed in partnership with the National Research Council, the Canada Foundation for Innovation, Canadian Institutes of Health Research and Natural Sciences and Engineering Research Council, expressed the concern that: 'no national data preservation organization exists, nor does Canada have any national data access strategy or policies. NCASRD participants expressed considerable concern about the loss of data, both as national assets and definitive longitudinal baselines for the measurement of changes over time' (2005: 2). The *NCSARD* report provides a comprehensive list of recommendations that include organizing a Data Force and the creation of Research Data Canada. None are specific to geospatial data and data preservation remains an issue.

Finally, the 2002 *SSHRC National Data Archive Consultation* was the National Consultation on Access to Scientific Data (*NSCARD's*) (Strong and Leach, 2005) predecessor. It primarily discusses data in the social sciences and humanities and the preservation of data created in the course of state-funded research projects. The *Consultation* identified important institutions, infrastructures, management frameworks, and data creators and calls for the creation of a national research data archive.

Some progress has been made, especially at GeoConnections and Research Data Canada, but deep integration of preservation as part of the life-cycle management of geospatial data has not yet manifested, nor have geospatial data become part of mainstream data and information policies and strategies. The new multistakeholder forum for open dialogue on issues and concerns that affect professional practice and activities in the geomatics sector, the Canadian Geomatics Community Round Table (2013) shows some promise at integration although there is no preservation working group. There is no national network of trusted digital repositories that can ingest geospatial, research, government or private sector-created data and LAC has undergone severe budget cuts. Overall, there are many recommendations, with much overlap, but limited implementation. Proactive archiving and a life-cycle approach to records management remain the best interim solutions.

Along with the initiatives above, the Committee on Data for Science and Technology *(CODATA)* Working Group on Archiving Scientific Data (2006) along with the CODATA Data At Risk Working Group have been holding symposia and workshops on the topic and the *Canadian National Committee for CODATA* are members of these. The Open Geospatial

Consortium (OGC) also has a *Data Preservation Working Group*, which aims 'to address technical and institutional challenges posed by data preservation, to interface with other OGC working groups that address technical areas that are affected by the data preservation problem, and to engage in outreach and communication with the preservation and archival information community'. Finally, the International Research Data Alliance may also help countries move forward on these issues.

21.3.1 Legislation, Directives, and Policies

As mentioned, the GeoConnections *Research Report* (Lauriault and Kennedy, 2011) and *Primer* (Lauriault et al., 2013) discuss legislation, directives, and policies in relation to geospatial data preservation. Tables 21.1 and 21.2 list those that apply to geospatial data and Table 21.3 includes NRCan legislation that explicitly mention geospatial data and information making these official government records that by law should be preserved.

Table 21.2 lists information management policies, directives, guidelines, and standards at the federal level related to the preservation of information and the primary responsibility lies with the TBS. The *Standard on Geospatial Data* is the only one, which addresses geospatial information domain. Its objective is 'to support stewardship and interoperability of information by ensuring that departments access, use and share geospatial data efficiently and effectively to support programme and service delivery'. Adopting this standard is part of a proactive archiving process and is good records management.

Table 21.3 lists acts and regulations that fall under the responsibility of the Minister of Natural Resources. The *Remote Sensing Space Systems Act* and the *Remote Sensing Space Systems*

TABLE 21.1 Overarching Legislation and Regulation That Apply to Geospatial Data

The Library and Archives of Canada Act (2004, c. 11)	Privacy Act (R.S., 1985, c. P-21)	Legal Deposit of Publications Regulations (SOR/2006-337)	Canada Evidence Act (R.S., 1985, c. C-5)
Copyright Act (R.S., 1985, c. C-42)	Personal Information Protection and Electronic Documents Act (2000, c. 5)	Privacy Regulations (SOR/83-508).	Access to Information Act (R.S., 1985, c. A-1)

TABLE 21.2 Information Policies, Guidelines and Standards

TBS Directive on Recordkeeping	TBS Standard for Electronic Documents and Records Management Solutions (EDRMS)	TBS Policy Framework for Information and Technology	TBS Multi-Institutional Disposition Authority (MIDA)
TBS Directive on Information Management Roles and Responsibilities	TBS Policy on Information Management	Policy on Management of Information Technology	LAC guidelines: local digital format registry (LDFR)
TBS Standard on Geospatial Data	TBS *Standard on Metadata*	Forthcoming Methodology associated with TBS Directive on Recordkeeping, which NRCan Records Management Group contributed	

TABLE 21.3 Acts and Regulation Governing Geospatial Data

Department of Natural Resources Act (1994, c. 41)	Resources and Technical Surveys Act (R.S., 1985, c. R-7)	Canada Lands Surveys Act (R.S., 1985, c. L-6)	Energy Monitoring Act (R.S., 1985, c. E–8)	Energy Efficiency Act (1992, c. 36)
Canada Oil and Gas Operations Act (R.S., 1985, c. O-7) and Canada Oil and Gas Geophysical Operations Regulations (SOR/96-117)	Forestry Act (R.S., 1985, c. F-30) and Regulations Respecting the Report on the State of Canada's Forests (SOR/95-479)	Remote Sensing Space Systems Act (2005, c. 45) and Remote Sensing Space Systems Regulations (SOR/2007-66)	Canada-Nova Scotia Offshore Petroleum Resources Accord Implementation Act (1988, c. 28)	National Energy Board Act (R.S., 1985, c. N-7) and National Energy Board Electricity Regulations (SOR/97-130)
Nuclear Energy Act (R.S., 1985, c. A-16)	Nuclear Fuel Waste Act (2002, c. 23)	Northern Pipeline Act (R.S., 1985, c. N-26)	Canada Petroleum Resources Act (1985, c. 36 (2nd Supp.))	Uranium Mines and Mills Regulations (SOR/2000-206)
Nuclear Safety and Control Act (1997, c. 9)	Canada-Newfoundland Atlantic Accord Implementation Act (1987, c. 3)		Charts and Nautical Publications Regulations (SOR/1995-149)	

Regulations were added since they impact how the Canada Centre for Remote Sensing (CCRS) manages its geospatial data sets, while the *Charts and Nautical Publications Regulations* addresses marine geospatial information. Acts and regulation mention raw data, how these are to be processed, the systems used to view and process the data and the use of cryptography. There is also mention of registrars and catalogs. In some there are specific references to archiving, deposition, data protection plans, the number of years data are to be kept and data curation. Finally, because these geospatial data sets are critical to the management of Canada's boundaries, the health and safety of Canadians, Canada's economy and the attestation of lands, these government records can all be called into evidence to resolve international and national disputes, or any other judicial proceeding. The Government of Canada produces geospatial data, databases, and maps in many other departments and agencies and each of these would have to adhere to acts and regulations for which they are responsible along with those in Tables 21.1 and 21.2.

21.4 CANADIAN GEOSPATIAL DATA PRESERVATION EXAMPLES

There are some geospatial data preservation strategies in Canada as seen in Table 21.4. The following introduces four of the most established. For a more detailed analysis that includes best practices and lessons learned, see the *GeoConnections Primer* (2013).

21.4.1 Ontario Geographic Information Archive Ministry of Natural Resources

The Ontario Geographic Information Archive (GIA) was established in 2009 to preserve snapshots of the entire data holdings of the Land Information Ontario data warehouse. Its

TABLE 21.4 Preservation and Data Management Initiatives

Earth Observation Data Management Services (EODMS),(NRCan & CCRS)	Ontario Geographic Information Archive (GIA), Ministry of Natural Resources (MNR) Land Information Ontario (LIO) Data portal	Canadian Polar Data Network, International Polar Year (IPY) Data Preservation,	Integrated Science Data Management (ISDM) Wave Data Archive, Department of Fisheries and Oceans (DFO)
Institute of Ocean Sciences (IOS/OSD) Data Archive, University of British Columbia	Lithoprobe Data Archive, NRCan, Geological Survey of Canada (GSC)	National WaveForm Archive (NWFA), NRCan, Earthquakes Canada	Geomagnetism Summary Plots: Archives, NRCan, GSC
National Climate Data and Information Archive, Environment Canada (EC) The Canadian Ice Service Archive (CISA), EC, Canadian Ice Service	The National Soil DataBase (NSDB), Agriculture and Agri-Food Canada (AAFC); Canadian Soil Information Service (CANSis)	National Water Data Archive, EC, Water Survey of Canada (WSC) w/ Archived Sediment Data	System of Agents for Forest Observation Research with Advanced Hierarchies (SAFORAH), Canadian Forest Service (CFS), University of Victoria and other academic and government partners.

objectives are to: 'preserve vital geospatial data records to comply with legislative and policy requirements; support Ministry of Natural Resources (MNR's) Information Management Strategy; preserve data of scientific and historical value for long term analysis; and better defend long-term decision-making around the protection of valuable resources' (Lauriault et al., 2013). The GIA has a retention strategy to keep annual snapshots for 10 years, after which every 5th year is preserved for 200 years. After 200 years, the records are to be transferred to the Archives of Ontario. The GIA has developed an excellent set of Guideline for Retiring and Retaining Geospatial Data Stored in the Land Information Ontario Warehouse (OMNR, 2010) among other policy documents, which ensures that all geospatial data moved from the LIO Warehouse to the GIA support the Ministry's mandate.

21.4.2 Earth Observation Data Management System, CCRS

The Earth Observation Data Management System (EODMS) is part of Natural Resources Canada's (NRCan) revitalization plan to improve Canada's earth observation (EO) satellite capacity and access to EO data, which was supported in the 2012 Federal Budget (Lauriault et al., 2013). The CCRS is procuring digital data archives and the GeoConnections program provided resources to develop a Canadian preservation strategy based on the Long Term Data Preservation (LTDP) guidelines (LTDP WG, 2012a,b). Once online, the EODMS will be capable of managing increased EO data collections within the Federal Committee on Geomatics and Earth Observation framework and its current EO data collection activities, which are unique in Canada. This will provide improved access to these data, facilitate the creation of EO-derived products and knowledge, as well as the integration of EO data with other scientific data. The

EODMS will also be able to accommodate the increasing volume of data generated in the future, such as those from upcoming RADARSAT Constellation Mission, Landsat Continuation Mission and Sentinel Missions. Furthermore, EODMS contributes to the CCRS Space EO Data Stewardship pillar set by the Director General in 2011 and it aligns with the following Government of Canada policies: Information Management, Operational Security Standard, Management of Information Technology Security, Web Usability, Accessibility and Common Look and Feel, Geospatial Standard, Standard on Metadata, Information Technology; and Web Interoperability. Finally, the EODMS is based on user requirements, long-term technology transition planning, was developed after a critical examination of the existing CCRC EO data archives and will conform to the best international standards and practices namely the *Open Archival Information System (OAIS) Reference Model*, *LTDP Guidelines*, and the *Trustworthy Repositories Audit and Certification (TRAC) Checklist*, which will be discussed shortly.

21.4.3 Integrated Science Data Management, Department of Fisheries and Oceans

The Department of Fisheries and Oceans (DFO) manages its data according to a departmental science strategy outlined in its *Management Policy for Scientific Data (2001)*. The Integrated Science Data Management (ISDM) functions as a national data centre for DFO data with preservation functions shared as appropriate with existing regional data centres and provides co-ordination among the centres to ensure that all data are properly managed (DFO, 2001). The *ISDM Wave Data Archive*, for example, integrates the preservation and archiving of data from multiple organizations into an operational information management environment and project budgets must clearly indicate the allocation of resources for data management and how data resources will be used and DFO Science and Oceans managers are the responsible agents for ensuring that data under their control are submitted. The policy ensures that data are quickly copied into a 'managed' environment where they are backed up and secured from accidental or circumstantial loss, and where the supporting metadata are linked with the data to preserve the long-term usefulness of the preserved data sets. The Information Management (IM) Policy adheres to the following principles:

1. DFO scientific data sets… are irreplaceable, and must be protected and managed to ensure long-term availability;
2. … it is essential that DFO Science/Oceans maintain responsibility for their quality control, management, archiving and dissemination; and
3. … all scientific data collected by the DFO must be migrated to a 'managed' archive immediately after the data have been processed.

In terms of archiving, all DFO scientific data must be managed, as part of an integrated system accessible through regional, zonal, and national data centres.

21.4.4 International Polar Year Data Preservation

The International Polar Year (IPY) project was a large three-year Arctic and Antarctic scientific programme, organized through the International Council for Science and the World Meteorological Organization. The IPY Data and Information Service was a global partnership of data centres, archives, and networks working to ensure the proper stewardship of IPY and related data. The IPY Data and Management Subcommittee developed a Data Policy (2008), which was endorsed by the Joint Committee in 2006 as well as a Data Strategy (IPYDIS, 2012). The policy states that:

> All IPY data must be archived in their simplest, useful form and be accompanied by a complete metadata description…it is the responsibility of individual IPY projects to make arrangements with long-term archives to ensure the preservation of their data. It must be recognized that data preservation and access should not be afterthoughts and need to be considered while data collection plans are developed.

The Canadian Polar Data Network (CPDN) established in 2010 is the infrastructure built to preserve IPY data with funding from the Canadian International Polar Year Program Office, Aboriginal Affairs and Northern Development Canada (CPDN, 2012). The operational archival system, like the EODMS, is based on the *OAIS Reference Model* Standard.

While there are few operating examples, these are starting points upon which to build a more comprehensive geospatial data and mapping preservation strategy. The data held in these, however, only represent but a very small fraction of the geospatial data produced. Only the GIA and EODMS are archives in the true sense of the term, while the other initiatives listed in Table 24.4 are examples of proactive archiving practices.

21.5 ESTABLISHING A GEOSPATIAL DATA PRESERVATION SYSTEM

It is beyond the scope of this Chapter to provide a detailed approach to the establishment of a geospatial data preservation system. The information just discussed is a useful place to start. In addition, the InterPARES 2 Project, a large and ongoing Social Science and Humanities Council of Canada research project based out of the University of British Columbia has conducted a number of case studies that examined geospatial data and a general study that investigated data portals and their preservation practices (Lauriault and Craig, 2007; Lauriault et al., 2008a,b). In addition, the InterPARES 2 project produced a number of important guidelines for digital record creators and preservers, and a number of other benchmark and policy documents. Furthermore, the US Geospatial Multistate Archive and Preservation Partnership provides access to essential tools such as business plans, scoping documents, self assessment templates, and a number of practical examples. Finally, the GeoConnections *Primer* (2013) provides instructions on how to establish a geospatial data archives.

The *Primer* is based on the InterPARES 2 Creator and Preserver guidelines (2007a,b), and the following standard frameworks: the *OAIS Reference Model* (ISO, 2012), the *Long Term Digital Preservation Guidelines* (LTDP WG, 2012a,b) and the *TRAC Checklist* (ISO, 2012), which are briefly described below. The *Primer* also includes definitions, legislation, policies, case studies, tools, examples, and best practices to help those who wish to get started with geospatial data preservation.

21.5.1 Reference Model for an OAIS

The Consultative Committee for Space Data Systems recognized that the rapid obsolescence of digital technologies was increasing the risk of being unable to restore, render, or interpret preserved information. The OAIS became ISO 14721:2012 Space data and information transfer systems – OAIS – Reference model. The *OAIS Reference Model* accommodates the highly distributed nature of digital information holdings and the need for the local implementation of effective policies and procedures supporting information preservation. The OAIS is the most commonly referred to general standard.

21.5.2 European LTDP Common Guidelines

The European Space Agency initiated a coordinated action to develop a common approach to the preservation of EO data with the goal to ensure the long-term preservation of all European (including Canadian) EO space data. The CCRS collaborates with the LTDP. The objective is to facilitate EO data accessibility and usability through the implementation of a cooperative and harmonized collective approach among EO space-data owners. The *LTDP Guidelines* (LTDP WG, 2012a,b) conform to the *OAIS Reference Model* and are specifically designed to meet the needs of EO data creators and preservers and their community of practices and systems. They are aligned with the *TRAC Checklist*. The EODMS design specifications, for example, are based on both the OAIS, the LTDP with the objective to becoming a trusted digital repository.

21.5.3 TRAC: Criteria and Checklist

The *TRAC Checklist* was created by the Online Computer Library Center (OCLC) and the Center for Research Libraries (CRL) to facilitate the objective evaluation of OAIS compliance (OCLC and CRL, 2007). It too has become an ISO Standard, 16363:2012 Space data and information transfer systems – Audit and certification of Trustworthy Digital Repositories (TDRs) (ISO, 2012). The *TRAC Checklist* is divided into three sections: Organizational Infrastructure; Digital Object (i.e. archival record or Preserved Data Content Definition) Management; and Technologies, Technical Infrastructure, and Security. The *TRAC Checklist* is particularly useful as it prompts evaluators and designers to look for concrete documents, processes, and components that are not specifically listed in the *OAIS Reference Model* or the *LTDP Framework*. The first certified TDR in Canada is The Ontario Council of University Libraries Scholars Portal and the documents associated with the *Checklist* are publicly available. Geospatial data are not currently being ingested into the Scholars Portal but will be.

21.6 CYBERCARTOGRAPHIC ATLASES AS 'ARCHIVES'

There are countless examples of culturally significant and technologically innovative data sets, maps, and atlases being created that would be archived if a preservation infrastructure existed. The GCRC has been actively engaged in the creation of cybercartographic atlases for over a decade and as seen in this volume has been mapping traditional knowledge in Canada's north. These atlases embody the collective memories (Josias, 2011; Nannelli, 2009) of those who have contributed to their creation and have become a means to record the historical, geographical, cultural, and scientific facts that have been transmitted orally for centuries. These atlases, in many instances, include data that are the first recordings of orally transmitted knowledge for which Elders and communities have authoritatively endorsed. The communities who have contributed to and authorized them regard these atlases as living 'archives' even though these do not constitute an archive in the institutional sense of the word.

The *Inuit Siku (sea ice) Atlas* (Chapter 14) for example was developed to respond to Inuit Elders' and hunters' expressions of interest: to share their knowledge with youth; see more Inuit knowledge and northern content in the northern education system; and to share their knowledge more broadly with scientists and the public. The *Siku Atlas* was compiled and developed to reflect the knowledge, stories, maps, language, and lessons shared through years of interviews, focus groups, sea ice trips, and workshops with local sea ice experts in Nunavut.

The Kitikmeot Heritage Society, Inuit Heritage Trust's Traditional Name Placing Project and the Gwich'in Social and Cultural Institute on the other hand (Chapters 15, 16, and 20) approached the GCRC to have their place name databases mapped along with their collections of audio and visual recordings of place names enunciated with narrated the stories by Elders. These place name atlases have become important knowledge transmission tools from elder to youth, are cultural geo-linguistic heritage preservation tools, have been incorporated into school curricula (Chapter 20), and are land occupancy records depicting the territorial extent of a community's land use and settlement. Furthermore, naming places is part of the infrastructure of experience (Dourish and Bell, 2005) and represents social relationships, kinship, historical events, and shared cultural memories (Chapter 16). These atlases are enabling local communities to re-place their histories onto the map and others to see space from a different cultural lens.

The *Cybercartographic Atlas of the Lake Huron Treaty Relationship Process* (Chapter 17) on the other hand is a retelling and remapping of the treaty process whereby historical archival records such as surveyor notebooks and diary entries, and numerous personal histories are geotranscribed (Pyne in Caquard et al., 2009) and aggregated into a database. This database of stories, archival records, treaties, and media are then mapped onto old and new maps. In addition, some historical maps were georectified according to contemporary map projections (Rumsey, 2013) and layered with the oral history of events accounted by community elders or with the geotranscribed stories from surveyor and explorer notes. These participatory and counter cartographies are performing a post-colonial mapping of Canada's treaty system and are part of the decolonization of representation.

Alternatively, the *Mapping Views from the North Atlas* (Chapter 13) is animating (Dwyer and Davies, 2010) the archives by having Aboriginal community members identify their ancestors in archival photos, annotate them with stories, and then retake the same photos in a similar setting using contemporary multimedia techniques.

These atlases are 'remapping' the official historical record by reflexively using western geospatial technologies, multimedia, and methods in such a ways so as not to recolonize (Pyne and Taylor, 2012). Furthermore, they are also providing new geonarratives (Caquard et al., 2009) to accompany colonial maps, making 'official' records of orally transmitted knowledge, and by doing so put into question the 'authenticity' and 'completeness' of the traditional archival treaty record. Also, by geotranscribing historical records, digitizing oral cultures, and repurposing old maps in new ways, cybercartographic atlases are giving each of these records a 'secondary provenance' (Podolsky Nordland, 2004), even though the provenance of the original record is provided, it becomes less important as the atlases allow for the reinterpretation of original record.[2]

Aboriginal social memory or traditional knowledge is now being recognized in the courts for evidential purposes (Sparke, 1998), to countervail archival sources (Frogner, 2010) and there is a growing recognition that archival records such as treaties, maps, and surveys were modernist legal devices of British and French colonialists. As Raymond Frogner (2010: 47) noted, these records represent 'sovereignty's positivization' and were legal fictions since they purported to have been created among equals, which was mostly not the case. The *Lake Huron Treaty Atlas*, in particular, puts into question the ideas of treaties as dispositive documents as the transaction did not necessarily involve willful participants (Frogner, 2010). While the *Mapping the Views of the North* (Chapter 13) enable descendants to enliven old black and white photos of Elders.

In these atlas examples, Aboriginal Elders and communities appraised the content of the maps; supplied, made and curated the records; annotated photos; and directed how these were to be cartographically rendered in collaboration with researchers and programmers at the GCRC. In a sense they are taking control of the remembering of their heritage (Smith, 2006) and have put into question the authenticity of what 'official' archival documents about these spaces purport to be and allow for the reimagining of their communities (Anderson, 1991) according to their worldviews. These atlases have become a community 'archive' created outside institutional archives with the use of 'authoritative' maps, methods and technologies, archival records and by georeferencing traditional knowledge from 'authoritative' community sources onto new and old maps.

21.7 CONCLUSION

The rescue and salvage of the CLI demonstrates that preservation as an afterthought is uncertain, will most likely yield only partial results, is expensive and time consuming, and is often not 100 per cent successful. The example reinforces the fact that maps and atlases designed in closed proprietary systems impede preservation, as does a lack of life-cycle management and that saving the data is not enough, as entire systems either need to be preserved or recreated to regain intended functionality and aesthetics. For every successful recovery, there are many data sets that have been permanently lost and many more that are at risk.

With regard to consultations, studies, reports and initiatives, few address geospatial data explicitly although the preservation issues related to research data are mostly transferable to a geospatial context. For the last decade, substantial studies have one after the other identified the issues and what needs to be done, but none of the key recommendations have been implemented and in some cases recommendations from the public have been ignored.

[2] Podolsky Nordland (2004) discusses transmedia shifts of pre-Gutenberg archival records and how these digitized records gain new life and acquire new layers of meaning. In the case of the Ac ko mok ki' map, it was argued that it be reinterpreted according to the Siksika world view and their cartographic conventions thus giving it a secondary provenance.

Some progress has been made with the creation of Research Data Canada and as seen in Figure 21.1 and data in 2013 are at the forefront of information policy discussions, but as of yet there is no concerted effort towards an integrated information management strategy, which includes geospatial data nor is there a digital data preservation infrastructure in place to ingest these.

Canada has in place excellent legislation, regulation, directives, policies, and standards guiding the preservation and management of its digital assets and it is clear that geospatial data are considered to be official records warranting their preservation. Concurrently, there are few dedicated resources set aside to implement them in the government bureaucracy in real terms, and as a result little exists in terms of data management, which is of great concern as many data sets by law are supposed to be preserved. It would seem that, as an interim strategy, data producers and managers should adopt a life-cycle approach to record management and practice proactive archiving and should build a business case for data preservation.

Some government and research institutions are preserving geospatial data and many are being proactively archived in data management systems. Even though these represent only a very small subset of the geospatial data produced, they demonstrate that it is possible to do so and provide concrete examples upon which to build. NRCan has also taken a proactive approach by producing research reports and primers on the topic and are beginning to implement the recommendations (e.g. EODMS).

Communities are looking to GCRC to help them incorporate their knowledge into the Nunaliit Cybercartographic Atlas Framework (Chapter 9) and to apply cybercartographic approaches to the management of their local and traditional knowledge. The atlases have become community-specific archives created outside of the traditional archival system but exemplify proactive archiving in their design. In addition, these communities are making records by recording oral knowledge, are contesting the validity of official treaty archival records, and are transmediating knowledge into 'official' records. In effect, communities are creating conditions for secondary provenance, and are controlling their own memories, which is part of the decolonization of representations. They are remapping colonial records with indigenous geonarratives and by doing so are reinterpreting old maps and legal processes according to an indigenous worldview. What is now needed is a digital archive within which to deposit these artefacts for long-term preservation.

Overall, we are losing digital data, maps, and atlases faster than we are producing them. We have also seen glimmers of hope in some initiatives, and there is tremendous will as witnessed in consultations. Some of the examined geospatial data-preservation initiatives, portals, and all the cybercartographic atlases are proactively building in preservation and records management practices. Geospatial data, maps, and atlases are part of Canada's scientific and cultural heritage, and ensuring that this knowledge is transferred across time and space is the responsibility of creators and preservers.

The chapter demonstrated that it is possible to preserve Canada's geospatial data assets as all of the right ingredients for doing so exist with the exception of national information strategy that includes preservation, the financial resources to build a preservation infrastructure, and the will to operationalize laws and policies. A cultural shift by the geomatics community is also required in order to recognize that their assets should be part of a mainstream information policy. This would be the recommended next steps.

References

Ahlgren, D., McDonald, J., 1981. The archival management of a geographic information system. Archivaria 13, 59–65.

Anderson, B., 1991. Imagined Communities. Verso, New York.

Canada Encyclopedia, 2012. Canada Land Inventory Definition. Available at: http://www.thecanadianencyclopedia.com/articles/canada-land-inventory.

Canadian Council on Social Development (CCSD), 2013. Community Data Program. Available at: http://communitydata-donneescommunautaires.ca/Home.

Canadian Geospatial Data Infrastructure (CGDI). Home Page. Available at: http://geoconnections.nrcan.gc.ca/home.

Canadian Polar Data Network (CPDN), 2012. CPDN Purpose. Accessed from: http://polardatanetwork.ca/?page_id=59.

Caquard, S., 2011. Cartography I: mapping narrative cartography. Progress in Human Geography. 10.1177/0309132511423796.

Caquard, S., Pyne, S., Igloliorte, H., Mierins, K., Hayes, A., Taylor, D.R.F., 2009. A 'living' atlas for geospatial storytelling: the cybercartographic atlas of indigenous perspectives and knowledge of the great lakes region. Cartographica 44 (2), 83–100.

CODATA, 2006. Preservation of and Access to Scientific and Technical Data in Developing Countries Task Group. Home Page Available at: http://www.codata.org/taskgroups/TGpreservation/ (accessed March 2011).

CODATA, 2013. Data at Risk Task Group. http://www.codata.org/taskgroups/TGdataatrisk/index.html.

Consultative Committee for Space Data Systems (CCSDS), 2012. Reference Model for an Open Archival Information System (OAIS). Available at: http://public.ccsds.org/publications/archive/650x0m2.pdf.

Department of Fisheries and Oceans (DFO), 2001. Management Policy for Scientific Data DFO Management Policy for Scientific Data. Available at: http://www.dfo-mpo.gc.ca/science/data-donnees/policy-politique-eng.htm.

Department of Fisheries and Oceans (DFO), 2013a. ISDM Wave Data Archive. Available at: http://www.meds-sdmm.dfo-mpo.gc.ca/isdm-gdsi/waves-vagues/index-eng.htm.

Department of Fisheries and Oceans (DFO), 2013b. Integrated Science Data Management (ISDM). Available at: http://www.meds-sdmm.dfo-mpo.gc.ca/isdm-gdsi/index-eng.html.

Digital Fishers. Available at: http://digitalfishers.net/.

Doorn, P., Tjalsma, H., 2007. Introduction: archiving research data. Archival Science 7, 1–20. 10.1007/s10502-007-9054-6.

Dourish, P., Bell, G., 2005. The infrastructure of experience and the experience of infrastructure: meaning and structure in everyday encounters with space. Environment and Planning B: Planning and Design 34, 414–430.

Dwyer, C., Davies, G., 2010. Qualitative methods III: animating archives, artful interventions and online environments. Progress in Human Geography 34 (88), 88–97.

Frogner, R., 2010. Innocent legal fiction: archival convention and the North Saanich Treaty of 1852. Archivaria 70, 45–94.

GeoConnections, 2005. Archiving, Management and Preservation of Geospatial Data: Summary Report and Recommendations. Produced by David L. Brown, Grace Welch and Christine Cullingworth as an initiative of the Policy Advisory Node.

GeoConnections, 2013. Canadian Geomatics Community Round Table. Available at: http://geoconnections.nrcan.gc.ca/1068.

Geogratis. Available at: http://geogratis.cgdi.gc.ca/geogratis/en/index.html

GeoMAPP, 2011. Geospatial Multistate Archive and Preservation Partnership. Available at: http://www.geomapp.net/.

Government of Canada, 1994. Natural Resources Act. Available at: http://laws-lois.justice.gc.ca/eng/acts/N-20.8/.

Government of Canada, 1995. Charts and Nautical Publications Regulations. Available at: http://laws-lois.justice.gc.ca/eng/regulations/SOR-95-149/.

Government of Canada, 2005. Remote Sensing Space Systems Act. Available at: http://laws-lois.justice.gc.ca/eng/acts/R-5.4/.

Government of Canada, 2007. Remote Sensing Space Systems Regulations. Available at: http://laws-lois.justice.gc.ca/eng/regulations/SOR-2007-66/FullText.html.

IASSIST, 2011. Canada's Open Data Pilot Project. Available at: http://www.iassistdata.org/resources/canadas-open-data-pilot-project.

Industry Canada (IC), 2010a. Digital Economy Consultation. Available at: http://www.digitaleconomy.gc.ca/eic/site/028.nsf/eng/h_00491.html.

Industry Canada (IC), 2010b. Improving Canada's Digital Advantage Strategies for Sustainable Prosperity: Consultation Paper on a Digital Economy Strategy for Canada. Available at: http://publications.gc.ca/collections/collection_2010/ic/Iu4-144-2010-eng.pdf.

Industry Canada (IC), 2010c. Summary of Key Findings. Available at: http://www.ic.gc.ca/eic/site/sittgateway-portailstit.nsf/eng/00055.html.

International Polar Year Data and Information Service (IPYDIS), 2012. International Polar Year Data and Information Service (IPYDIS), Copies of Key Documents are Available from Mark A. Parsons. http://parsonsm@nsidc.org/parsonsm@nsidc.org.

InterPARES 2, 2007a. Creator Guidelines: Making and Maintaining Digital Materials: Guidelines for Individuals. Available at: http://www.interpares.org/public_documents/ip2%28pub%29creator_guidelines_booklet.pdf.

InterPARES 2, 2007b. Preserver Guidelines – Preserving Digital Records: Guidelines for Organizations. Available at: http://www.interpares.org/ip2/display_file.cfm?doc=ip2%28pub%29preserver_guidelines_booklet.pdf.

IPY Data and Management Subcommittee, 2008. Data Policy. Available at: http://classic.ipy.org/Subcommittees/final_ipy_data_policy.pdf.

ISO, 2003. ISO 14721:2003, OAIS Reference Model Space Data and Information Transfer Systems – Open Archival Information System – Reference Model. Available at: http://www.iso.org/iso/catalogue_detail.htm?csnumber=24683.

ISO, 2012. ISO 16363:2012, Space Data and Information Transfer Systems – Audit and Certification of Trustworthy Digital Repositories. Available at: http://www.iso.org/iso/iso_catalogue/catalogue_tc/catalogue_detail.htm?csnumber=56510.

Josias, A., 2011. Toward an understanding of archives as a feature of collective memory. Archival Science 11, 95–112. 10.1007/s10502-011-9136-3.

Lauriault, T.P., 2012. TBS Open Data and Open Government Consultation Response. datalibre.ca. Available at: http://datalibre.ca/2012/01/16/tbs-open-data-and-open-government-consultation-response/.

Lauriault, T.P., Craig, B.L., 2007. General Study 10-Preservation Practices of Scientific Data Portals. Final Report InterPARES 2.

Lauriault, T.P., Hackett, Y., 2005. CS06 Cybercartographic Atlas of Antarctica Case Study. Available at: http://www.interpares.org/ip2/ip2_case_studies.cfm?study=5 (accessed 18.08.12.).

Lauriault, T.P., Kennedy, E., 2011. TA 2: Final Report: Geospatial Data Archiving and Preservation. As Part of the Science & Technology Policy Research and Analysis Resource Team Prepared for: NRCan, GeoConnections Operational Framework Team by Hicklings, Arthur and Low (HAL).

Lauriault, T.P., Craig, B.L., Pulsifer, P.L., Taylor, D.R.F., 2008a. Today's data are part of tomorrow's research: archival issues in the sciences. Archivaria 64, 123–179.

Lauriault, Tracey P., Hackett, Yvette, Taylor, D.R.F., 2008b. InterPARES 2, case study 06, cybercartographic atlas of Antarctica is discussed in Part 1-case and general studies in the artistic, scientific and governmental sectors focus task force report. In: Duranti, L., Preston, R. (Eds.), International Research on Permanent Authentic Records in Electronic Systems (InterPARES) 2: Experiential, Interactive and Dynamic Records.

Lauriault, T.P., Kennedy, E., Hackett, Y., 2013. Geospatial Data Preservation and Archiving Primer. As Part of the Science & Technology Policy Research and Analysis Resource Team Prepared for: NRCan, GeoConnections Operational Framework Team by Hicklings, Arthur and Low (HAL).

Library and Archives Canada (LAC), 2007. Canadian Digital Information Strategy (CDIS). Available at: http://www.lac-bac.gc.ca/obj/012033/f2/012033-1000-e.pdf.

LTDP Working Group, 2012a. Long Term Data Preservation Earth Observation Preserved Data Set Content LTDP/PDSC. Available at: http://earth.esa.int/gscb/ltdp/LTDP_PDSC_4.0.pdf.

LTDP Working Group, 2012b. Long Term Preservation of Earth Observation Space Data: European LTDP Common Guidelines. Available from: http://earth.esa.int/gscb/ltdp/EuropeanLTDPCommonGuidelines_Issue2.0.pdf.

MacDonald, J., Shearer, K., 2006. Toward a National Digital Information Strategy: Mapping the Current Situation in Canada, Ottawa: Library and Archives Canada Summary: http://www.lac-bac.gc.ca/digital-initiatives/012018-3200-e.html and full Final Report: http://www.collectionscanada.gc.ca/obj/012018/f2/012018-3200-e.pdf.

Morris, R.J., 1991. History and computing: expansion and achievements. Social Science Computer Review 9 (2), 215–230.

Nannelli, E., 2009. Records of the Commission for Reception, Truth, and Reconciliation in Timor-Leste. Archival Science 9, 29–41. 10.1007/s10502-009-9103-4.

Nordland, L.P., 2004. The concept of "secondary provenance": re-interpreting Ac ko mok ki's map as evolving text. Archivaria 58, 147–159.

OCLC and CRL, 2007. Trustworthy Repositories Audit & Certification: Criteria and Checklist. Version 1.0. Available at: www.crl.edu/sites/default/files/attachments/pages/trac_0.pdf.

Ontario Council of University Libraries (OCUL), 2011. Trusted Digital Repository. Available at: http://www.ocul.on.ca/node/97.

Ontario Ministry of Natural Resources (OMNR), 2010. Guide Line for Retiring and Retaining Geospatial Data Stored in the Land Information Ontario Warehouse. Province of Ontario.

Open Geospatial Consortium (OGC), Data Preservation WG. Available at: http://www.opengeospatial.org/projects/groups/preservwg.

Pyne, S., Taylor, D.R.F., 2012. Mapping indigenous perspectives in the making of the cybercartographic atlas of the Lake Huron Treaty relationships process: a performative approach in a reconciliation context. Cartographica 47 (2), 92–104.

RAPLIQ, 2013. Regroupement activists pour l'inclusion Québec (RAPLIQ) The Montréal Accessible. Available at: http://montrealaccessible.ca/.

Research Data Alliance (RDA). Available at: https://rd-alliance.org/node.

Research Data Canada (RDC), 2008. Stewardship of Research Data in Canada: A Gap Analysis. Available by contacting: Bronwen.Woods@nrc-cnrc.gc.ca.

Research Data Strategy (RDC), 2011. Mapping the Data Landscape: Report of the 2011 Canadian Research Data Summit. Available at: http://rds-sdr.cisti-icist.nrc-cnrc.gc.ca/eng/events/data_summit_2011.html.

Research Data Strategy (RDC), 2013. Research Data Strategy Home Page. Available at: http://rds-sdr.cisti-icist.nrc-cnrc.gc.ca/eng/.

Rumsey, D., 2013. Map Collection Have Been Geo-rectified to Be Viewed along in Google Map and Google Earth. Available at: http://rumsey.geogarage.com/gmaps.html.

Scassa, T., 2013. Legal issues with volunteered geographic information. The Canadian Geographer 57, 1–10. 10.1111/j.1541-0064.2012.00444.x.

Schut, Peter, 2000. Back from the Brink: The Story of the Remarkable Resurrection of the Canada Land Inventory Data. Available at: http://web.archive.org/web/20011104160045/http://www.igs.net/~schut/cli.html.

Smith, L.J., 2006. Uses of Heritage. Taylor & Francis.

Social Planning Council of Ottawa (SPCO), 2013. Community Information and Mapping System (CIMS). Available at: http://www.cims-scic.ca/ (accessed 29.08.13.).

Social Science and Humanities Research Council (SSHRC), 2002. Final Report: National Consultation on Research Data Archiving, Building Infrastructure for Access to and Preservation of Research Data. Available at: http://www.sshrc-crsh.gc.ca/about-au_sujet/publications/da_finalreport_e.pdf.

Sparke, M., 1998. A map that roared and an original atlas: canada, cartography and the narration of nation. Annals of the Association of American Geographer 88 (3), 463–495.

Statistics Canada, 2010. Surveying and Mapping Services Bulletin. Catalogue no. 63-254-X. Available at: http://www5.statcan.gc.ca/bsolc/olc-cel/olc-cel?catno=63-254-XIE&lang=eng#formatdisp.

Strong, F.D., Leach, P.B., 2005. The Final Report of the National Consultation on Access to Scientific Data, (NCASRD). Government of Canada, Ottawa. Available by contacting: Bronwen.Woods@nrc-cnrc.gc.ca.

Treasury Board Secretariat (TBS), 2009. Standard on Geospatial Data. Available at: http://www.tbs-sct.gc.ca/pol/doc-eng.aspx?id=16553&section=text.

Treasury Board Secretariat (TBS), 2011a. Open Government Advisory Panel. Available at: http://www.open.gc.ca/open-ouvert/bio-bio-eng.asp.

Treasury Board Secretariat (TBS), 2011b. December 6, 2011 to January 16, 2012, Open Government Consultation Report. Available at: http://www.open.gc.ca/consult/wwh-cr-eng.asp.

Treasury Board Secretariat (TBS), 2012. Open Government Plan. April 12, 2012. Available at: http://www.tbs-sct.gc.ca/media/nr-cp/2012/0412-eng.asp#.

Treasury Board Secretariat (TBS), 2013a. Open Data Roundtable Report. Available at: http://data.gc.ca/eng/open-data-roundtables-summary-report.

Treasury Board Secretariat (TBS), 2013b. Open Data Portal. Available at: http://data.gc.ca/.

Trottier, O., 2012. Email Communication: Natural Resources Canada, Centre for Topographic Information on 28 August 2012.

WeHub, Water and Environmental Hub (WeHub) project. Available at: http://www.waterenvironmentalhub.ca/about/project.

Wilson, C., O'Neil, R.A., 2009. GeoGratis: a Canadian geospatial data infrastructure component that visualizes and delivers free geospatial data sets. Proceedings of the International Cartographic Association Conference, Ottawa, Available from: http://icaci.org/publications/.

CHAPTER

22

Conclusion and the Future of Cybercartography

D. R. Fraser Taylor[1], *Tracey P. Lauriault*[2,3]

[1]Geomatics and Cartographic Research Centre (GCRC), Carleton University, Ottawa, Ontario, Canada, [2]National Institute for Regional and Spatial Analysis (NIRSA), National University of Ireland at Maynooth, Maynooth, Co. Kildare, Republic of Ireland, [3]Member of the Geomatics and Cartographic Research Centre (GCRC), Carleton University, Ottawa, Ontario, Canada

OUTLINE

22.1 INTRODUCTION

The 21 chapters in this book illustrate the progress that cybercartography has made both in theory and in practice since the first edition was published in 2005. These will not be further described here except to point to the impressive work on indigenous mapping and traditional knowledge with First Nations and Inuit communities reflected in the various atlases described in this volume. In addition, the work on legal and ethical issues relating to the representation of traditional knowledge makes a very important contribution to the field. Issues on individual and collective data ownership, informed consent, intellectual property, privacy, confidentiality, and others have not been adequately considered in cartography in general and in particular

Developments in the Theory and Practice of Cybercartography, Second Edition, ISSN 1363-0814
http://dx.doi.org/10.1016/B978-0-444-62713-1.00019-2

by those involved in indigenous mapping. The importance of cybercartography as a means of preserving the cultural heritage of indigenous people should also be noted. The map has always been more than a material artefact but the strong impact of technology in recent decades has tended to concentrate on this aspect and to give less attention to the important cognitive aspects of cartography as well as to the map as a social construct. Critical cartography has drawn attention to this latter aspect but the theoretical aspects of the arguments made have not always been followed by concrete and useful applications. Cybercartography attempts to combine all three aspects of mapping and to present these in ways that give voice to people who are often 'invisible' in the technological maps of the twenty-first century. Cybercartography is a holistic concept that will continue to advance as a result of the iterative interaction between theory and practice as demonstrated by many of the chapters in this book.

Despite the progress made many challenges remain, one of which is how to integrate cybercartography more effectively with other developments in the emerging 'Age of Location' as society become increasingly location based.

22.2 THE AGE OF LOCATION

Location is increasingly central to almost all aspects of daily life. As Lawler has recently observed, 'The "where" dimension is one of the most natural, powerful, insightful and intuitive ways to explore the growing world of data and services' (Lawler, 2013: 92). The extent of the impact has recently been quantified in a study commissioned by Google entitled *What is the Economic Impact of GeoServices* (Oxera Consulting Ltd, 2013). The direct effects were estimated at US $150–$270 billion generated annually and US $113 billion value added. To give some context this was estimated as 0.2% of the global annual Gross National Product (GNP) of US $70 trillion. A second study, which concentrated solely on the USA (Boston Consulting Group Ltd, 2012) estimated that in 2011 the geospatial industry in the USA, of which mapping is a part, generated US $73 billion in revenues and involved 500,000 high-wage jobs.

Over 80% of existing digital databases on a wide variety of subject matters have a locational component and, as a result, the concept of 'spatially enabling society' (Rajabifard and Coleman, 2012) has emerged as an important research topic, especially over the last decade. With the advent of services such as Google Maps, Google Earth, Bing, Foursquare, and others, society as a whole is heavily involved in the creation and use of location-based services. Furthermore, complex algorithms are being created to mine and fuse data collected from multiple social media sources and shopping outlets and credit card companies' databases, under the rubric of big data, and these are increasingly being spatialized by the private sector to geoprofile locations and identify and target niche markets. In other cases, the same process is used as a health surveillance and early warning system, such as Google's Flu trends. In other instances, social media applications like Twitter are having their content qualitatively analysed and represented into thematic maps often lacking methodological rigour but nonetheless are used to assess how Twitter users feel or think, and where they do so. In addition, open data are being normalized by many governments, and interestingly most of the datasets found in the state-run open-data portals are geospatial while state-funded App contests or civil society-led Hackathons produce mobile applications that almost always include a spatial component. Also, most mobile devices such as phones and iPads capture the location of devices users. Finally, the increase of geosurveillance as exemplified by the US PRISM scandal is another politically controversial aspect of a geoenabled society. Geo-applications and location-based devices are central to daily life and there are social, legal, and technical issues associated with this phenomenon.

Maps and mapping, including cybercartography, have an opportunity to be a central facet of the age of location both as a means of the integration and linkage of disparate datasets and the display of the new information created as a result of these linkages, and as discussed in this volume, there are legal and ethical implications that must be considered. Cartography in general, and cybercartography in particular, cannot afford to be isolated from other location-based technologies, applications, and societal issues. As David Schell (2013) has pointed out 'geospatial technology has become a property of information processing itself, flowing round domain after domain like oil in a huge machine, looking to each to be integral with the service and information models that drive it' (112).

22.3 CYBERCARTOGRAPHY AND SPATIAL DATA INFRASTRUCTURES

The term 'spatial data infrastructure' as Onsrud points out (Onsrud, 2007: 1) is largely self-explanatory but when applied in practice the concept is quite complex and contains many elements. As their name suggests, such infrastructures are the means of integrating data by location to facilitate spatial data discovery, evaluation and applications including decision-making. As Onsrud points out, they include geographic data, metadata, framework, services, clearinghouses, standards, partnerships, education, and communication. Many spatial data infrastructures are being developed by governments as part of electronic governance and are in the main technological 'top-down' management structures but as Erik de Man points out, 'information infrastructures are embedded within social systems. They are multifaceted and face similar challenges and dilemmas in their development, adoption, and application' (Erik de Man, 2007: 33). Technical and bureaucratic issues dominate in the application of spatial data infrastructures and they do not always fit well with socioeconomic and cultural realities at the local level. Erik de Man also argues that such infrastructures tend to assume that they are separate from information management in more general terms.

Cybercartography is a 'bottom-up' concept as argued in Chapter 1. In the case of Inuit communities in Nunavut the Canadian Geospatial Data Infrastructure (CGDI) and the emerging Arctic Spatial Data Infrastructure have little meaning as they are government systems designed to help manage what central governments see as Canadian (in the case of the Federal Government), territorial (in the case of the Government of Nunavut) and circumpolar (in the case of Arctic Spatial Data Infrastructure (SDI) stakeholder nations), space, and resources.

Cybercartography, as the chapters of this book illustrate, is a very different kind of spatial infrastructure. It uses location to link disparate datasets and has many of the characteristics of a spatial data infrastructure but serves a very different purpose and very different uses dominated by community wishes and community choices. It is a technological construct using many of the same geospatial technologies as 'mainstream' spatial data infrastructures but it is primarily a social and cognitive construct rather than a technical one, although the Nunaliit Cybercartographic Atlas Framework and the creation of a network of nodes in the North, National Snow and Ice Data Center (NSIDC), and the Geomatics and Cartographic Research Centre (GCRC) are technical manifestations (see Chapter 9). Its use and value does, however, intersect with those of the formal spatial data infrastructures. Key socioeconomic issues for the Canadian North, for example, are climate changes leading to sea ice shrinkage and rapid environmental change. Local and traditional knowledge is of great value in understanding these issues. A major challenge for cybercartography is to find ways of integrating traditional knowledge with other forms of knowledge that currently dominate the structure of the Canadian Spatial Data Infrastructure (CGDI) and the emerging Arctic Spatial Data Infrastructure (ASDI). Research at the GCRC is already underway to respond to this challenge. An important element in this is involvement in a new network called Connect North. The overall goal of Connect North is the preservation and sharing of traditional knowledge and linking this more effectively with scientific research. The partners involved in Connect North, in addition to the GCRC are:

- The Canadian Cryospheric Information Network (CCIN) centred in the Department of Geography and Environmental Management at the University of Waterloo (http://www.ccin.ca/home/), which has two decades of experience in data management and discovery and a comprehensive repository of research data and information from across the Canadian Arctic in the form of the Polar Data Catalogue co-founded with ArcticNet. CCIN is a member of the Canadian Polar Data Network.
- The Centre d'études nordiques (CEN) is a centre of excellence for northern research and training composed of nine universities in Quebec. CEN has many decades of experience in working with northern communities and has formal linkage to universities across Canada and abroad through collaborative projects such as ArcticNet. The CEN has recently developed a field science program in the Kativik School Commission curriculum with 14 Inuit villages and took a leadership role in the first ArcticNet Integrated Regional Impact Study.
- Inuit Tapirit Kanatami (ITK) represents the Inuit of Canada and is interested in connecting Inuit to the national and international information management movement. ITK has been involved with GCRC for some years in the creation of our atlases. They have established an Inuit Knowledge Centre (Inuit Quajisarvingat) to bridge the gap between Inuit knowledge and western science.
- ARCTICconnexion, based at the Université du Québec at Rimouski works with the scientific community to make Arctic research teams more aware of Inuit social and cultural values and equip them with key tools for the development of research projects grounded in a collaborative approach.
- The Nasivvik Centre for Inuit Health and Changing Environment is a multidisciplinary research and training centre funded by the Canadian Institutions of Health Research – Institute of Aboriginal Peoples Health. It is based at the Universities of Laval and Trent and has important health data.

The ConnectNorth network is a realization that there is a pressing need to link the information and data for the large number of data management related projects on going in Canada. Cybercartography can be an important means of doing this, especially if this initiative can be linked to the emerging Arctic Spatial Data Infrastructure (ASDI). The network is an illustration of the need for greater 'human interoperability' as a starting point for technical integration and interoperability of datasets. In addition, there is a need for greater 'legal and policy interoperability' as alluded to in Chapters 18 and 19 with respect to traditional knowledge and with the integration of multinational data.

Change in the Arctic is inevitable and irreversible but it is possible to develop mitigation strategies that will lessen the negative impacts on local people. Integrating indigenous knowledge in an effective fashion with other knowledge systems could help to achieve this. For example, if an ice breaker has to be used to clear a navigation channel to facilitate resource exploitation the timing and the route could be modified to reduce the impact on the walrus hunt or other community activities. The melting of the ice also means longer travel routes requiring more fuel, which necessitates a closer examination of the environmental and economic impact of the melting of the ice. Cybercartographic atlases can, and do, present different information on the same topic as well as different interpretations of the same factual data, which is rarely possible in a standard spatial data infrastructure context, which tends to present 'authentic', 'authoritative', and 'objective' data. The Inuit judge the authenticity and authority of information by the reputation of the Elders who provide it rather than by the 'accuracy' of the information as determined by a government spatial data infrastructure as reported in Chapter 14. This was also briefly discussed in Chapter 21, where data in state archives are found in records that were created according to the norms of a western colonial legal system that maybe authentic, but may not accurately represent the nature of the 'agreements' between colonizer and colonized.

These are, however, not completely incompatible views. Cybercartographic atlases can include both. At present, spatial data infrastructures do not, but they can do so and the challenge of including traditional knowledge as well as other forms of 'crowd sourced' or 'volunteered geographic information' should be met. Cybercartographic atlases also enable the representation of multiple 'authoritative' views.

CentroGEO in Mexico City has developed a map-based tool to help resolve spatial conflicts that is called STRABO (Luscombe and Reyes, 2004). This is a spatial version of the well-known Delphi technique. Participants enter their views on issues such as the location of a school or a highway on a computer-generated map. Each map is then amalgamated by the STRABO computer program and the results are viewed by participants as a group that discusses the reasons for the location each individual choice. A subsequent round of map drawing proceeds and the results are again discussed. The process goes on until consensus is reached although some conflicts may remain insoluble.

At first sight Inuit and First Nations views may appear to be incommensurate with those of government, western science and western legal systems, but cybercartographic atlases can act as reconciliation mechanisms as in the case of the Lake Huron Treaty Atlas (Chapter 17) and as discussed in Chapter 19.

The narrow technological approach being taken by spatial data infrastructures such as the Arctic Spatial Data Infrastructure (http://arctic-sdi.org/) – for example, could be considerably enriched by including the traditional knowledge, which indigenous peoples can provide. Spatial data infrastructures should not simply serve the purpose of government. They should also serve the needs of the people and find ways to include their perceptions and narratives.

Cybercartography and the work done over the years at the GCRC can help to make this happen. The reality of power imbalances suggest that the Inuit and other Arctic peoples, despite the limited self-government that territories such as Nunavut enjoy, are likely to be overwhelmed by the processes of resource implementation and navigation through the Northwest Passage. Giving them voice in the knowledge infrastructure may help mitigate some of the adverse effects, and, if the inevitable resource exploitation is to take place, maximize the economic advantages to local people.

22.4 UTILIZING GEOGRAPHIC INFORMATION MANAGEMENT

Geographic information management tools such as cybercartography and spatial data infrastructures are dependent upon reliable access to data and data sharing. Both are technical as well as policy issues. They are dependent on a wide range of factors, the most important of which is the willingness of data stewards to share their data and to make them as open and available as possible. The technical capabilities for doing this are available through various means of making different datasets interoperable and the development of standards and specifications such as those developed by organizations such as the International Standards Organization and the Open Geospatial Consortium. The policy and legal issues are more complex. Although there is a growing trend towards open government this is by no means universal and the challenges of 'human interoperability' are more difficult to resolve than those of technical interoperability. For example, the trend towards open data under the banner of open government may not always be in the interest of indigenous people, as much of traditional knowledge is about sensitive environmental areas, sacred places, hunting and trapping sites or endangered species. Making these more open may put these places and wildlife at risk.

Preferred technical solutions such as the use of the cloud to store increasing amounts of 'big data' centrally may not be the best solution for indigenous communities such as the Inuit with limited bandwidth and technical capacity. In addition, when their data are shared in the cloud or on the Internet communities lose control and ownership of the data. The sharing of data, until recently, has been based on trusted personal relationships often built up over many years of personal contact, which is the case with many of the atlases discussed in the book. This is often given through verbal or written assent. As discussed in Chapters 18 and 19, the situation changes dramatically in the Internet era, while Chapter 9 discusses some technological methods to address these issues. If Inuit communities are to make effective use of the power of geographic information management the means of doing so must be readily available in each of the communities themselves. As this book was in the final preparation stage, the GCRC was awarded a grant by the Canadian Foundation of Innovation and the Ontario Leading Edge Fund of over Canadian $1 million to build a distributed data management system for northern communities to help address this and implementation of the new system will begin in late 2013.

22.5 RESPONDING TO THE GROWING INDIVIDUALIZATION OF MAPPING TECHNOLOGIES

Even in indigenous communities in the Canadian North and despite the technical constraints, the cell phone and other Global Positioning System-enabled personal devices are becoming more popular. In society in general, such devices are now ubiquitous and are transforming how individuals obtain information, including maps. The technology is now also being used to input information as people create, or customize, their own maps (Chapters 15, 6 and 7). Cybercartographic atlases can already be displayed on mobile devices and, as described in Chapter 9, a prototype application has been developed for the iPad to allow the input of information in a variety of formats to the Nunaliit platform used to create the atlases.

These are, however, simply the first steps in responding to the challenges of dealing with the individual, both as a map user and a map creator. Cybercartography has not been designed for a small screen and the design challenges involved are substantial. Visual displays demand simplicity if they are to be effective but the use of multimedia supplements such as sound and augmented reality need to be built into the design in order to increase comprehension. The technical means of doing so are well developed but the design elements, including the important user interface design, pose many challenges. The Nunaliit Cybercartographic Atlas Framework continues to evolve and has been developed to ingest information in a wide variety of formats but more work will be required to deal with input by individuals.

Increasing individualization also poses additional legal and ethical issues and intensifies others. Some of these have been describe in Chapters, 4, 18, and 19. Copyright, intellectual property, data ownership, consent and legal liability all take on new importance. This is in addition to the outstanding issues in dealing with the legal and ethical aspects of mapping indigenous knowledge. We are exploring the use of template licences to govern the use of traditional knowledge in cybercartography (Scassa et al., 2013; CIPPIC, 2013) but many issues still remain to be resolved.

Georg Gartner (Chapter 7) and his colleagues at the Technical University of Vienna are involved in ground-breaking research on individual mapping and mobile devices. The concept of EMOMAP is designed to gather peoples' responses to their immediate environment regarding issues such as comfort, safety, diversity, and emotions that are used to create a 'heat map' overlain on an OpenStreetMap coverage (Hodson, 2013). These interesting new developments, however, illustrate some of the design challenges outlined above as well as legal and ethical challenges. If an individual maps the area outside a restaurant, for example, as 'unsafe' has the restaurant owner the right to sue for damages caused by lost business? Alternatively, citizen science, the 'Internet of things', and 'sensor web' are growing trends where official and nonofficial scientists are using individual and integrated sensor technologies to monitor, survey and map wildlife, the environment and events, sometimes in near real time.

22.6 PRESERVING CULTURAL HERITAGE

There is no future without the past and Lauriault and Taylor have presented some of the challenges in Chapter 20. Although advances have been made, the challenge of effectively archiving and preserving cartographic atlases, especially the interactive nature of these atlases, remains. The atlases, especially those dealing with traditional knowledge, are an important artefact of the cultural heritage of the indigenous groups involved. The atlases present this information in its cognitive and societal context without which (as was argued in Chapter 1) the knowledge loses it full meaning. The challenge of doing so remains.

22.7 SCALING UP OF CYBERCARTOGRAPHY

We are entering an era of what has been called 'big data' as large amounts of data are becoming available. The applications described in this book deal with relatively small datasets. Can cybercartography be 'scaled up' to be useful in a much larger data environment?

We are beginning to experiment with the use of cybercartography to analyse and display health data at the national scale including very large datasets. In theory, there should be no problem in scaling up cybercartography but these new applications may reveal new challenges.

22.8 CONCLUSION

This book has presented an interesting set of chapters on the theory and practice of cybercartography with a number of innovative applications in the field of indigenous mapping. Cybercartography continues to evolve in an iterative fashion and each new application areas bring new insights and developments. The book presents an overview of where this holistic concept is in 2013 and how far it has come since the first edition of this book was published in 2005. Cybercartography will continue to evolve in response to both societal and technological change.

References

Boston Consulting Group Ltd, 2012. Putting the US Geospatial Services Industry on the Map. Report prepared for Google, Boston. p. 11.

CIPPIC and GCRC, 2013. Traditional Knowledge Licensing: Final consultation report.

Erik de Man, W.H., 2007. Are spatial data infrastructures special. In: Onsrud, H. (Ed.), Reseach and Theory in Advancing Spatial Data Infrastructure Concept, ESRI Press, Redlands, California, pp. 33–54.

Hodson, M., 29 June 2013. Maps that know you. New Scientist.

Lawler, M., 2013. Where is the future? The future is "where". Geospatial World 6, 92–94.

Luscombe, B.W., Reyes, C., 2004. Building consensus in environmental decision making: a methodology integrating GIS tools and structured communication. Paper Read at the 29th Annual Conference of the National Association of Environmental Professionals (NAEP), Portland, Oregon, April.

Onsrud, H. (Ed.), 2007. Research and Theory in Advancing Spatial Data Infrastructure Concepts, ESRI Press, Redlands, California.

Oxera Consulting Ltd, 2013. What Is the Economic Impact of Geo Services Prepared for Google. Oxera Consulting Ltd, Oxford and Brussels, p. 32.

Rajabifard, A., Coleman, D. (Eds.), 2012. Spatially Enabling Government, Industry and Citizens, GSDI Association Press, Needham.

Scassa, T., Taylor, D.R.F., Lauriault, T.P., Engler, N., 2013. Ethical mapping of traditional knowledge through template licences. Presentation to the Associate of American Geographers Conference, April 13.

Schell, D., 2013. A cultural development called geospatial industry. Geospatial World 6, 112–113.

Index

Note: Page numbers followed by "f" and "t" indicate figures and tables, respectively

H

I

O